Computer Integrated Manufacturing Systems:
Selected Readings

COMPUTER INTEGRATED MANUFACTURING SYSTEMS: *SELECTED READINGS*

Edited by

John W. Nazemetz,

William E. Hammer, Jr.,

and

Randall P. Sadowski

Industrial Engineering and Management Press
Institute of Industrial Engineers

Industrial Engineering and Management Press
25 Technology Park/Atlanta, Norcross, Georgia 30092

© 1985 by the Institute of Industrial Engineers. All rights reserved.

Published 1985
Printed in the United States of America

ISBN 0-89806-066-4

No part of the original material appearing in this publication may be reproduced in any form without permission in writing from the publisher. Articles previously published and appearing in the publication may be reproduced only in conformance with the copyright provisions of the original publisher.

CONTENTS

PREFACE..xi

I. INTRODUCTION AND OVERVIEW

Editorial Overview..1

History and Definition

History of Computer Use in Manufacturing Shows Major Need Now Is for Integration
Randall P. Sadowski
Industrial Engineering, March 1984..3

The Role of Robotics and Flexible Manufacturing in Production Automation
Mikell P. Groover
1982 Fall Industrial Engineering Conference Proceedings..8

Predictions

A Machine Tool Manufacturer's View of the Factory Environment of the Future
Karl B. Schultz
1982 Fall Industrial Engineering Conference Proceedings......................................16

Factory of the Future/IE of the Future
William R. Welter
1984 Annual International Industrial Engineering Conference Proceedings..........22

Placing CIMS in Perspective

Change the System or Change the Environment
Hal Mather
1984 Fall Industrial Engineering Conference Proceedings......................................27

CIM — Target or Philosophy By-Product?
Jason L. Smith
1984 Fall Industrial Engineering Conference Proceedings......................................32

CIMS in Perspective: Costs, Benefits, Timing, Payback Periods Are Outlined
Roger G. Willis and *Kevin H. Sullivan*
Industrial Engineering, February 1984..35

The Economics of Computer Integrated Manufacturing
Jack Meredith
1984 Fall Industrial Engineering Conference Proceedings......................................42

The Social Issues

Productivity Benefits of Automation Should Offset Work Force Dislocation Problems
Mikell P. Groover, John E. Hughes, Jr., and *Nicholas G. Odrey*
Industrial Engineering, April 1984...47

II. MANUFACTURING COMPONENTS, STRUCTURE, AND TECHNOLOGIES

Editorial Overview..55

Computer-Aided Technologies

Computerized Facilities Planning and Design: Sorting Out the Options Available Now
H. Lee Hales
Industrial Engineering, May 1984..57

Production and Order Planning in the FMS Environment
W. Bruce Webster
1984 Fall Industrial Engineering Conference Proceedings......................................63

Implementing a Computer Process Planning System Based on a Group Technology Classification and Coding Scheme
Joseph Tulkoff
1984 Fall Industrial Engineering Conference Proceedings......................................72

Support Systems

Approaches to the Scheduling Problem in FMS
S.C. Sarin and E.M. Dar-El
1984 Fall Industrial Engineering Conference Proceedings......................................76

Operations Research and Computer-Integrated Manufacturing Systems
William E. Biles and Magd E. Zohdi
1984 Fall Industrial Engineering Conference Proceedings......................................87

Automated Material Handling

Modular Integrated Material Handling System Facilitates Automation Process
Neil Glenney
Industrial Engineering, November 1981...92

Robotic Vehicles Will Perform Tasks Ranging From Product Retrieval to Sub-Assembly Work in Factory of Future
Robert E. Smith
Industrial Engineering, September 1983...98

Automated Controls

Automated Visual Inspection Systems Can Boost Quality Control Affordably
John W. Artley
Industrial Engineering, December 1982...104

Machine Vision — The Link Between Fixed and Flexible Automation
David Banks
1983 Fall Industrial Engineering Conference Proceedings....................................108

Automatic Identification and Bar Code Technology in the Manufacturing and Distribution Environment
Edmund P. Andersson
1983 Fall Industrial Engineering Conference Proceedings....................................112

Manufacturing Technologies

Cellular Manufacturing Systems Reduce Setup Time, Make Small Lot Production Economical
J.T. Black
Industrial Engineering, November 1983..120

Cellular Manufacturing Becomes Philosophy of Management at Components Facility
William J. Dumolien and *William P. Santen*
Industrial Engineering, November 1983..131

Well Planned Coding, Classification System Offers Company-Wide Synergistic Benefits
Glenn C. Dunlap and *Craig R. Hirlinger*
Industrial Engineering, November 1983..135

III. ISSUES AFFECTING DEVELOPMENT AND ACHIEVEMENTS OF CIMS

Editorial Overview..143

Importance of Integration

Systems Integration Is a Mandatory Component in Achieving an Optimum Systems Environment
Thomas R. Cornell
1983 Annual Industrial Engineering Conference Proceedings..145

Design and Software Considerations

Criteria for the Selection of a CIMS Data Base System
John C. Windsor and *Chadwick H. Nestman*
1984 Fall Industrial Engineering Conference Proceedings..154

Designers, Users and Managers Share Responsibility for Ensuring CIM System Integrity
Harley O. Britton and *William E. Hammer, Jr.*
Industrial Engineering, May 1984..162

Building the Bridge Between CAD/CAM/MIS
C.W. Devaney
1984 Fall Industrial Engineering Conference Proceedings..168

Distributed Computing in the Manufacturing Environment
David G. Hewitt
1982 Fall Industrial Engineering Conference Proceedings..173

Achieving Integrated Manufacturing Systems
Michael J. Daugherty
1984 Fall Industrial Engineering Conference Proceedings..180

Ten 'Cardinal Sins' to Avoid in Selecting Software for Manufacturing Management Control Systems
Robert L. French
Industrial Engineering, April 1982..186

Human Factors

Human Factors Problems in Automated Manufacturing and Design
Martin G. Helander
1983 Fall Industrial Engineering Conference Proceedings...188

IEs Must Address Automation's Effects on Worker Motivation and Capability
Kelvin Cross
Industrial Engineering, August 1983...194

Impact on Manufacturing Support Functions

The Impact of Automation on Work Measurement
Richard L. Shell
1982 Fall Industrial Engineering Conference Proceedings...197

Computerized Work Measurement and Process-Planning — Vital Ingredients in a Computer-Integrated Manufacturing (CIM) System
Kjell B. Zandin
1983 Annual Industrial Engineering Conference Proceedings...203

Management Concerns

Flexible Manufacturing Systems: Combining Elements to Lower Costs, Add Flexibility
H. Thomas Klahorst
Industrial Engineering, November 1981...206

Management Looking at CIMS Must Deal Effectively with These 'Issues and Realities'
Robert L. French
Industrial Engineering, August 1984...209

IV. PLANNING AND APPLICATIONS

Editorial Overview..217

Systems and Concepts

How to Define and Plan a System for Integrated Closed Loop Manufacturing Control
Edward J. Anstead
Industrial Engineering, September 1983..219

CIM Project May Be Doomed to Failure if Key Building Blocks Are Missing
Stan Manchuk
Industrial Engineering, July 1984..222

The Development of a Computer-Integrated Manufacturing System: A Modular Approach
Scott E. Lee and Charles T. Mosier
1984 Annual International Industrial Engineering Conference Proceedings...229

Implementing Flexible Automation
F.E. Harkrider
1984 Fall Industrial Engineering Conference Proceedings...237

Computer Integrated Manufacturing Systems for Small Firms

How Small Firms Can Approach, Benefit from Computer-Integrated Manufacturing Systems
H. Lee Hales
Industrial Engineering, June 1984.. 246

Assessing Economic Attractiveness of FMS Applications in Small Batch Manufacturing
Daniel P. Salomon and John E. Biegel
Industrial Engineering, June 1984.. 252

Case Studies

Automating Existing Facilities: GE Modernizes Dishwasher, Transportation Equipment Plants
Frank T. Curtin
Industrial Engineering, September 1983.. 258

Technology Requirements to Support Aerospace Plant Modernization
John Huber
1984 Annual International Industrial Engineering Conference Proceedings.......................... 262

Pratt and Whitney's $200 Million Factory Showcase
E.F. Cudworth
Industrial Engineering, October 1984.. 271

Production Planning and Control at an Aeronautical Depot: An Analysis of Stochastic Scheduling
John. S.W. Fargher, Jr.
1984 Fall Industrial Engineering Conference Proceedings.. 276

The Evaluation of a Flexible Manufacturing System — A Case Study
Gary R. Redmond
1983 Annual Industrial Engineering Conference Proceedings... 288

V. INDEX.. 295

PREFACE

As early as 1981 the Institute of Industrial Engineers began to anticipate the need to provide its members with information on the subject of computer integrated manufacturing systems (CIMS). In order to educate the industrial engineer on CIMS, three divisions within the Institute, Computer and Information Systems, Manufacturing Systems, and Production and Inventory Control, joined forces. This cooperative effort has resulted in three tangible products to date: the two-year CIMS series published in *Industrial Engineering* magazine, the co-sponsorship of technical sessions at the 1984 Fall IIE Conference, and this collection of readings. All three efforts have taken a substantial amount of planning and coordination by members of the three divisions. We hope the results have been worth the effort and will help the practicing IE stay current with today's technology.

CIMS has been defined as the ability to control all phases of the manufacturing system using computers, from planning—through design—to shipping. CIMS also implies the ability to take each system and subsystem that presently operates independently and create a single system which integrates all operations. This integration means that information could be shared by each area, thus making planning and control easier and more efficient. Rather than optimizing individual components, which often results in "islands of automation," the ultimate CIMS will optimize the operation as a whole.

While it is generally accepted that no true CIMS are in existence today, there are systems which already have in place many of the components required for CIMS. The ideal CIMS will consist of a careful integration of hardware, software, and human resources. The hardware is clearly in the most advanced state. This is evident by the abundance of highly sophisticated, computer-controlled machine tools, sensors, and material handling equipment. The continued advancement of the state-of-the-art in computers is also contributing to the development of CIMS by offering faster, cheaper, smaller, and more powerful computing capabilities. These two factors are the primary reasons for the development of flexible manufacturing systems (FMS). Much of the required software also exists, although there is much disagreement as to how it should be linked and which tools are best. The existence of FMS demonstrates the ability of the software to successfully link multiple elements into complex systems. Although the increasing sophistication of FMS is aiding the development of CIMS, several vital pieces of CIMS are still missing. The ability to directly link CAD and CAM systems relies on the development of computer-aided process planning (CAPP). CAPP is still in the research stage, except in a few isolated cases. Effective resource planning which would allow the development of a practical shop floor scheduling capability is a topic of great discussion, but little progress has been made in recent years toward achieving this goal.

In establishing the guidelines for the selection of papers to be included in this volume, the editors first divided the areas of CIMS to be treated into four main sections. The first section develops an historical perspective on computer integrated manufacturing systems and provides a glimpse into their future. The use of CIMS as an operating philosophy, as well as the economic and social impact of CIMS use is also discussed in this section.

The second section covers the area of manufacturing components, structure, and technologies.

The articles and conference papers selected for the third section address many of the issues and factors which affect the development and implementation of CIMS. Both technical issues and management concerns are discussed.

The final section focuses on the planning aspects and applications of computer integrated manufacturing systems. It covers the basic building blocks required for the development of CIMS and suggests methods for implementation. Five case studies of systems in use today are also included.

During the paper selection process we became very conscious of the fact that many aspects of CIMS have not been covered and need to be addressed in future articles. We hope that you, the readers of this collection, will research and write papers and articles on the topics that still need to be examined.

John W. Nazemetz
William E. Hammer, Jr.
Randall P. Sadowski

I. Introduction and Overview

History and Definition

Predictions

Placing CIMS in Perspective

The Social Issues

EDITORIAL OVERVIEW of Section I

Comprehension and appreciation of the major trends that are transforming our society is essential for the professional who aspires to prosper in these turbulent times. During the next few years we will witness fundamental structural changes in the way we carry out our manufacturing activities. We will begin to realize the integration of our design, process planning, scheduling, and control functions using computer technology. The readings in this section were selected to provide perspective on the trends that have taken us to the present state of the manufacturing art, as well as to raise some of the questions that must be answered as we proceed.

In the first subsection, Randall Sadowski and Mikell Groover provide a brief history of recent developments in the integration of manufacturing functions. They discuss developments from the precomputer control era through computerization and development of a classification for the major trends in fixed and flexible automation. These papers provide an historical anchor for evaluation of the techniques and applications that will be discussed in later sections.

The second subsection builds on this background and tries to forecast the future. During almost any discussion on the future of CIMS, questions are raised regarding vendors' plans, capabilities, and concepts. Cincinnati Milacron is always included among the vendors driving CIMS technology and Karl Schultz provides a summary of what this manufacturer sees as the future of CIMS. William Welter also provides his view of the "factory of the future" and includes in his discussion the implications of the use of this technology for the industrial engineering profession.

In the third subsection Hal Mather and Jason Smith raise some questions as to the validity of and prerequisites for CIMS implementation. Both authors discuss the necessity of getting a production system under control before attempting to integrate manufacturing functions, noting that substantial benefits can be obtained from these efforts. Roger Willis, Kevin Sullivan, and Jack Meredith look at the economics of a true CIMS and the problems of justifying and assessing the synergistic effect of interlocking subsystems.

As a fitting finale to this first section, the social implications of CIMS implementation are discussed. Since industrial engineering is the only engineering discipline to specifically consider the human being in the workplace, our displacement of him/her must be considered and assessed.

This section is intended as a brief overview and, as such, all aspects and issues cannot be raised and examined in-depth. The most notable omission is a discussion of systems which represent the culmination of efforts to integrate our manufacturing systems. As will be seen in the remaining sections, the subsystems are in various stages of development and integration.

History Of Computer Use In Manufacturing Shows Major Need Now Is For Integration

By Randall P. Sadowski
Purdue University

The earliest applications of computers in industry and manufacturing-related activities were primarily in the administrative and finance areas. These initial applications were learning experiences which required projects that involved easily identifiable and tangible savings.

Administrative cost reduction or cost avoidance systems provided such projects. They also provided relatively straightforward applications of well documented procedures. In the manufacturing area, on the other hand, procedures were not clearly developed or documented.

These administrative applications also required calculations which were consistent and compatible with then-available hardware and software capabilities. Administrative calculations such as payroll, general ledger, etc., provided well defined, repetitive computations that were relatively trivial and lent themselves directly to serial or sequential processing. This serial processing was a key attribute, since the early computers were incapable of totally random access of large volumes of data.

Before computers could be successfully applied in the manufacturing area, technical capabilities would have to advance to a stage at which computers could handle large volumes of data in a non-sequential fashion, and the disciplines and approaches of manufacturing would have to become much better defined.

Operationally, it made sense that initial applications would be in the administrative area, with its well defined, sequential disciplines. These applications were often in small departments and were initially not integrated together. The computers of the 1950s were relatively slow and expensive, and the cost of using a computer in these early stages of development limited its use in most business environments.

Organizationally, it was natural that the newly created data processing and systems departments would report directly to finance or administration in the early days when applications were limited to the business areas. However, this structure has caused much grief in recent years in many industries trying to implement successful computer-integrated manufacturing systems. Fundamentally, it became apparent in the early use of computers that it was much easier to succeed in administrative applications than it was in the ill-defined manufacturing area.

The punched card era

The punched card provided a reusable form of data which could be analyzed and manipulated with unit record equipment (counting machines, collators, sorters, key punches, etc.). It became widely accepted and used in administrative and financial applications during the 1950s and '60s. Although these bear little resemblance to computer implementations as we think of them today, they represent the origins of today's computer systems.

The early use of punched cards in manufacturing was predominantly seen in their inclusion in job or order packets for material requisition, labor reporting and job tracking. These cards were often key punched from manufacturing paper documents such as routings and parts listings. The punched cards were also used for simple inventory accounting systems.

Limited attempts were made to use punched cards for bill of material explosions. However, by and large this proved to be technically infeasible from a time and effort standpoint.

During the same time, computers were being introduced into the engineering areas of manufacturing for companies with heavy engineering requirements. This was partially due to the introduction of FORTRAN and the availability of lower-priced engineering computers. A mainstay of that period was the classic IBM 1620.

Although the computers available at that time were compatible with engineering activities, the evolution of computers discouraged their use in the logistics side of manufacturing. The available computers were principally binary word machines with limited input/output capabilities. They were primarily interfaced with printers, card readers and magnetic tapes, which were ideally suited for the sequential-type operations previously mentioned. In addition, the languages available at that time were simply not suitable for the types of tasks required in the manufacturing environment.

The production and inventory control environment at that time required large data processing type jobs, not engineering oriented jobs. Thus the primary obstacles to successful implementation of manufacturing systems were the hardware technology, inadequate languages and the inability to perform random access of data.

Pre-computer controls

Gradually the early manual data collection systems evolved to allow for numeric data entry devices in production control or dispatching departments. This was achieved by linking key punch output devices in these areas to the centralized facility. Although such systems were not truly utilizing the computer, they did prove beneficial in dynamic job tracking and centralized dispatching activities.

There were even attempts by some facilities, primarily long lead-time build-to-order durable goods manufacturers, to manually explode engineering bills of materials for component manufacturing and purchasing requirements. The result was often called the "quarterly schedule" because the time required to manually explode the bill of materials precluded the possibility of doing it more often.

Some companies utilized non-computer punched card equipment to develop an alternative approach to exploding bills of materials by simply calculating the gross requirements. The level-by-level approach which resulted in the total netting of requirements was not practical on this non-computer type equipment.

During this time, the initial concepts for manufacturing systems were being developed by long-lead-time type users. Unfortunately, short-lead-time manufacturers didn't have the time or computation capability to produce the required manufacturing control information. Such manufacturers often utilized manual order point or min/max systems to replenish individual components rather than utilizing the bill of material relationship requirements.

One successful early computer approach to inventory management grew out of techniques developed by distribution companies to control distribution of finished goods inventories. Such inventories were being effectively managed using statistical inventory control or scientific order-point concepts.

> **" There is a distinct lack of executive leadership which has the foresight to look beyond a CIMS as a purely technical challenge and view it as the management challenge that it really is. "**

Inventory and management replenishment systems were developed which utilized ABC analysis, EOQs, order-point concepts, forecasting with exponential smoothing and statistical techniques for dealing with safety stocks. Such systems were readily adaptable to the period's computers and provided an exciting new use in inventory management which actually worked for finished goods or distribution inventories.

Unfortunately, the technique was so desirable that many consultants and manufacturers attempted to apply it to discrete component manufacturing inventories, with often disastrous results. It became apparent that although a technique was needed, it was simply the wrong tool for these kinds of environments.

Early computerized systems

The late fifties and early sixties provided, for the first time, relatively low-cost computing capabilities which allowed the concepts of data processing to emerge. These new systems arrived with extremely fast and capable input/output devices and commercially oriented or non-engineering programming languages. No longer limited to sequential processing, they opened the field for computerized manufacturing applications.

The first manufacturing application that evolved was the preparation of shop documents and paperwork on pre-punched cards for job, labor and material reporting. During this time a limited number of companies were performing an explosion of the bill of materials using sequential tape systems. This frequently required many, many passes of reading the tape.

With the development of direct access storage, such systems were often converted to take advantage of this capability to perform a direct explosion of the bill of materials. The addition of this new direct access storage device suddenly brought adequate and reasonably priced data processing equipment to a large number of manufacturers.

Although the power of the computer had finally arrived at the point at which it could be used in the manufacturing environment, there was little, if any, software available.

The late sixties saw the emergence of the minicomputer, produced by manufacturers such as Digital Equipment and Hewlett-Packard. Although the minicomputers lacked applications software, they were widely accepted by the engineering and technical community for their computing capability and were quickly adopted for a wide range of engineering and process monitoring and control applications.

These were predominantly stand-alone and well justified islands of automation that emerged in factories nationwide. They were predominantly in the process and flow manufacturing industries (petroleum, chemicals, pharmaceuticals, consumer goods, etc.).

During this phase, the computerization of manufacturing systems was slow primarily because of the lack of software. Many companies attempted to write their own software and proceeded to implement such systems. The primary application areas were inventory accounting, requirements planning (subsequently called MRP), and in some advanced systems, operations scheduling and capacity requirements planning.

The COPICS concept

During the 1960s, IBM undertook the development of an overall concept for a production and inventory control system. This concept was announced in the late 1960s and subsequently published by the company in the early 1970s in eight paperback volumes.

It is important to recognize that the development of the communications oriented production information and Control System (COPICS) concept involved no software. However, it did provide a rather detailed view of the data flow in a manufacturing organization with an integrated system consisting of sales forecasting, engineering data control, inventory control, requirements planning, purchasing, operations scheduling, shop floor control, etc. In this sense it made a major contribution to the subsequent development of computer-integrated manufacturing systems.

This concept provided the outline and defined the boundaries which allowed the intelligent development of software for the control of manufacturing systems. The actual software that emerged over the next couple of years included the bill-of-materials processor program (BOMP), inventory control, requirements planning systems and rudimentary capacity planning systems.

The BOMP was a breakthrough in data storage and access concepts (it used a chain file technique). Although this concept represented a significant breakthrough for dealing with the bill of materials, it is definitely not a data base system by today's standards.

These software systems gained immediate acceptance and subsequent widespread use in many types of manufacturing operations. Although the requirements planning system was only one section of the original COPICS concept, the systems being implemented were referred to as MRP (material requirements planning) systems.

The decade of the seventies saw the emergence of the COBOL language and an explosion of knowledge of MRP systems, fueled by some successful systems and a cadre of manufacturing consultants. Many of these consultants were prominent in the MRP crusade which occurred during the 1970s.

This crusade resulted in a large number of actual implementations of MRP systems and a vast amount of experience being gained, primarily through failures. The age of the computer had arrived, and manufacturers were rushing to take advantage of its capability. Unfortunately, many of these MRP implementations were subsequently determined to be failures.

During this same period, the growth of the minicomputer was exponential. However, it was primarily in stand-alone islands of automation which would ultimately have to be incorporated together in order to achieve a truly functional computer-integrated manufacturing system.

The initial computerization of production and inventory control systems consisted of limited application software offerings by only a relatively small number of vendors. Many companies based their systems on available BOMP or BOMP-like software, utilizing some of the inventory accounting and requirements planning software available, and then developed and wrote a significant amount of home-grown or in-house software to supplement that which was available. The COPICS concept was widely used as a blueprint or architecture for what was now being called manufacturing systems.

At this stage, manufacturing systems for most companies did not include design engineering functions or plant automation. The few exceptions tended to occur in a small number of leading edge manufacturing companies and in highly automated process plants which were very flow shop oriented.

Explosion of computerization

The second half of the 1970s and the early 1980s saw a truly dramatic increase in the development, acceptance and implementation of computerized production and inventory control systems. Unfortunately, not all of these systems ultimately proved to be successful.

This sudden explosion of computerized manufacturing systems can be traced directly to the development of new technology (hardware and software) and a recognition by industry that such systems were vital to the success of their companies. Computers were now easily affordable by most manufacturing companies. The price/performance curve of components was experiencing approximately a 20% annual improvement, which was compounding over time.

Computer hardware and software vendors recognized and targeted manufacturing as a significant marketing opportunity for subsequent sales. Simultaneously, the consulting community and selected computer vendors were beginning to spread the gospel of manufacturing systems. This consisted of a rather intense phase of teaching, telling, selling, consulting and publishing of the benefits of computerized manufacturing systems.

The number of commercially available computerized systems increased from relatively few in the early 1970s to a great many in the late '70s and early '80s.

A survey of commercially available production management systems published by CAM-I in 1981 listed 283 different computerized systems available. The majority of these systems were designed and marketed by software and system houses rather than computer manufacturers.

Although much has been publicized regarding the high cost of computerized manufacturing systems, most of the systems listed in that report were under $50,000. However, this does not include the hardware cost or the cost of any software modifications required to interface the software to the company files or to provide additional capabilities not currently available.

Interestingly, most packages do not appear to be generic in nature, as the number of users for each system is often relatively small.

This sudden commercial availability of large numbers of computerized manufacturing systems was complemented by the emergence of an extensive array of available computer hardware and software, particularly from 1980 on. At the same time, the attractive computer price/performance reduction was fueling a similar explosion of computing applications in engineering design and plant automation.

The acquisition and implementation of material requirements planning systems became the objective of virtually every manufacturer in the United States. The only exceptions were the pure job-shop fabricators who had no bills of materials and were primarily interested in detailed job-shop operations scheduling and capacity management to support their efforts to satisfy specific customer orders.

Accompanying the implementation of a large number of systems during the 1970s was the observation by many individuals that many of these systems were ultimately considered to be failures. The alarming failure rate caused practitioners, consultants and academicians to attempt to isolate the reasons for it.

What evolved from this examination was a new understanding of what a manufacturing or production and inventory control system was. It illuminated the need to focus more attention upstream from the traditional MRP on the master production schedule and the production plan and downstream on purchasing, capacity management and shop floor control techniques.

It also showed the need for timely and accurate feedback within the system itself. Additionally, this new understanding put renewed emphasis on the bill of materials and on inventory accuracy.

The implications for the direction of computerized manufacturing were significant. It became apparent that a truly integrated manufacturing system was required which necessitated an overall data base management system. The need for significant data communications and a data communications management system, as well as the desirability of on-line interactive systems rather than traditional batch modes, could also be seen.

The implications for management that resulted from this self-analysis were possibly even more significant. In most companies, a variety of non-integrated manufacturing systems existed. In order to create a truly integrated system, these existing systems needed to develop closer, more disciplined ties to marketing, engineering and physical distribution elements. Achievement of such a goal implied more computerization, rather than less, in that ideally there would be a company-wide integrated system with the capability to communicate successfully using a company-wide data base and data communications management system.

As selective chief executive officers reviewed the computerization that had evolved over recent years, a consistent picture emerged. Most companies found that they had multiple hardware and software systems scattered throughout their facilities, with little if any interaction and little if any compatibility.

It also became apparent that the once sought-after islands of automation defied integration, primarily because they had never been designed with integration in mind. Finally, it became all too apparent that the numerous sources of data that existed often contained completely inconsistent and contradictory information.

The picture that began to emerge was that the existing manufacturing and financial systems were totally inadequate to allow companies to successfully compete in the modern world. At the same time, U.S. manufacturing was faced with increased pressure from foreign competition, government regulations, increasing cost of capital and consumerism, which provided a rather bleak picture for the future.

One reason for this inadequacy may be that historically, many executives, though competent, have been terrified by technology and have assumed that data processing and information systems were purely technical activities. Such executives often fundamentally abdicated their responsibility to the computer technicians, who were, however technically competent, ill-prepared and ill-placed in the organization to take an effective leadership role.

Formulation of CIMS

As a result of these realizations and subsequent analysis, it has become apparent in the last couple of years that computer-integrated manufacturing systems are clearly the direction required by today's manufacturing environment. The continued availability of powerful computing capability and modest cost will help CIMS become an attainable objective.

Although the concept has been somewhat defined, neither a total systems architecture nor the requisite integration software is yet available. Many problems still remain to be resolved.

For example, it has become apparent that engineering and manufacturing systems need to be integrated. This has focused attention on the need to create the ability to manage the differences that exist between various types of engineering data and the subsequent manufacturing data derived from them.

The solutions being pursued currently would interface these dissimilar systems so that data could be communicated between them. At a minimum, successful implementation will require that a standard communications software system be developed.

Although there are numerous examples of integrated data base systems for production and inventory control environments, very few of these have been successfully interfaced with engineering systems to provide for automatic generation of bill of material data; and fewer still include the CAD/CAM geometry systems which would allow automation of the downstream process planning and execution functions. These problems and many more will have to be resolved before the truly computer-integrated manufacturing system emerges.

Although the need for such integrated systems has been clearly defined, there still are significant barriers to their successful development and implementation. There is a distinct lack of executive leadership which has the foresight to look beyond a CIMS as a purely technical challenge and view it as the management challenge that it really is.

It has also become apparent that the current operating style of most manufacturing companies will make it very difficult to justify the time and expenditure required to successfully develop and implement such systems. Typically, their planning horizons are far too short, and traditional piecemeal budgetary techniques do not lend themselves to such long-term massive projects.

It is also difficult to accurately estimate the benefits that one would attain from a truly integrated system. This suggests that management will have to play a key role in championing the development and implementation of future systems.

Future implications

A review of the progress toward computerized manufacturing over the last 30 years reveals that phenomenal strides have been made toward the development of the ultimate manufacturing control system. Going from the punched cards of the 1950s to the concepts for a truly computer-integrated manufacturing system of the early 1980s represents phenomenal leaps in technology and our ability to control our manufacturing environments.

A cautious assessment of our current status reveals that the concepts for a CIM system have been developed and discussed, but our current state is not too different from that of the late 1960s following the announcement of IBM's PICS concepts. The direction now appears to be clear, but the means to achieve the final objective still elude the practitioner and theoretician alike.

Although the obstacles may be many and large, the picture that is emerging appears to be bright. The problem has been clearly defined and an objective clearly stated. Many of the concepts that are needed to satisfactorily meet the objective have been developed, and many more of the remaining problems are currently being discussed and resolved.

It is clear that the future of manufacturing in the United States depends upon the successful development of such systems, and on whether the foresight of management and the expertise of our technical community allow the achievement of these objectives in the foreseeable future. It is to be hoped that the many mistakes made during the past 30 years of development and implementation of computerized manufacturing systems will not be repeated in development of the new computer-integrated systems. IE

☐ *The author wishes to thank Mike Kutcher, chairman, IBM Corporate Automation Council, for valuable discussions on the subject of computer use in industry and Jim Clark, consultant, manufacturing industries marketing, IBM, for providing extensive background information that served as a basis for this article.*

Randall P. Sadowski is an associate professor in the school of industrial engineering at Purdue University. He holds BSME and MSME degrees from Ohio University and a PhD from Purdue University. Previously he was on the faculty of the University of Massachusetts. His teaching and research interests are in the design and control of production and manufacturing systems.

Reprinted from the 1982 Fall Industrial Engineering Conference Proceedings.

The Role of Robotics and Flexible Manufacturing in Production Automation

Mikell P. Groover
Professor of Industrial Engineering
Lehigh University
Bethlehem, Pennsylvania

Introduction

I believe there was a time when a machinist working in a factory could take pride and satisfaction in his job. To have such a job often required a rather intensive apprentice training program. The job itself required a certain level of skill and intelligence in reading blueprints, laying out the job, and operating the mechanical production equipment. Production had a relatively high manual content and there was a more direct relationship between a person's efforts and the resulting output. At the end of a shift, the worker could have a sense of accomplishment for his contributions. Certainly not all production jobs were as rewarding as portrayed here. In fact, much of the factory production work was menial and repetitive. However, in the uncomplicated, unsophisticated world of those times, manual production workers possessed the self-respect which derives from the importance and recognition attached to their work by others.

Today, conditions have changed in the factory. There is a greater variety of processes performed, the machinery is more automated, computer systems are widely used, and the management of production has become more complex. Although the activities which take place in a factory today would seem to be far more exciting from a technological standpoint, the regard for manual production labor has definitely declined. Owing to a variety of sociological and economic reasons, individuals are inclined to shun factory work if they can. The glamour of a college education, the apparent disdain among many persons for doing manual labor, the growth of the service industries, the unrest created by union activities, and other factors have all contributed to a reduced social esteem for factory production work.

It is the force of these social and economic factors, as much as the rush of technology, that is moving us towards more automation in manufacturing. The number of direct factory workers is decreasing just as the proportion of farm workers has decreased significantly over the last century. The current proportion of the labor force engaged in direct manufacturing is about 20%. In 1947, the corresponding figure was 30%. By the turn of the century, the proportion will no doubt be significantly lower. One estimate is that the percentage will be a mere 2% (7).

The objective of this paper is to examine some of the trends and advances in the technology of automation, with particular emphasis on the role of robotics and flexible manufacturing. These are the systems that will gradually take over for manual labor in the operation of tomorrow's factories.

Automation and Types of Production

Automation has been defined (3) as "the technology concerned with the application of complex mechanical, electronic, and computer-based systems in the operation and control of production." The initial efforts at automation were in mass production (e.g., the automobile industry) where large numbers of products were manufactured. This type of automation is sometimes described as "fixed automation" because the sequence of production operations is fixed by the equipment configuration. Production rates achieved by this type of automation are very high. However, the problem is that this equipment is inflexible in the sense that it cannot be used for producing more than one product type. In many cases, no product variations can be tolerated. The equipment is designed to produce one item only, and when that item is no longer made, the equipment either becomes obsolete or requires a major changeover to adapt it to a new product.

Because of the inflexibility of fixed automation, attempts were made to design equipment which could be conveniently rearranged for new product situations. One of the first examples of these attempts was the development of numerical control (NC) around 1951. NC allows the machine tool to be programmed for a particular production part. Accordingly, the name "programmable automation" is often applied to this general type of automation. The distinguishing feature of programmable automation is that the equipment is designed to be flexible. Because of this flexibility, programmable automation can be used for relatively small production lot sizes. For example, the typical range of batch sizes for NC is from several dozen parts to several hundred parts.

The manufacturing industries are normally associated with the production of discrete items. Although there are a variety of production techniques involved, three general types of production can be distinguished:
1. Mass production
2. Batch production
3. Job shop production

Mass production involves the dedicated production of large quantities of a single item. There is little or no model variation in the pure case of mass production. Batch production is the manufacture of medium lot sizes (typical lot sizes might range from hundreds to thousands) of the same product. Finally, job shop production deals with low quantities of specialized product. The quantities range from one-of-a-kind to perhaps several dozen and the product is often customized and technologically complex. The relationship among the three types is illustrated in Figure 1.

These production types are defined here with no regard to automation. Of course, attempts have been made to automate the operations in each of the three cases. Fixed automation is normally associated with the mass production type, and programmable automation is associated with batch production and job shop production. The cost reduction and productivity improvement resulting from the application of automation technology has been most pronounced in the case of mass production. It has been estimated (2) that a V-8 automobile engine block would cost roughly $3000 if conventional machine tools and techniques were used to do the machining. The same engine block costs approximately $30 to machine on an automated transfer line. The improvements resulting from automation in batch and job shop production are also significant, but not nearly so dramatic as in this example of mass production automation. The job shop has been the most difficult type of production to automate because of the low quantities and associated economics involved.

Some interesting statistical data were published in the October 1980 issue of American Machinist magazine (1). These data are presented in pie chart form in Figures 2, 3, and 4. They show how time is spent in a typical production shop for each of the three cases of mass, batch and job shop production. All three cases involve metal cutting operations. It is interesting that the proportion of time spent with the tool actually engaged in productive work is relatively low for the three types. For the transfer line (mass production), representing the highest level of automation, the effective utilization of the equipment is only 22%. The utilization is much lower for batch production (8%) and job shop production (6%). By comparison, in a survey conducted for a previous article in Industrial Engineering magazine (5), the utilization of equipment in the continuous chemical and petroleum industries was found to be as high as 96%. This greater level of efficiency is a reflection of a much higher level of automation, process integration, and computer control found in those industries. It also helps to explain why these industries are typically quite profitable and why the associated products can be sold at relatively low prices. Even at $1.25 per gallon, the price of gasoline is cheap compared to most other manufactured substances.

My argument concerning these figures is, perhaps, a statement of the obvious. There is a close correlation between the level of automation and the utilization of the production process. A higher level of automation leads to a higher level of utilization. There are two reasons for this. First, management has a large investment in production processes that are highly automated, and they are inclined to protect that investment by using the equipment as much as possible. Indeed, the unit cost of production is reduced with a higher level of utilization because the high equipment cost is spread over a larger number of units. This is in contrast with manually driven operations where the unit cost tends to remain the same (and sometimes even increases) as the output is increased.

The second reason why greater utilization is associated with automation is that highly automated processes require little or no tending by human operators. In the ideal, the automated process could be operated continuously, 24 hours per day, with no human attention. The day shift might be used to organize the process (e.g., prepare raw materials, tooling, routine maintenance, etc.) so that production could continue unattended during the two overnight shifts. With a fully automated process, management need not be concerned over the many problems associated with a three shift manual operation.

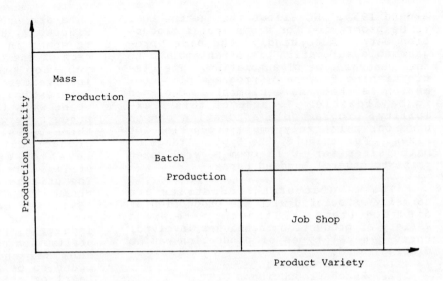

Figure 1 - The Three Production Types arranged according to Production Quantity and Product Variety.

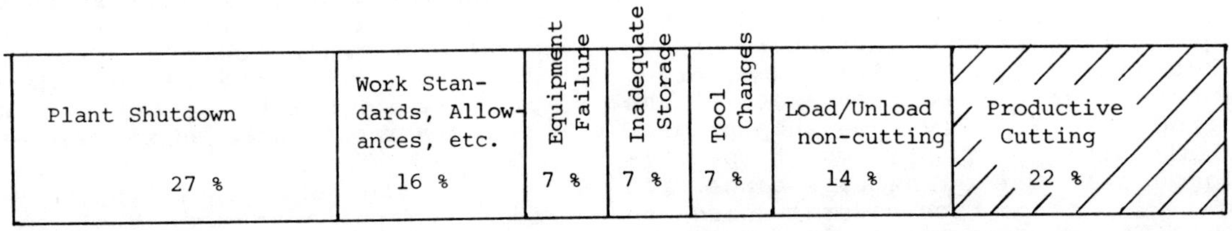

Mass
Production
Batch
Production
Job Shop
Production Quantity
Product Variety

Plant Shutdown
27 %
Work Standards, Allowances, etc.
16 %
Equipment Failure
7 %
Inadequate Storage
7 %
Tool Changes
7 %
Load/Unload non-cutting
14 %
Productive Cutting
22 %

Figure 2 - Time Allocation in High Volume (Transfer Line) Production

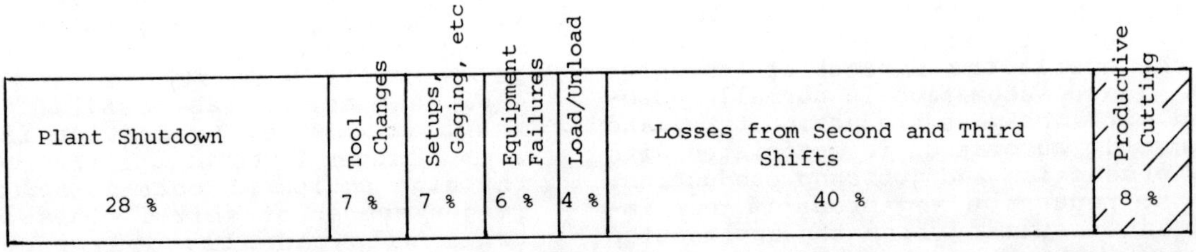

Plant Shutdown
28 %
Tool Changes
7 %
Setups, Gaging, etc.
7 %
Equipment Failures
6 %
Load/Unload
4 %
Losses from Second and Third Shifts
40 %
Productive Cutting
8 %

Figure 3 - Time Allocation in Mid-Vloume (Batch) Production

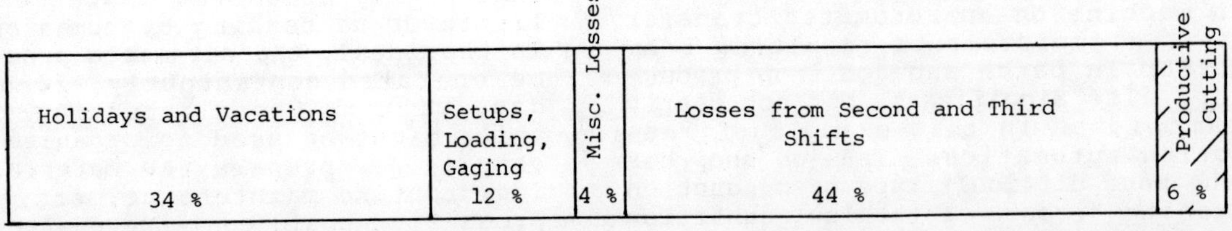

Holidays and Vacations
34 %
Setups, Loading, Gaging
12 %
Misc. Losses
4 %
Losses from Second and Third Shifts
44 %
Productive Cutting
6 %

Figure 4 - Time Allocation in Job Shop Manufacturing

On the one hand, management is permitted to run automated processes at higher utilization levels because they are unencumbered by labor requirements. On the other hand, management is forced towards higher utilization with automated equipment because of the greater investment associated with this equipment.

The logical consequences of higher utilization are economies of scale, greater production capacity, and lower prices. These are the kinds of benefits which management strives to achieve in manufacturing.

Automation for Low Quantity Production

One of the trends in manufacturing involves the continuing efforts to introduce more automation to batch and job shop operations. Figure 5 presents a listing of automation achievements in the three production categories. This list tends to show that more achievements have been made in automating high production operations relative to low production. In the past, it has been technologically and economically easier to automate operations to produce large quantities of items. The systems involved were largely mechanical in nature and a high investment could be justified because of the large quantities. It has been much more difficult to automate production operations for low quantities. Automated systems based mostly on mechanical components are suitable for repetitive operations but not for situations where the motion sequence must be frequently altered to accommodate new products. In other words, mass production automation reflects a high level of achievement in terms of mechanical complexity; however, it is quite inadequate in terms of the kinds of features required for automating low quantity production. The following are the features needed in automated equipment for job shop and batch operation:
1. Quick changeover of production jobs.
2. Adaptable to a variety of workparts.
3. Capable of performing the process under program control.
4. Easily programmable.
5. Capability for off-line programming.
6. Capability to save the programs.

In nearly all batch and job shop situations, the production equipment must be set up and fixtured for the particular job. This setup procedure can consume a significant amount of time. If the production lot size is small, the setup time can figure prominently in the piece cost. One of the objectives in automating this kind of operation would be to provide for a fast and convenient way to changeover the equipment from one production job to the next. A second feature of the equipment is that it should be possible to accommodate a variety of different workpart configurations. This variation in product configuration is a characteristic of batch and job shop manufacturing, and any automated equipment used for these types of production must be designed to be adaptable. Practically speaking, this would mean that the equipment could accept a limited range of products. For example, a machine tool system might be capable of handling parts up to a 15-inch cube in size.

Other important features of batch and job shop automation are concerned with programming. The program contains the step-by-step instructions for accomplishing the production operations on a given workpart. The equipment must be capable of performing the process sequence completely under program control, with no human assistance. Today, numerical control represents an example of this kind of program control. However, most shops require a human operator to be in attendance at the NC machine (in many cases, one operator tends two or three machine tools). In the ideal automated system, no human attendance would be required.

The programming features would include ease and convenience of programming using a powerful and versatile language such as IBM's AML (A Manufacturing Language) or the McDonnell Douglas MCL (Machine Control Language) developed under U.S. Air Force sponsorship. The most effective use of these computer languages would involve off-line programming where the control program could be written without making use of the production equipment itself. This form of programming is used in NC part programming. Its advantage is that a program can be developed for a new part while the NC machine is working on other jobs. Under ideal circumstances the new program could be downloaded from the computer to the machine tool through direct connection (similar to DNC in numerical control). Another programming feature which seems easily within grasp is the capability to save the program for future use. This capability exists today in NC and it could easily be incorporated into the design of new automated equipment for small batch operations. It is anticipated that future program storage media would be other than the one-inch wide punched tape used so commonly today in NC.

Robots and Flexible Manufacturing Systems

Many of the features required for automation of low quantity production are possessed by today's robots and flexible manufacturing systems. I have previously

Figure 5 - Achievements in Automation for the Three Production Types

Production Type	Automation Achievement
Mass Production	Automated transfer lines and dial indexing machines Partially and fully automated assembly systems Industrial robots for parts handling, spot welding, etc. Computer production monitoring Finely tuned production scheduling Integrated automated material handling systems
Batch Production	Numerical Control, Direct Numerical Control Computer Numerical Control Flexible Manufacturing Systems Material Requirements Planning Shop Floor Control
Job Shop Production	Numerical Control Computer Numerical Control Material Requirements Planning

Figure 6 - Application Characteristics of Numerical Control, Flexible Manufacturing Systems, and Transfer Lines

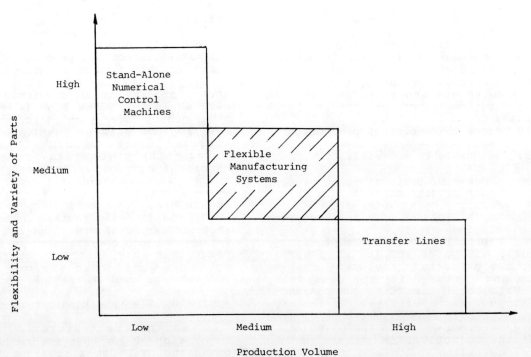

defined a robot as a general-purpose programmable machine possessing certain anthropormorphic characteristics (3). This definition seems especially appropriate for our discussion of programmable automation for low quantity production. An industrial robot represents the type of automated equipment which will play a large role in future batch production systems. If methods can be developed for fast, off-line computer-automated programming of robots, these machines will also have an important function in job shop production as well.

The term flexible manufacturing system, or FMS, is the name given by Kearney & Trecker to its computer-integrated manufacturing lines. Since Kearney & Trecker is the market leader for this type of equipment, the term flexible manufacturing system has become almost generic. An FMS is a group of processing stations (the stations are typically NC machines) connected together by an automated workpart handling system and operated under computer control (3). The FMS is designed to efficiently produce parts in the mid-volume, mid-variety range as indicated in Figure 6. It incorporates many automated features to deal with this production range, including automatic tool changing, automatic parts handling, downloading of NC programs from a control computer, and automatic monitoring and reporting of system performance.

There is great interest among manufacturing people in robotics and flexible manufacturing systems because these technologies represent the most promising approaches to automation in low and medium quantity production. Flexible manufacturing systems can take a variety of design forms and are not necessarily limited to the machining equipment of the type marketed by Kearney & Trecker and other machine tool builders. The flexible manufacturing approach is appropriate for many applications including assembly operations, electronics components, printing, garment making, various metalworking processes, and others. Also, current day flexible manufacturing systems for metal cutting operations are typically used for mid-volume production. The trend in future systems will probably be towards lower and lower quantities per batch. The general objectives of systems designed for flexible manufacturing are:
1. To approach the efficiencies and economies of scale normally associated with mass production, and
2. To maintain the flexibility required for small and medium lot size production of a variety of different parts.

Trends in Small Quantity Production Automation

There are several trends and developments which are gradually permitting us to realize these objectives in automated manufacturing systems. It is doubtful that we will ever achieve the high production rates of dedicated fixed automation, or that the flexible machines will be as versatile as we might desire. However, these are the directions in which this type of automation is moving. I will present a concise listing of these trends and developments, and then proceed to discuss them. The trends in the technology of flexible manufacturing automation are:
1. Greater use of the computer for process control.
2. Greater use of management information and communication systems in manufacturing.
3. Development of computer-automated programming routines for production equipment.
4. Improved data input techniques and human-machine interfaces.
5. Improved machine breakdown analysis and diagnostic systems.
6. Greater use of vision systems and other noncontact inspection techniques.
7. More use of 100% inspection.
8. Greater use of robots as components in flexible manufacturing systems.

The first of these trends is well-known. Computer-aided manufacturing (CAM) is the term applied to the use of computers in production applications. However, not all CAM applications are for process control. Indirect applications such as process planning, computerized scheduling, material requirements planning, and computer-assisted NC part programming are included within the scope of CAM. In process control, computer applications are becoming more common and more sophisticated. These applications often involve the use of microprocessors to bring intelligence and control to the particular location in the process where it is needed. This has sometimes been referred to by the term distributed computer systems. In manufacturing, distributed systems are made up of local "intelligent" work cells connected to other computers in a hierarchical arrangement. Various levels in the hierarchy have different responsibilities but the local computer performs the process control function, receiving instructions and programs from the levels above.

A second recognizable trend in small quantity production is the greater use of management information and communication systems. Generally, there is an inverse relationship between the volume of production and the need for communication of complex information to the shop floor. For job shop manufacturing, the need for this communication is greater than for high production because each new job requires a new set of instructions. Traditionally, these instructions have been communicated by means of paper documents (engineering drawings, assembly lists, route sheets, etc.). Increasingly in the future, computer-based systems will be utilized to convey instructions to the shop, either to direct labor personnel in printed documents or via CRT terminals, or in the form of programs to flexible automated machines. The information system would be driven from a common data base for both design and manufacturing.

There are two directions in which information would flow in this computerized communication system. Not only would information flow down to the shop, but also performance data would be collected from the shop to be used in the preparation of production records and reports. This two way communication and information system has sometimes been referred to by the term "shop floor control." The trend in this area is towards much greater use of computer systems in shop floor control than is presently the case. It is very possible that some of the greatest increases in productivity in batch and especially job shop manufacturing over the next decade will result from improvements in management information and communication systems.

The third trend is the evolution of computer-automated routines for generating job programs for programmable equipment. A good example of this is the automated NC part programming routines developed for CAD/CAM systems. These techniques involve the use of interactive graphics as an aid to the human user in the part programming procedure. In its most productive form, the routines use the part geometry data developed during design on the CAD system, thus saving the programmer the time and effort to reconstruct the part definition. Most of the turnkey CAD/CAM systems available today allow the user to define the tool path for machining using high-level commands to the system. The user receives almost immediate verification of the correctness of the program as it is being entered. Many of the turnkey systems have software which will automatically generate the tool path for common situations such as profile milling, pocketing, and sculptured surfaces. The output can be an APT listing or the CLFILE (cutter location file) that can be used to generate the actual NC tape.

Further enhancement of these automatic part programming packages and similar developments in robotic programming will shorten the lead time and reduce the human involvement in preparing the particular job programs for production. Ultimately, if this type of automatic part programming can be perfected, it should be possible to economically produce one-of-a-kind products starting with the design data as the only input to the computer system.

It may require a great deal of developmental effort before automatic part programming routines are available to cover all situations. In the meantime, the programming of the machines will be done by human operators. A fourth trend that relates again to computer technology consists of the improvements being made in data and program input methods and in human-computer interfaces. The use of voice recognition systems for data input and "user friendly" computer systems are two examples of this trend. Voice input systems have been used in numerical control part programming. Savings in programming time of up to 50% are claimed for voice NC systems due to the quicker exchange of data between the programmer and the system. User friendly computer systems make it easier to do the programming for individuals not accustomed to programming. These systems lead the user through the programming or data input task by interrogating and interacting with the user.

A fifth trend which is becoming more common in computer numerical control (CNC) systems is self-diagnosis of malfunctions and breakdowns of the production equipment. These systems are designed to perform two valuable services. The first is to analyze a machine breakdown when it occurs and to identify the component(s) which failed. This allows the repair crew to quickly replace the component without spending a great deal of time trying to figure out what went wrong. The second function of the self-diagnostic system is to analyze the continuing performance of the equipment to detect clues which might indicate iminent failure of a component. This type of analysis might help to avoid an actual breakdown and resulting damage to the machine or workpart by allowing the suspicious machine component to be replaced before it fails.

These self-diagnostic capabilities will become increasingly important in the future as fully automated production systems become more prominent. As the production workload gradually shifts away from manual labor towards greater use of automated machines, the problem of maintenance and service becomes more critical.

Another growing trend in automation is the increased use of vision systems and other noncontact inspection techniques. This trend currently applies more to high production automation than to low quantity production. However, it is anticipated that improvements in the technology will allow these techniques to be applied to a greater variety of production situations. The advantage of noncontact inspection, including vision, is its inherent faster speed. It is not necessary to stop the part, position it, and physically probe the part with noncontact methods. Instead, the inspection procedure, driven by the high speed data processing capability of a dedicated computer system, can be completed in a fraction of a second. Current technology requires the system to be taught what to inspect for by using an actual workpiece during the teach routine. This procedure is quite compatible with high volume production since the inspection system, once taught, can repeat the process over and over during production. However, for low quantities of product, the teaching procedure becomes economically impractical. Future technological developments will allow the part design data to be downloaded to the inspection system without the need for a physical part. This would make it economically feasible to inspect one-of-a-kind products.

One of the probable consequences of the noncontact inspection techniques will be that production parts are inspected on a 100% basis rather than on a statistical sampling basis so commonly used at present. The implications of 100% automated inspection are perhaps more significant for high production because it permits feedback control to be employed to compensate for undesireable shifts and out-of-tolerance conditions in the manufacturing process. These kinds of adjustments would not be possible if only one product were made. However, under present practice, manual inspection often results in lengthy production delays. Automated, on-line inspection would tend to remove the inspection bottleneck. Hence, manufacturing lead time would be reduced, a longstanding objective in batch and job shop production.

The final trend in the list is the growth in the use of robots as components in flexible manufacturing systems. As improvements are made in robotics, these machines will become increasingly versatile at dealing with the variations common to low volume production. Assembly operations seem to represent an area where robots hold great potential for future applicability. The kinds of variations typically encountered in assembly include:

- Variations in position of components
- Variations in orientation of parts
- Defective components
- Different products assembled on the same line.

Several research projects have been devoted to the batch assembly problem. What is required to solve this problem is a flexible assembly system. The term adaptable-programmable assembly system (APAS) is sometimes used to describe this type of flexible manufacturing line. Robot arms constitute an important ingredient in these systems.

Technological developments in the field of robotics are tending to make these machines more humanlike. Better programming techniques, increased intelligence and memory, and improved sensors will all tend to enhance the capabilities of robots to accomplish flexible manufacturing.

References

1. Ashburn, A., Hatschek, R., and Schaffer, G., "Machine-Tool Technology," Special Report 726, American Machinist, October 1980, pp. 105-128.

2. Cook, N. H., "Computer-Managed Parts Manufacture," Scientific American, February, 1975, pp. 22-29.

3. Groover, M. P., Automation, Production Systems, and Computer-Aided Manufacturing, Prentice-Hall, Inc., Englewood Cliffs, NJ, 1980.

4. Groover, M. P., and Zimmers, E. W., Jr., "Automated Factories in the Year 2000," Industrial Engineering, November, 1980, pp. 34-43.

5. Groover, M. P., and Hughes, J. E., Jr., "A Strategy for Job Shop Automation," Industrial Engineering, November, 1981, pp. 66-76.

6. Groover, M. P., and Zimmers, E. W., Jr., CAD/CAM: Computer-Aided Design and Manufacturing, to be published by Prentice-Hall, Inc., 1983.

7. Merchant, M. E., "The Inexorable Push for Automated Production," Production Engineering, January, 1977, pp. 44-49.

Biographical Sketch

Mikell P. Groover, Ph.D., P.E., is Professor of Industrial Engineering at Lehigh University. He is Director of the Manufacturing Processes Laboratory and Co-Director of the Robotics Laboratory at Lehigh. He is past Director of the Manufacturing Systems Division and currently Vice President of Region II of the IIE.

Reprinted from the 1982 Fall Industrial Engineering Conference Proceedings.

A Machine Tool Manufacturer's View of the Factory Environment of the Future

Karl B. Schultz
Manager
N.C. Systems Development
Cincinnati Milacron

ABSTRACT

Over the last decade, all of our manufacturing capabilities have changed at a rapid rate. Even so, never before in the history of the United States have so many papers been published on the necessity to further revitalize the manufacturing industries and increase their productivity. This paper will examine the factors which contribute to the productivity in the manufacturing sector, the technical developments and management skills to create the factories of the next decade, and a snapshot of what a factory of the future might be.

INTRODUCTION

As Americans we have grown accustomed to having manufactured things available to us at reasonable prices. It seems to many of us that they have always been there, and it's almost a right, rather than a privilege to have them. The availability of these goods separate us from people of different cultures. They give us freedom from extreme physical labor, freedom of time to do higher level things.

For the last four decades, the U.S. has had an undisputed leadership position in industry. No other country produced as much or exported as much. We became a very wealthy nation with the highest standard of living in the world. Obviously, a nation that others were envious of. The temptation when things are rolling this well is to extract from the productive sector to pay for other needs. Mistakes are easy to cover because the strength of the economy would cover them. We became very short term in our thinking. We want it now! Young people want now what it took their parents a lifetime to accumulate! Business people in the U.S. want immediate return on investment now!

Things are changing a little. U.S. industry, previously unchallenged, has been successfully attacked by a nation with long term economic goals. The lack of change in U.S. industry, and the rapid rise of the industrial challenge, particularly from Japan, has sent shock waves through our economy. A once strong positive balance of foreign trade has eroded to virtually no market at all for many industries outside the U.S. boundaries. Many are still in disbelief, and some think it is a nightmare that will go away in the morning. There is a strong movement toward trade barriers and isolationism to try to protect our current industrial way of life.

The parts or workpieces that comprise the mechanical things around us are manufactured using varying processes which are influenced by the part's design and characteristics, its size, the quantity to be produced, and difficulty of manufacture to name a few. Two general classifications of manufacture are currently employed: Mass production and job-lot production. Mass production is typified by the wide use of automatic transfer lines in the automobile industry. Industries of the U.S. were pioneers in the use of transfer line technology and have made a science out of it. We have taught the rest of the world. Today, transfer lines are a common answer where a high volume of a specific product part is required. They are arranged for high efficiency, and usually are very inflexible. If there is a part design change, major revisions in the setup or major changes in the line itself are required, and are costly and time consuming.

Workpieces which qualify for mass production account for not more than 25% of the total value of metalworking production in all industrialized metalworking nations. One half of the remaining 75% is produced in job lots of less than 50 workpieces. We are also aware that the number of these parts are increasing each year and that their complexity and requirement for accuracy are also increasing. Most of these parts are produced on stand-alone equipment, some semi-automated with NC or CNC but much of it conventional. It is important in these shops to have a high degree of flexibility. The reduction of setup time is critical. The skill of the worker must be high to handle the wide range of parts he sees. Machining efficiency is much lower than with transfer line technology, but flexibility is high and the time required for new setups is relatively low.

A wide range of parts falls in the mid-volume job-lot class; lot quantities are too small for mass production techniques, but total work content per lot is sufficient to require more automation than traditional small job-lot production. Today, economic and social pressures are bringing about the need for more involvement in integrated manufacturing systems to handle this mid-volume requirement.

ECONOMIC CONSIDERATIONS

Charles F. Carter (1) has revealed that, in a conventional job shop environment, 95% of the time a workpiece is sitting in queue awaiting action from someone. The workpiece is just tieing up expensive floor space and contributing to high in-process inventory. Of the remaining 5%, in which the workpiece is on the machine, the machine tool is actually in cut less than 30% of the time. The rest of the machine time is spent positioning, loading, gaging, idle time, etc.

Much of our installed capacity is not used. Let us assume the theoretical capacity of any machine tool is 100%. In an extensive survey covering a large cross-section of machine tools in many plants, Kuhnert (2) found that 34% of the available time was not utilized due to holidays and vacations, 44% because of incomplete use of second and third shifts, 2% for just idle time, 12% for setup, loading, gaging, etc., and 2% for cutting conditions (Figure 1). This left only 6% of the total available machine time as the production fraction. For just 6% of the time the machine was theoretically available, it was actually doing what you wanted it to do to produce parts. What a potential this situation offers for increased productivity! We <u>must</u> improve on this.

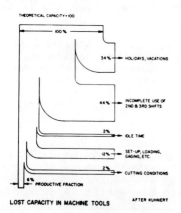

(FIGURE 1)

Using figures supplied by the U.S. Department of Labor and the National Machine Tool Builders Association, we can see that the United States is lagging behind in the annual percentage change of growth in productivity. Japan was far and away the leader with a 10.5% change in annual growth between 1967 and 1978. The United States, while still the leader in manufacturing capability, grew by only 2.6%. Japan invested approximately 40% of their GNP in equipment and facilities to improve productivity during the period 1960-1977 (Figure 2). This resulted in a productivity increase of 175%. In contrast, 15% of GNP investment for the U.S. resulted in a productivity gain of less than 10%. As a result of our limited investment in capital equipment, the aging of our metal-working equipment is getting the best of us. Using 1978 figures, Japan has an installed machine base of which 60% is newer than 10 years in contrast to 31% for the United States. In 1980, for the first time, Japan surpassed the U.S. in automobile production. This was a real shock, we had virtually believed it was our destiny to be always number one in automobile production.

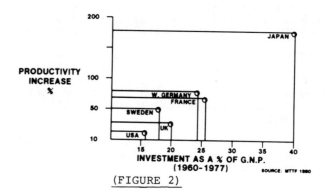

(FIGURE 2)

SOCIAL CONSIDERATIONS

Years ago, the popular view of a farm was that of a family unit working long hours for very little. The word "farmer" often had negative overtones of an uneducated laborer. At the time Cincinnati Milacron was founded in 1884, 50% of the U.S. population were farmers. By the mid 1930's that number had dropped to 20%, and today, less than 3.7% of our population provides food for not only the U.S., but exports to a large portion of the world. Today's farmer is a professional businessman who has a huge capital investment in his facilities and equipment and is automated to the hilt. He must do this to compete.

In 1947, about 30% of all people employed in the U.S. worked in manufacturing. By the end of 1980, that number had dropped to

22%. By the end of the century, The Rand Corporation forecasts that perhaps 2% will be engaged in manufacturing. The 2% figure seems low, but the trend is clear, it shows that fewer and fewer people will be producing more and more of our manufactured products. It is the same thing we saw in the agriculture industry. Fewer and fewer people wish to be a part of the industry under the same conditions and technologies that have existed. A higher style of life is desired.

Dr. Eugene Merchant (3) argues that "The long term economic force at work in industrialized countries is the continuing need to reduce the cost of creating real wealth. The service sector of a nation's economy being non-wealth producing depends for its support on the principal wealth producing sector -- the manufacturing sector. Thus, increases in the standard of living, quality of life, employment, and the general economic well being of a country stem directly from decreases in the cost of wealth production." If we want the quality of life that we have grown to expect, we must improve our ability to produce the wealth that supports it all. We must do more to improve our manufacturing capability. This will require more than just new equipment. It means a sharp departure from the way we have traditionally done things.

The output per manhour has risen slightly in the last 30 years. But the hourly compensation of each of our employees has risen sharply. This produces a unit labor cost which looks quite dramatic when plotted on a graph.

If we wish to remain competitive in the world marketplaces, we must find ways of reducing this unit labor cost.

It is interesting to observe that all of the mechanical and electrical technologies to build an integrated system exist in today's world and have for some time. NC machine tools, part transport, computer hardware and associated operating system software have all been functioning well in various industries and applications independently for years. The advances in computer technology and the current ability to obtain a lot of computer power at reasonable cost has helped make integrated systems a reality. The "glue" provided by the system supplier is made up of interfaces and controlling application software so all of the components work together in an integrated fashion.

At the 1982 International Machine Tool Show in Chicago, several FMS type systems fitting all the characteristics discussed were shown by various vendors including Milacron. At the 1980 Show Cincinnati Milacron demonstrated a VARIABLE MISSION manufacturing system (Figure 3). The observer saw the high spindle utilization of the machines, the

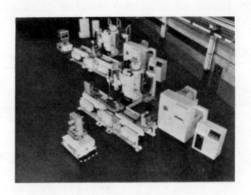

(FIGURE 3)

motion of the transport system and the tool management system. All of this worked in an unmanned environment. What was not easily visible was the hierarchy of computer software control that allowed it to operate so effectively under the many conditions affecting it. The development of modular software for integrated systems is a major, maybe the major integrated systems effort at Cincinnati Milacron. The same is probably true for most systems suppliers. While the cost of computer power is dropping, the cost of software and its maintenance is rising. It is critical for the future of companies who may use systems as well as system producers to employ a hierarchy of software tools which will grow and change to suit a company's changing needs. The hierarchy of system software is the tool used by management to communicate with the system and to control it. It must be flexible enough to fit into a user's methods of management. The communication with integrated systems is where new management capabilities and ideas must be developed.

A properly configured integrated system will provide the capability to produce as it was designed. Any influences which affect this productivity will become apparent quickly because of computerization, and management must be able to respond quickly. System management's ability to plan and respond correctly and quickly will play the largest role in the productivity of any integrated system. Proper selection and training of system managers is vital.

The integrated system must be designed in a fashion to allow two forms of flexibility: the flexibility to manufacture the range of parts for which the system is intended, and the flexibility to change, adapt, and grow to accommodate the changing influences on the system users. These changing influences can be changes in a user's product requirements or changes in the technologies of the components.

INTEGRATED (VARIABLE MISSION) MANUFACTURING SYSTEMS

Integrated manufacturing systems have been described by some as the panacea for the world's manufacturing problems. An integrated, or VARIABLE MISSION as we call those we build at Milacron, manufacturing system is at least several CNC machine tools connected by automated workpiece handling devices, the entire system being integrated under the control of a hierarchy of computerized functions with man as the administrator. The potential exists for significant increases in productivity and reductions in workpiece in-queue time. The prime target for integrated systems is the mid-range quantity lots although high quantity and low quantity job-lot work can be accommodated. Integrated systems are in themselves not a panacea. Properly managed, they can be highly productive and very economical, but there are many considerations.

Systems are capital intensive. Although they are designed to operate under light manning conditions, they do call for various support services. They require a high degree of skill from both technical and management people. The learning curve must be considered, and there must be a way of predicting performance. I will not discuss details of justification of an integrated manufacturing system except to say that they have been designed to alleviate the social and economic problems that we have just discussed. Justification methods used by most companies do not account for many of the so called "intangible" costs associated with these social and economic problems. Integrated systems will reduce direct labor, they will increase labor productivity, they will reduce floor space and in-process inventory, and they will improve your product flexibility along with many other things. The justification, however, will be unique to a specific user's applications.

The steps leading to the decision to install an integrated manufacturing system require a long term view of one's corporate plans and strategies and specifically product strategies. Questions need to be answered: What are the types of products to be produced? What parts will be manufactured? What are their sizes, attributes, tolerances required, quantities to be produced? Attention must be paid to manufacturing philosophies. What controls the input of parts to a system, or in other words what "pushes" the system? What will be required of the output, or what are the "pulling" forces? What importance is placed on minimizing in-process inventory? What are the tooling and fixturing problems? Questions concerning the life of the products and associated parts to be manufactured on a system must be considered to determine the degree of flexibility required. How will the social and economic forces we have discussed affect this specific application? It could well be that a system is required because of outside influences such as machine tool operators not being available in the labor market or one's competition reducing their unit cost. The questions asked initially have more to do with manufacturing strategies and scheduling than the specifics of the system machine tools and part transport components. An integrated system is a management tool which is part of, and must fit a company's management scheme.

Within the mid-volume automation segment, there are three separate and distinct automation types:

1. CELL SYSTEMS

Cell systems most closely resemble a stand-alone machine. Two lathes and a robot or a machining center with a carrousel work changer would be examples. They are typically designed to handle a confined family of parts, often loose rather than palletized, in a predefined sequence. The sequence is loaded into a programmable controller, and since the sequence is fixed no scheduling or traffic control is required. A customer would want a cell system if he ran reasonably sized job lots and either did not mind the set-up time required to change from one job to the next, or because of job lot size did not have to. Turning tends to fit well in the cell environment because the parts usually must be handled loose, and they also tend to have lower cycle times and come in higher quantities. The technology of holding and handling loose parts also tends to force turned parts into the cell environment. There are, however, many factors which determine whether a cell or other form of automation is to be used, and generalizations may be suspect. Many people look with great favor on the cell as a general systems solution because it does appear as a stand-alone machine, does not depend on computer science to drive it, and appears to have comparatively little impact on the user organizations supporting it as they stand today. A cell system can provide improved spindle utilization and is best suited for reasonably sized batches to be run in sequence. Cells do not adequately address the in-process inventory problem and the just-in-time philosophy currently popular.

2. FMS OR VMM TYPE SYSTEMS

Cell systems have a lot of machine content and limited computer control. VMM systems have both machine content and computer control. The object of the computer control is to reduce in-process inventory by entering parts into the system on an as-needed basis while getting the best possible utilization of the machine tools and tooling. This means that a wide mix of workpieces and tooling will be moving through the system at any point in time. It's this "random" capability that requires the computer control. Unlike the Cellular System, a VMM system is a factory in itself complete with scheduling capability. The computer does far more than drive machines, it provides systems management. This is the prime distinguishing feature between cells and VMM. The user gets not only better spindle utilization, he gets lower in-process inventory.

The machine tools, and in some cases the material handling devices, may look the same in cells and FMS systems, but the applications for the systems are different based mainly on quantity or parts required and on manufacturing philosophy. A VMM system is designed to handle the lower volume end of mid-volume manufacturing. Many different part types run simultaneously on the system, and the user is striving for a limited manned operation, high spindle utilization, and lower in-process inventory.

3. FACTORY DATA AUTOMATION

Factory Data Automation is the information and management control vehicle in the blue collar segment of the factory. Those interested in factory data automation are vitally interested in improving thruput in the shop and reducing inventories. They are working to reduce the 95% of the time the part is sitting. Their interest is in their manufacturing process as a whole: What to do when! This includes discrete scheduling, tool allocation and planning, tool crib set-up, automatic down loading of tool and set-up data, down loading of N.C. Data (DNC), management and maintenance reports to name a few. Conventional and N.C. machines, cells, and VMM systems can all be linked in a hierarchy with factory data automation. Prime emphasis is on the control and management of the system as a whole.

As a system vendor we are acutely aware of these requirements. Cincinnati Milacron has employed a philosophy coined FM3 for the development of its VARIABLE MISSION manufacturing systems. FM3 stands for flexibility, modularity, maintainability, and manageability. Simply stated, the FM3 philosophy means that the mechanical components, computer hardware, and computer software were all selected or designed to meet the requirement of the second flexibility need: the need to adjust to changing influences.

There has always been a cost penalty associated with flexibility. It's always easier to visualize and design for a specific well defined requirement than it is to try to visualize and incorporate all contingencies into the design. While from an initial design point of view it might appear cheaper to go for a specific, or "dedicated" system, it might well be very expensive in the long run if the system cannot be economically adapted to meet changing needs. This has an effect on corporate management as a whole. Decisions are made or not made based on their manufacturing department's ability to respond. Many opportunities are lost due to the inability to adapt.

Specifically what does a specific user need to do to get involved with integrated systems? Integrated systems have champions. People with foresight and the knowledge, ability, and patience to bring them to reality. A user needs a champion to insure an orderly and successful installation. There is still the aura of magic surrounding computerized systems, and many people have anxiety problems when dealing with them. Good people often see their entire careers passing in front of their eyes when a system is mentioned. Proper selection of a dedicated champion and systems team is vital.

Where do systems apply? One of the first tasks is to review the requirements and the future effects of the economic and social trends affecting the production to fulfill the requirements. Integrated systems as mentioned earlier fit well in mid-volume manufacturing requirements where the need to reduce labor costs for both production and set-up exists. The person studying this must understand the effect a system will have on the processes feeding the system with raw material and the processes downstream accepting the output. He must also plan for interfaces with management control systems and provide proper support systems, personnel, and tools.

Build a relationship with a vendor. Once an area and a basic approach is planned, a study of what the integrated system vendors have to offer must be made. We all have our specific attributes and will be glad to discuss them. A basic decision then comes to the surface: Should the company develop its own integrated system? Chances are the study showed that there are pieces of every system that's desirable, and some software engineer is suggesting that given one year, a few bucks, and a cot to sleep on, he can have the whole thing. There can be plusses to doing one's own, but in just about all of the cases I have seen, even the most sophisticated users build themselves into narrow boxes eventually. It is best to choose a vendor with a reputation in the integrated systems field. Take advantage of their experience across a spectrum of applications. This relationship becomes a marriage and just like any marriage involves differences of opinion and a few squabbles, but also provides a good relationship which leads to a solid long term approach which can adapt to the pressures of change.

Integrated systems vendors will work with a

potential customer to configure a system to meet his needs. They will review the information provided by the customer and provide proposals to meet the needs described. Cincinnati Milacron, for example, provides three levels of proposals. The first, or "Concept" proposal is quite basic, provides a system layout and is designed to be a talking paper to see if the vendor and potential customer are thinking alike. Several discussions are held at the concept level and much insight is gained by both vendor and customer.

The second, or "Preliminary" proposal gets into more detail. It provides a detailed layout, tooling and process studies, details on system operation and facilities. It also includes simulations of the system to give both the customer and vendor assurance the system will perform as configured. An implementation plan is sometimes provided, and if a phased installation is proposed, some detail on how that is to be accomplished is given. A "preliminary" proposal gives a fairly detailed description of the workings of the system and a very good estimate of costs. After discussions, the customer and vendor will make final adjustments and move on to the last step.

The "Final" proposal involves much of the work of the preliminary plus any adjustments made. It becomes a contract between vendor and customer on what each is expecting, how and when each phase of the installation is to occur, who is responsible for what, training, post installation service, and the financial terms of the contract.

During the development of these proposals, both the customer and vendor gain tremendous insight into the application. New ideas are developed and tested. It is important that the customer's systems team be closely involved. Much learning is taking place at this time. It takes a lot of time and effort from both parties to put together studies of this type. A serious sincere effort on the part of vendor and customer is required.

Integrated systems vendors have a lot of information and guidance available because of previous installations. This can provide great assistance to the customer. Customers should use it wisely, and they own the system. A successful installation is when the customer accepts the system as his, and uses his capabilities and capabilities available to him to make it work for him and his company.

As we look into the future, we see much development work taking place on elements of integrated systems. The entire fields of sensors, machine and tooling diagnostics, systems scheduling, tool management, system modeling and simulation among others are alive with activity.

As we move through the 1980's, integrated systems will invade the mid-volume manufacturing arena. The way we do things will have to change to accommodate this automation, and this automation will be required to stay in business. Proper planning now will allow a company to take advantage of these capabilities economically as we proceed through this decade.

Forty-two years ago, Mr. Frederick V. Geier, then President of Cincinnati Milling Machine Company said, "Our position is different from that of ten years ago. We have grown larger. We have a great name to live up to. This next decade will test us day by day." This statement is relevant to each of your companies as well as mine, and it is just as relevant today, at the dawn of the eighties, as it was in 1940. This decade will present challenges far greater than we have experienced in the past. It will test and change the way we manage our manufacturing processes. Integrated manufacturing systems will be at the heart of this change and we must be prepared to use them to our best advantage.

REFERENCES

1. Carter, C.F., Jr. "Trends in Machine Tool Development and Application", Proceedings of the Second International Conference on Product Development and Manufacturing Technology, University of Strathclyde; Macdonald, London (1972).

2. Kuhnert, H. "Study on Efficiency as Seen from the Consumer Point of View" (in German); ICM-Festschrift, VDW, September, 1970.

3. Merchant, M.E., "Economic, Social and Environmental Considerations for Computer-Automated Manufacturing Systems", Proceedings of CIRP Seminars on Manufacturing Systems, Vol. 4, PP. 247-256 (1975).

BIOGRAPHICAL SKETCH

Karl B. Schultz, Manager, N.C. Systems Development, Cincinnati Milacron.
Mr. Schultz is currently involved in the Development and Systems Design of Milacron's Variable Mission Manufacturing Systems. He is a member of the NC Society and recognized as a certified NC manager, and is a member of SME and recognized as a certified manufacturing engineer.

Factory of the Future/IE of the Future

William R. Welter
Management Consulting Division
The Austin Company

ABSTRACT

The purpose of this paper is to describe the manufacturing strategy and major components of the factory of the future, then to investigate the implications for the men and women who will be called upon to implement those concepts. For the purpose of this paper, the factory of the future is simply that which we believe will exist in the mid-1990s.

The dire straits of the present manufacturing community and the challenge of the future have received considerable publicity. In fact, the "factory of the future" has been promoted as a panacea for America's economic ills. This paper looks at and beyond the factory of the future and presents an answer to the question "So what?"

More specifically, this paper addresses three subjects:

- the organizational mix of factory of the future -- <u>what</u> is coming

- manufacturing strategy -- <u>how</u> it will be achieved

- the future role of the industrial engineer -- one of many <u>who</u> will be affected either positively (through heavy involvement) or negatively (by being left behind).

There are three objectives for this paper: to present a brief description of each of the three subjects, to show the evolving relationship among the three subjects, and to suggest ways we as industrial engineers can expand our areas of expertise to meet the challenges of the factory of the future.

FACTORY OF THE FUTURE

Describing the factory of the future is like making any other forecast -- it's bound to be wrong simply because it deals with the unknown. For the purposes of this paper, however, we'll consider the factory of the future as a composite of the factories that will exist ten years from now. Ten years was chosen because it is far enough into the future to plan and implement major changes, yet it's timely enough to avoid a science fiction connotation. In the factory of the future we will see:

- "lights out" factories

- factories essentially the same as today

- factories with dramatically different

  -- production technologies, such as robotics, FMS, and lasers

  -- material handling and storage methods, such as ASRS and wire guide vehicles

  -- information integration and use, such as CAD, CAM, CIM, and MRP II. The ability to act as a single system depends on the ability of all subsystems to share data. Shared data is the integration "glue" that holds the factory of the future together.

What does "factory" in factory of the future encompass? Only the shop floor? Engineering? What else? In other words, what's the scope of the factory of the future? The United States Air Force ICAM project addressed this subject and defined a factory as "a set of functions that can be logically grouped." (1) The functional groupings that they chose were:

- management
- customer liason and service
- product planning and production planning

- resource procurement
- product production
- logistics support. (2)

As shown in Figure 1, the scope of the Air Force certainly goes beyond what many of us consider as "the factory." That's because the ICAM project examined the business of manufacturing, not just the physical process. Furthermore, as a result, Figure 1 shows that <u>what</u> is done in a factory will not change -- the <u>how</u> will.

- We have to be aware of changes in the mix of tasks. For example, we traditionally thought that low cost, high product performance, and quality were mutually exclusive. The automotive industry learned the error of this assumption at the hands of foreign automakers.

- All companies have a manufacturing strategy, whether they realize it or not.

FACTORY OF THE FUTURE FRAMEWORK

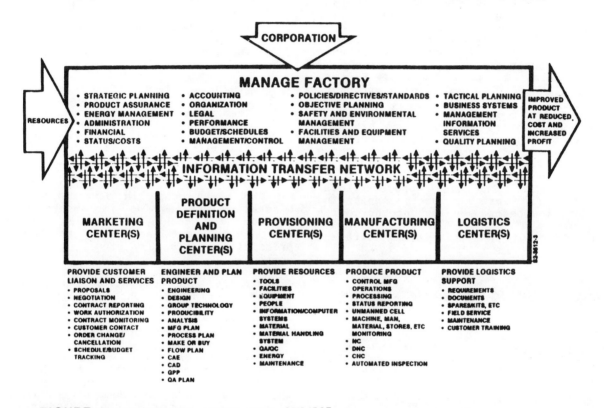

FIGURE 1 SOURCE: AIR FORCE ICAM PROJECT

But why change? Why invest the hard work to go from today's factories to the factory of the future? The obvious answer is that manufacturing tasks are changing. A recent article in <u>Harvard Business Review</u> states that ". . . the tasks of a manufacturing system . . . can be defined in terms of requirements for cost, product flexibility, volume flexibility, product performance, and product consistency." (3) The article then presents a case for tying manufacturing policies to product strategies.

Although the article focuses on companies operating in the international environment, it raises points relevant to all manufacturers:

MANUFACTURING STRATEGY

What is manufacturing strategy? The word "strategy" has its origins in ancient Greece and simply meant "the art of the general." Centuries later, the Prussian General Carl von Clausewitz, considered by many to be one of the great students of the art of war, wrote: "Strategy forms the plan of the war, and to this end it links together the series of acts which are to lead to the final decision: that is to say, it makes the plans for the separate campaigns and regulates the combats to be fought in each." (4)

Are plans the strategy? No. Are the "series of acts" the strategy? No. Simply put, strategy is the scheme or general approach that's used to win. From a business point of view, strategy is the way a company decides to compete. At a another level, manufacturing strategy is the manner in which manufacturing operations support the business in competition. This sounds simple, but as von Clausewitz said: "In strategy everything is very simple, but not on that account very easy." Why? Because companies are complex systems, not simple organisms. Consider, for instance, a company that has decided to use the strategy of low prices to compete. To support this strategy, the manufacturing division has to be the low-cost producer, or the company has to be willing to forego profit. But what general approach should be used? What quality levels? What transportation costs? What inventory levels? What are the trade-offs? Obviously these questions cannot be answered here. The point is simply that an approach is needed to determine a strategy.

But first, a goal is needed. It would do no good for the manufacturing division to focus on high-quality levels if cost was the chosen method of competition. As shown in Figure 2, determining the mission of manufacturing is part of the manufacturing audit process. The audit process identifies where you are and what you're trying to accomplish before you lay out the course of action. The audit is the essential first phase of a manufacturing strategy project.

MANUFACTURING STRATEGIC PLANNING

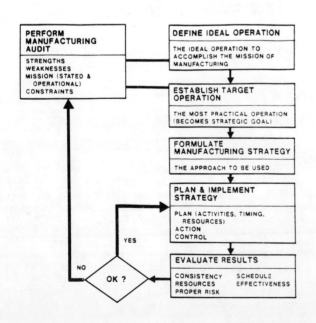

FIGURE 2

The next phase in developing a manufacturing strategy involves the iterative processes of:

- defining an ideal operation -- doing the research to determine the state-of-the-art equipment, processes and management approaches available. This step involves examining competitors, suppliers, and research processes.

- establishing a target -- looking at what's the best manufacturing operation you can design, taking all constraints -- financial, managerial, and operational -- into consideration. This target becomes the strategic manufacturing goal.

- formulating a strategy -- designing the best approach to reach this goal, considering prime cost drivers and other financial factors. Note: This is not a linear process; it's iterative and closely related to the previous steps.

The third phase in developing a manufacturing strategy is implementation and includes the tasks of planning (deciding how to implement the chosen strategy and prioritizing necessary steps), action, and control.

Finally, an evaluation phase is required. During this phase, progress toward the target operation is examined. If the target operation has not been reached, then the reasons why are investigated.

Unfortunately, few companies have an ongoing process of manufacturing strategic planning. This may have been acceptable in the '50s and '60s when companies could sell all that the factories could produce; however, this same process in the '70s and early '80s sounded the death knell for many companies.

<u>THE INDUSTRIAL ENGINEER</u>

The role of the industrial engineer is difficult to describe because of the variety of duties assigned. Therefore, for the purposes of this paper, we'll use the latest <u>Handbook of Industrial Engineers</u> as a reference. The handbook defined these functions for which IE managers are responsible:

- facilities planning and design
- methods engineering
- work systems design
- production engineering
- management information and control systems

- organizational analysis and design
- economic analysis
- operations research
- work measurement.
- wage administration
- quality assurance. (5)

In the same book, industrial engineering was defined as "concerned with the design, improvement and installation of integrated systems of people, materials, equipment, and energy." (6) If this definition is accurate, then the relationship with the factory of the future and manufacturing strategy should be self-evident.

INTERACTION

The relationships among factory of the future, manufacturing strategy, and industrial engineering could almost take on the aura of a truism. After all, if something is going to happen, it should be planned and someone should be charged with bringing it about. However, unlike some truisms, this one will not happen naturally.

Manufacturing management can approach the factory of the future in different ways. They can approach it from the point of view of "business as usual" and just let it evolve. But then the goal may never be reached, or if it is reached, it will happen after much of the competition.

The opposite approach is the one that I hope will be taken by the American manufacturing community: to plan and implement the factory of the future in an effective and timely manner. This planning process, then, will tie manufacturing strategy to the factory of the future.

But what about the industrial engineering department? What is their relationship with the factory of the future? The answer lies with the basic requirements needed to construct and implement strategy. Strategy requires the ability to view and understand the manufacturing operation from the top down. In other words, to understand the "big picture."

None of us are naive enough to think that the industrial engineers will be solely responsible for the factory of the future; that responsibility certainly lies with company management. However, company management has neither the time nor the full range of expertise needed to accomplish the task. In other words, they will need to delegate much of the work. And much of that work may go to the company's industrial engineers -- if they have the expertise needed to help management. If the expertise is there, they will be heavily involved. If not, they won't. It's as simple as that.

IMPLICATIONS

The factory of the future is going to happen. We can be sure of that. Successful companies won't get that way by accident. We can be sure of that. And things are going to change for the industrial engineer. We can be sure of that.

The question before us now is "How do I make the industrial engineering function more valuable to my company as it addresses the challenge of the factory of the future?" The answer might best be found by doing some contigency planning in the very beginning and addressing the question of "What can go wrong?" Typical causes of strategy failures include the following:

- inaccurate situation assesment
- unforseen change in the environment
- inadequate resources for the job.

Understanding the causes of failure can help you avoid it. For example, when dealing with a situation assessment, determine what the mission of manufacturing is at your company. Determine what is the one thing that manufacturing has to be best at in terms of cost, quality flexibility, and consistency to support your company's competitive strategy.

When confronted with changes in the environment, consider the following excerpt from a recent issue of *Harvard Business Review*:

> "Conventional wisdom has it that manufacturing equipment: is directly and exclusively appli-cable to limited sections of the production process; has capabilities that are known and stable; produces benefits that are best judged by engineers and supervisors in affected operations; and makes a contribution to process efficiency and cost reduction that can be closely measured at the outset. None of this wisdom applies to CAM." (7)

If the author is correct (I believe that he is), then the work that many industrial engineers do in the area of new equipment justification may be altered significantly.

Finally, as an example of a needed resource, consider the growing need for group technology (GT) expertise in manufacturing companies. The use of GT will affect design engineering, layout, machine tool purchasing, planning and processing, and purchasing. Someone has to know enough about GT to assist in these areas and point out the interrelationships. Should it be the industrial engineer? My answer is "Why not?"

RECOMMENDATIONS

Factories of the future will not need specialists as much as knowledgeable generalists who can:

- solve problems
- explain results
- learn by experience
- determine relevance.

To become a knowledgeable generalist:

- Continue your education. Many of us, for example, know far less about the uses of computers than do new industrial engineering graduates. Learn about the new technologies that are evolving.

- Challenge your assumptions about how manufacturing operates. Things surely have changed since you've been doing your job.

- Expand your view of the company. Become knowledgeable about your customers and competitors, the mission of manufacturing, and how your company competes.

While it's a fact that manufacturing is changing, we can either be at the leading edge of our profession, or we can be left behind. The choice is ours.

WILLIAM R. WELTER

William R. Welter is a Manager in the Production and Operations Group, Management Consulting Division, The Austin Company.

He received a B.S. degree in Industrial Engineering from the University of Illinois and an M.B.A. from DePaul University. He is certified in production and inventory management, through examination, by the American Production and Inventory Control Society.

Mr. Welter joined the Management Consulting Division from Sealed Air Corporation where he was Manager of Technical Services. In this role, he was responsible for providing manufacturing expertise to support the company's efforts in automation at customer sites.

Mr. Welter was with the management consulting services group at Ernst & Whinney for four years. During this time he assisted companies in a variety of industries to improve performance in the area of manufacturing controls. He previously worked with Uarco, Inc. and Stewart Warner Electronics in various engineering and production positions.

Mr. Welter is an active member of the American Production and Inventory Control Society, IIE, and Technology Transfer Society.

BIBLIOGRAPHY

(1) "ICAM Conceptual Design for Computer Integrated Manufacturing, Task B Establishement of the FOF Framework." Needs Analysis Document. Wright-Patterson Air Force Base: ICAM Program Office.

(2) Ibid.

(3) Stobaugh, Robert and Telasio, Piero. "Match Manufacturing Policies and Product Strategy." Harvard Business Review, March-April 1983.

(4) Leonard, Roger Ashley. A Short Guide to Clausewitz On War. New York: Capricorn Books, 1968.

(5) Salvendy, Gabriel. Handbook of Industrial Engineering. New York: John Wiley & Sons, 1982.

(6) Ibid.

(7) Gold, Bela. "CAM Sets New Rules for Production." Harvard Business Review, November-December 1982.

Reprinted from the 1984 Fall Industrial Engineering Conference Proceedings.

Change the System or Change the Environment

Hal Mather
Hal Mather, Inc.
Atlanta, Georgia

ABSTRACT

The idea that complex systems can help us control the complex world of manufacturing has largely been a failure. The real successes are few and far between.

A better idea is to simplify the manufacturing environment. Better planning, improved layout, designs that are not only producible but also give low inventories and high customer service, faster changeovers, better supplier relationships, fewer upsets, and better risk/reward measurement systems, have far more potential for real and lasting improvements.

INTRODUCTION

Software and hardware for planning and control of manufacturing have been installed at a phenomenal rate over the past 15 years and this rate shows no signs of slackening. Ever more complex systems are being developed and marketed as the solution to the perceived conflict among high customer service, low inventories, and efficient production. The truth is the money invested in manufacturing control systems to date is far in excess of the benefits.

Could it be we have focused on the wrong solution? Instead of developing complex systems to manage a complex environment, shouldn't we have simplified the environment? For those companies that have done this, the systems now needed to manage are also simple. The synergistic relationship of a simpler environment and simple, effective systems has provided enormous payback for the companies that have taken this approach.

If this idea is introduced carefully into a plant, it can instill a way of life in the work force to attack and solve problems, resulting in higher and higher levels of improvement. Let's develop this idea further by looking at some key environmental problems and their solution.

ENVIRONMENTAL OPPORTUNITIES

Layout

A company installed a computerized inventory system in the late 1960's. They added sub-systems and expanded hardware over the next 12 years until they had close to a state-of-the-art manufacturing control system. Benefits were considerable but problems remained.

After a thorough evaluation of the environment they decided to change the factory layout. Two years later, factory inventories and scrap rates are down almost 50%, productivity is up over 30%. A throughput time reduction of over 50% has improved their ability to meet marketplace changes. The information needed to run the plant is minimal with the new layout. The system is simpler and easier to control.

The layout change was from a functional grouping of like machines and processes into a product grouping or process flow layout. Parts are now made on a production line under one supervisor's direction. Detailed scheduling, needed with functional machine groups, has almost disappeared. Responsibility and accountability are very clear.

Take a tour of your plant. Start at the receiving dock and follow the flow of production until you reach the shipping dock. If the flow seems logical and efficient, then your layout is probably not the problem. But if it's like most plants you will think the person conducting you on the tour is either lost, drunk, or enjoying a huge joke at your expense. Straightening out this maze could be the best business improvement program you'll ever start.

The objective of every manufacturer is to receive the correct materials from vendors and quickly move them to the shipping dock, transforming them along the way. The quicker and more efficient this process, the higher a company's return on investment.

A good analogy is travelling by automobile. If you want to move quickly from point A to point B by road, you always look for the nearest highway. Traffic moves safely and fast provided the road's capacity is not overloaded. Little in the way of "system" is needed to control your movement or that of those around you.

The worst road system to move you quickly from point A to point B is a downtown city center. Traffic moves in every direction; complex systems of one way streets, stop lights and police try to cope, with limited results. Controlling the stop lights with computers provides some relief but traffic still doesn't move as quickly and safely as on a highway.

Do you have highways linking receiving and shipping in your company or a downtown city center? Attacking your layout to get highways would not only improve your product flow but also simplify the control process.

Changeovers

The cornerstone of the Just-in-Time philosophy is fast changeovers. For example, Buick has recently changed some set-up times in one of their metal stamping plants from 6 hours to 18 minutes. This change has reduced inventories at the same time as flexibility has increased. Many companies have made similar gains, the amazing thing is, without huge capital investments. A better engineered changeover process is often all that is required for dramatic reductions in this nonproductive activity. All new tools and machines must be bought with short changeover times as one of the specifications. This is rarely done but is a fundamental environmental change.

The impact this change has on the total product flow is not readily apparent. But large batches, run because of long changeovers, cause uneven work loads on machines. Long queues of work are a natural result, there to absorb the peaks and valleys in work flow caused by the batching logic. Reducing changeover times allows all batch sizes to reduce, resulting in a remarkable improvement in the flow of material. Faster throughput with lower queues gives higher flexibility and lower inventories, and there's no production efficiency loss.

Using the highway analogy again, erratic flows occur at rush hour time, causing traffic jams and delaying us all from reaching our destination. This happens even though there is huge excess capacity during the day or at night.

Large batches flowing through a plant are equivalent to rush hour traffic. They demand huge amounts of capacity in a short period, causing queues and delaying all other products.

Smoothing this flow out with many small batches will stop this "pig through the python" syndrome. Reduced inventories, and higher reliability of deliveries will be the result.

Master Production Schedules

A master production schedule (MPS) is a future plan of what a plant should make, how many and when. It is the short range company game plan. It should be a realistic plan, meaning the factory and vendors must have the capability to execute the plan.

But many times, the comparison of what needs to be made and a plant's capability to produce it, is not done or is done poorly. An overloaded MPS is the result.

Resources now have more work to do than they are capable of. This is the most damaging environment possible to get low inventories, good customer service and a productive plant.

To explain why, assume we have four workcenters, each with 250 standard hours per week capability, but with a list of needed jobs of 300 standard hours per week. These four workcenters are fabrication workcenters, all delivering parts to an assembly area.

All supervisors have choices to make. They can only produce 250 standard hours of work per week but are asked for 300 standard hours worth of different parts. It's obvious that 50 standard hours of parts won't get made each week. What may not be quite so obvious is the formal system of priorities is now impotent. And the manual informal system started to try and get matched sets of parts into assembly cannot possibly succeed. Hot lists, shortage meetings, and expeditors cannot keep up with all the combinations and permutations of parts, engineering changes, master schedule revisions, scrap, late vendors, etc.

Each supervisor generates his 250 standard hours but they are unmatched sets of parts in assembly because the pressure to produce forces the supervisors to "cherry pick" the easiest jobs that give them most output. Expeditors are added, more supervisory time is spent chasing shortages instead of managing, costs and inventories increase, and assembly slows down. Output to the customer drops at the same time as inventories grow.

This is analogous to an overloaded highway where all traffic crawls along at 5 mph. An emergency vehicle can always get through, but not very quickly, and with the penalty of all other vehicles moving aside and waiting longer. Few people reach their destination on time, tempers are high, and fender benders are common.

The master schedule has to be realistic. An excellent MPS is one the factory exceeds regularly by a very small margin. A poor MPS is one the factory fails to achieve by any margin.

The MPS also has to be balanced. By balanced, I mean it has to provide a level amount of work each week to all work centers. An unbalanced MPS means erratic loads on work centers just as with the large batch sizes mentioned earlier.

For example, our task this month is 200 product A, 100 product B and 100 product C. We have the capacity to produce 100 per week of any of these products.

A normal MPS would show 100 product A in week 1, another 100 A's in week 2, 100 product B in week 3 and 100 product C in week 4. This would be done for efficiency, changeover and learning curve reasons. But as these are different products with different components using different routings and different standards, the demands on the feeding work centers will not be level.

A better balanced MPS would be 50 product A, 25 product B and 25 product C each week. Of course this can only be an efficient plan if changeover times have been reduced as described earlier. The optimum is one A, one B, one A, one C, one A, one B...etc. If there are no lot sizes underneath the MPS then demands on all the resources would be exactly level.

Design

Conflicts between inventories, customer service, and efficient production are often created in the basic design of a product. This is because the engineer's objective is to design a product to perform a function at an acceptable cost. Nowhere do we also specify "with minimum inventories and maximum customer service".

A company in Europe designed a new line of consumer electronic products. Six bare printed circuit boards could be made into 22 different stuffed boards. These 22 boards could in turn be assembled into 56 finished products. Some printed circuit board components had lead times of 10 months. It took 2 months to assemble the printed circuit boards, 1 month to assemble the finished products and 3 months to distribute them to the end consumer. The design outperformed the competition and could be produced to sell at an acceptable price with satisfactory profit margins.

Now the bad news. The long lead time components were unique to the 22 finished boards, so predictions at this level of detail had to be made 16 months (3 months distribution, 1 month assembly, 2 months PCB assembly, and 10 months vendor lead time) into the future to order these components. You're right, they had excess inventories of some and severe shortages of the others. The first assembly operation on the printed circuit boards added the parts unique to each configuration. So decisions of which printed circuit board to assemble had to be made six months before the consumer bought. A similar condition existed for final product assembly. Finished goods inventories were excessively high and unbalanced, with many backorders. Retailers reported sales of competitive products because this company's products were not available. Factory inventories were also high with much confusion and extra costs caused by expediting to the latest crisis. Targetted costs were being exceeded almost everywhere.

A review of the design from a planning and scheduling viewpoint showed that the long lead time purchased components, unique to each printed circuit board, could have been made common to the six bare boards. A specification change could also have reduced the component lead times from 10 months to 2 months. The common parts could have been assembled to the printed circuit boards first, with the uniqueness left until the last moment. The same was possible for the finished goods. There would have been a calculated cost penalty, which the reduction in inventories and improvement in customer service would have paid for in a matter of months.

There's a saying, "If you build a better mousetrap, people will beat a path to your door". This is undoubtedly true, but if you don't have the one they want, they won't come back. "Uniqueness at the last minute" must become a catch phrase for all design and industrial engineers and we must learn to rate designs and the process on this capability.

Supplier Relationships

A supplier is a work center not under your roof. Therefore, the same problems of overstated master schedules, poor layout, and erratic flow occur for your supplier as for you. Late deliveries, long lead times and strained relationships are the inevitable result.

We also play off one supplier against another, trying to get lower prices. This adversarial position, common to many customer/supplier relationships, rarely gives the best results. Neither does multisourcing from several vendors.

A division of a large, multinational corporation decided recently to reduce their approved supplier list by 75%. Their reasoning is this will put more business in fewer hands, simplifying communications problems. With more business going to fewer vendors, a more stable and level flow of business will be requested. Of course, the vendors must be willing to allocate increased capacity to this customer, otherwise this is useless.

Early results prove that the small cadre of suppliers are more flexible to this division's needs. Quality has improved and on time delivery performance is better. Suppliers are helping in cost reduction and design improvements. A better working relationship now exists between customer and vendors, sure to result in more improvements in the future.

Your suppliers must be viewed as an extension of your business. This statement is often made but rarely carried out successfully. Bad vendors usually deliver to bad customers. It's important to get your own house in order before approaching vendors to get better. Select the best method of working with them to improve the total business, not just get today's lowest price.

Disturbances

No company can function well with many surprise upsets. Systems can cope but people cannot. Too many changes overload the supervisors and indirect people needed to implement these revised plans. We rarely consider indirect or supervisory capacity, but these people have capacity limitations, just as direct labor people do.

You have three ways of using this capacity - to develop good plans, execute good plans or make changes to the original plans. It is obvious the first two give major payback, so reduce the disturbances that force resources into a change mode. Having to make changes is an admission of failure, either in plans or execution. Poor quality, late vendor deliveries, short term engineering changes, machine breakdowns, bad records and so on, are all disrupting influences. Attack these problems and reduce their incidence. The capacity you will free up to concentrate on better planning and execution will pay you back many times over.

Measurements

"People do what you inspect, not what you expect", is one of my favorite sayings. Our traditional "inspection" methods, the accounting system, may stimulate efficient production. The problem is, it gives kudos regardless of the need for the items being produced. The customer may not want what is being made or we may already have plenty of it in stock.

Our measurement system needs a complete overhaul. It must drive us towards good business performance. Currently it drives us towards high stocks and poor service because of its bias to rewarding low purchase prices and efficiency. Traditional accounting systems don't even have a comprehensive measurement system to help us focus on the real ways to control inventory and improve customer service. All we get is a statement of actuals after the fact.

The financial score card must become secondary to a set of business performance indicators. Credit should be given for low stocks and high customer service at the same time as we get efficient production and low purchase prices. A company's improved return on assets will show the logic of this change.

Some suggested measurements are:-

Distance travelled - Measure the actual distance products move from your receiving dock to the shipping dock. This will be a good measure of the quality of your layout. Reduce this distance and see the benefits roll in.

Average changeover time - calculate the average length of time spent on changeovers. A changeover is defined as the time between when a machine or work center finishes the last piece of one batch and starts to produce the first good part of the next batch. Pressuring this average down will smooth out production and reduce batch inventories.

P:D ratio - P is the procurement and production cycle for a product, sometimes called the aggregate, stacked, cumulative, or critical path lead time. D is the length of time between when a customer orders the product and expects delivery. It's obvious the ratio should be 1 or less to reduce speculation in the planning system. Measure and reduce this ratio for real improvements.

SUMMARY

We have looked to systems to improve our operations, especially reduce inventories, improve customer service and increase productivity, but they haven't done it. Some improvements have been made, at considerable cost, but the real opportunities remain. An attack on the environmental problems listed, plus others, will result in bigger gains at less cost. Identify your major problem and get it solved this year. Watch the resulting improvement in operations and see the results on the bottom line.

HAL MATHER
Hal Mather, Inc.
P. O. Box 20161
Atlanta, GA 30325

Mr. Mather is President of HAL MATHER, INC. ATLANTA, GA., an International Management Consulting and Education Company. Since 1973 he has been working with all types of industrial concerns on ways to improve their Business Planning and Control. Recent assignments have taken him throughout NORTH AMERICA, EUROPE, the FAR EAST, AUSTRALASIA and SOUTH AFRICA.

Mr. Mather is renowned for his personal consulting. He has worked with both large and small companies and has stimulated many successful projects with enormous payback.

He is a dynamic lecturer and educator. He conducts private courses for companies, specific to their needs, and conducts public courses in association with GEORGE PLOSSL EDUCATIONAL SERVICES, INC.

Mr. Mather is a prolific author. His many articles have appeared in a number of magazines, among them the HARVARD BUSINESS REVIEW, and he has been quoted in FORTUNE MAGAZINE. His two books, "BILLS OF MATERIALS, RECIPES AND FORMULATIONS" and "HOW TO REALLY MANAGE INVENTORIES", are classics in the field.

He has been certified at the fellow level by the AMERICAN PRODUCTION AND INVENTORY CONTROL SOCIETY, is a member of the INSTITUTION OF MECHANICAL ENGINEERS (U.K.), and is a senior member of the COMPUTER AND AUTOMATED SYSTEMS ASSOCIATION of the SOCIETY OF MANUFACTURING ENGINEERS. He is listed in WHO'S WHO IN THE SOUTH and WHO'S WHO IN INDUSTRY AND FINANCE.

Reprinted from the 1984 Fall Industrial Engineering Conference Proceedings.

CIM — Target or Philosophy By-Product?

Jason L. Smith

John Fluke Mfg. Co., Inc.

ABSTRACT

CIM can be a by-product from implementation of key productivity improvement philosophies, but should not be a primary target.

The U.S. has for decades savored technological advantages over international competition. As one consequence, we have not recognized the quiet entrenchment of several burdensome conditions that negatively affect our productivity. These conditions are so burdensome, in fact, that they can no longer be neglected and must be diminished or the U.S. risks losing its position as a world economic leader. These conditions should surprise no productivity student of 1984 and include at least:

- minimal management job-related training

- minimal worker involvement in decision making

- U.S. Government interference in free enterprize

- profit emphasis instead of emphasis upon lower price and improved quality

- rigid, overly specialized organizations

- decline of employee self-motivation and job pride

As the competitive countries of the world such as Japan, Korea, Brazil, Malaysia and Taiwan develop comparable technological products, sophisticated manufacturing, and strongly motivated employees; the question of America's very manufacturing survival arises.

For the U.S. to retain its competitive world position, these burdensome conditions have to be replaced by instituting specific corrective philosophies. These key philosophies all target "people involvement and education". The U.S. productivity solution is through people.

Many influential leaders of the U.S. government and industry are only giving these corrective philosophies lip service. They are looking for the quick and easy solutions. They are attempting to "buy" their solution by purchasing CIM, computers, CAD/CAM, robots, AS/AR, automation, and state of the art technology. They don't realize that there is no "short-cut solution" whatever the price. They are searching for the "golden calf on the wrong mountain".

The burdensome conditions listed were either caused by or allowed to develop by the U.S. Government and U.S. industrial managers. Today, the very same type of leaders that caused the problems are saying to the American people; "We have the answers - we can buy a solution". Such thinking will solve nothing. The real answer is to embark upon a long, slow, difficult journey involving the education and participation of all employees so as to firmly implement the needed philosophies. There are no short cuts, and only the concerted, educated efforts of all employees can solve the productivity problems.

Without involving all employees and targeting these new philosophies, CIM efforts are doomed to hinder productivity improvement. The very reason that CIM approaches have been successful in a few companies is because the key philosophies were alive and working well enough to support such a complex system as CIM implementation.

CIM can be the logical result following implementation of key philosophies, but should not be persued as a primary solution target. In other words: CIM can only be the result of a very comprehensive company strategy and not a purchaseable, off-the-shelf hardware/software product.

Unless we become smarter at managing human resources, CIM is destined to be described in history as "another good idea whose time never came".

What Are These Key Philosophies?

Hopefully, by now there is someone who is wondering "what are these key philosophies that should define the primary U.S. company productivity improvement strategy?"

Productivity Philosophies are:

maximizing employee pride

minimizing organizational layering

focusing on the best possible quality of work

involving workers extensively in decision making

training and educating both managers and workers

planning for maximum, long-range productivity, not profit

simplifying systems and procedures so they are understandable

integrating all department workers as a team (e.g. product marketing with sales with design with manufacturing engineering with workers)

CIM Complexity

CIM is a centralized company strategy consistent with the specific philosophies presented. CIM is, in the most general sense, a colousal logarithm specifying company operating interrelationships amongst most levels and most departments. This logarithm is therefore extremely complex and might involve millions of lines of computer software.

The orchestrator of such a mathematical model must be a very well-rounded, executive generalist who can understand the business from bottom to top. He must create a vision so emotional and with so much internal PR that nothing can stop progress.

The computer system required to operate CIM must include, as key elements:

    shared data bases
    automatic data gathering
    networking (computer communications)

Computer sub-systems that are each a small part of a CIM system include:

    MRP
    AS/AR
    ROBOTS
    CAD/CAM
    NC machines
    CAQC (inspection)
    AI (auto-insertion))
    CAPP (process planning)

This CIM model must be custom developed for a particular company. Few U.S. companies are today in a position to attempt CIM development. They are victims of the unproductive conditions identified earlier. The typical outward manifestations of these conditions are very complex company operations with minimal interdepartment understanding and cooperation. In such companies, it is almost impossible to describe the step-by-step procedures through which a product production undergoes from purchasing through shipping. Such is the case as these procedures involve many hundreds of steps requiring scores of days to complete. Also, these procedures vary from product to product, from time frame to time frame, and from employee group to employee group.

Obviously, the central problem in such companies (that are by the way, typical in the U.S. today) is that if the operations are so complex and generally not understandable, then how can they be described by a CIM computer model? Of course they can't! Any attempt to do so will end unprofitably until the operations are first simplified by adapting the key corrective philosophies described earlier.

While the key philosophies are in the process of implementation, a company will probably undergo many, many drastic changes:

    The employee skills will change. Education and training of all employees will become one of the most important goals. Much of middle management will no longer be needed.

    Communications of all kinds amongst all interfacing company departments will improve by orders of magnitude. Communications will be direct from worker to worker neglecting traditional and inefficient "chains of command". As examples, design engineers will be talking directly with assemblers and salesmen. MIS "ivory tower" people will have to learn behind the scenes shop paper and discuss issues with dozens of workers from all occupations.

Work flow, as well as material flow, will become more understandable, and faster, and have smaller batch sizes.

The activities and data input from workers will have a stronger and more direct influence upon all company operating decisions.

Employees as the natural result of their education, training, and burdens of influence will develop more old-fashioned self-motivation, responsibility for one's own actions, and a strong sense of work pride.

When most such key philosophies become a way of life or the operating strategy of the company, productivity improvement will be a result. Sometimes the necessary communications will be informal and verbal; sometimes it will be formal involving dozens of filled in forms; and sometimes it will result in the CIM factory of the future. One main point here is that CIM is not the necessary result of a more productive U.S. factory but just one of a range of possibilities.

A frequently cited and apparently successful large U.S. company whose productivity efforts did end up in a real CIM facility is the Ingersoll Milling Machine Co. of Rockford, Ill. Their CIM tied together:

    CAD
    accounting
    purchasing
    NC fabrication
    costing efforts
    labor reporting
    master scheduling
    bill of materials
    inventory control
    shop floor control

Only one computer data base was used. The computer was an immense IBM 3033U mainframe supporting 121 terminals and 30 CRT's. The computer memory included 16 million bits of core memory with on-line disk storage of 6 billion characters. The computer process rate had to be fast to be effective - a remarkable 5.5 million bits per second.

The computer software logarithm essentially mathematically described the operations, control, and planning of the factory.

This example should point out quite clearly that the factory had to first be simple to understand and extremely effective at communicating before the CIM computer was plugged in. In other words, the factory was productive before CIM. Only then could CIM be implemented economically.

Also, this example should demonstrate that CIM had to be orchestrated from the executive level. The obstacles and resistance during CIM factory preparation, software development, and implementation could only have been overcome with an almost religious vision that had to be pursued every hour, of every day, by every single person associated with the factory. This executive orchestrator could probably have become another Billy Graham if his vision was only oriented a little differently. Nothing less than this type of inspirational, dedicated, consuming leadership could inspire effective CIM implementation.

Jason Smith is the Instrumentation Division Methods Manager at John Fluke Mfg. Co., Inc. Mr. Smith currently develops Divisional manufacturing philosophies and strategies. He is the Fluke National Association of Suggestion Systems representative as well as a senior member of AIIE and a Puget Sound Chapter Board officer. Mr. Smith is an experienced manufacturing consultant and has just completed writing a text book entitled <u>Learning Curve and Your Factory</u>.

CIMS In Perspective: Costs, Benefits, Timing, Payback Periods Are Outlined

By Roger G. Willis
and Kevin H. Sullivan
Arthur Andersen & Co.

Imagine a factory that machines parts, fabricates metals and assembles items with very few people operating the machines. Imagine an environment where parts are manufactured without the traditional long setup times and long runs involving huge supplies of components. In this factory, parts are automatically counted, inspected and moved.

Even five years ago this would have seemed like a setting for an Isaac Asimov novel. But these factories exist today in Smyrna, TN; Rockford, IL; Campbellsville, KY; and Waterloo, IA. What's their secret?

The driving force behind all these factories is a concept called computer-integrated manufacturing systems, or CIMS. For many manufacturing executives, CIMS represents a chance to beat rising manufacturing costs and cut-throat competition. CIMS provides the tools they need to develop a competitive advantage.

With much of the necessary technology available, today's managers face the challenge of putting it all together and making it work—of planning and implementing a CIMS strategy that will turn a competitive advantage into long-term health for their organizations.

As with traditional manufacturing approaches, the purpose of a CIMS is to transform product designs and materials into salable goods at a minimum cost in the shortest possible period of time. But unlike traditional manufacturing philosophies, CIMS builds on the premise that management should work to optimize the whole business process rather than individual functions or elements. And, unlike traditional approaches, the process in a CIMS begins with the design of a product and ends with the production of that product. With CIMS, the customary split between the design and manufacturing functions disappears.

In a traditional manufacturing environment, the layout of plants usually groups like processes and machines together as is shown in Figure 1. This requires extra time to move materials and schedule machines. A CIMS, however, generally uses flow-through manufacturing concepts. (See Figure 2.) With a flow-through design, machines are arranged in sequence by operation.

These arrangements are called manufacturing cells. The plant includes a number of independent manufacturing cells, and each of these cells includes the machines required to produce or assemble the part or product with a minimum amount of human intervention.

A totally automated machining cell is called a flexible manufacturing system (FMS). These systems generally include robots, programmable controllers, direct numerically controlled machine tools and automated material handling systems. Similar automated systems can be established for assembly operations.

A CIMS also differs from a traditional manufacturing system in the dynamic role the computer plays in the CIMS manufacturing process. Computer-integrated manufacturing systems are supported by a network of computer systems tied together by a single set of integrated data bases.

Using the information in these data bases, a CIM system can direct manufacturing activities, record results and maintain accurate data. Therefore, the computer sys-

Implementation of a CIM system will entail major changes in the jobs of all levels of workers in a company. (Photo courtesy of IBM)

tems and data bases are arranged to support the various functions of a CIMS as shown in Figure 3.

Implementing a computer-integrated manufacturing system involves a dramatic change in manufacturing philosophy. Because this change will affect your entire organization, it is important that you be prepared for the transition and the problems that may arise.

Four important areas must be considered when implementing a CIMS: hardware and equipment, personnel requirements, implementation costs and management commitment.

Hardware and equipment

The heart of a CIMS is the integration of design, manufacturing, distribution and financial functions into a coherent system supported by computers. This integrated network is comprised of three basic building blocks: computers, data bases and programmable controllers. The relationship between these building blocks is illustrated by the factory information loop in Figure 4.

Computer-integrated manufacturing systems usually include two major data bases. The manufacturing data base collects data from the process automation computer and process controller. The data are used to report shop activities and to plan and control future activities. The product data base is used by programmable equipment to perform manufacturing activities such as machining, inspecting and counting.

Typically, the eight key functions of a CIMS are distributed across several computers to maximize efficiency and cost effectiveness and to minimize the risk of computer downtime. The relationship between these functions and the computer hardware is shown in Figure 3.

Four of the eight CIMS functions—production scheduling and control, maintenance scheduling and control, distribution management, and finance and accounting—are performed on the host mainframe. Master production schedules, for example, are developed on the mainframe computer and transferred periodically to the process automation computer, which schedules and directs the flow of goods through the work stations and cells. This transfer of information is known as "downlink" or "download."

Drafting and design, process automation, process control, and material handling and storage are usually performed on satellite computers. The drafting and design functions (CAD) usually require a mainframe or large minicomputer. The design and process information generated by the CAD/CAM functions is stored on the product data base. The CAM information is used to give operating instructions to the machinery and to route parts.

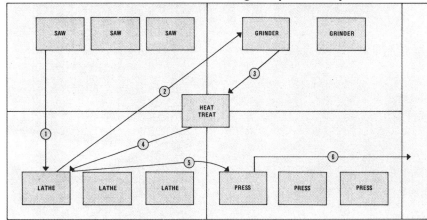

Figure 1: Traditional Manufacturing Layout (numbers indicate sequence of operations for a part). Long travel distances can cause large in-process queues.

The process automation function uses a minicomputer or mainframe to collect data, schedule manufacturing events and communicate these events to the manufacturing data base. This communication is commonly called "uplink" or "upload." Special sensing devices are used to collect this information.

Laser scanners, for example, are used to read the bar coded labels on containers and skids to identify part/location coordinates. This technology is much more efficient, reliable and timely than move tickets or other paper-oriented identification systems.

The process control function uses programmable controllers and sensing devices to control operations. These sensing devices will collect data that can be used for analysis. Coordinate measuring machines can be used to inspect machined parts and thereby control quality. If a serious problem occurs, the process controller can notify an operator of the problem and shut down the operation.

Programmable controllers are also used to perform numerically controlled functions such as machining, fabrication and inspection. In the past, special programming skills were required to use this equipment.

The current trend, however, is towards the development of more user-friendly equipment and programming languages by manufacturers.

Material handling and storage systems can store and retrieve parts, assemblies and finished goods. These mechanized storage systems often serve as buffers between various manufacturing cells. For example, there may be a storage depot between the machinery and assembly functions. The mechanized system would retrieve items from the depot as required for assembly operations. To operate efficiently, a computer system is used to maintain information and optimize storage and retrieval activities.

With the trend towards flow manufacturing concepts, there will be less need for automated storage and retrieval systems for in-process parts. However, these systems are applicable in a distribution environment where finished goods inventory is maintained.

Personnel requirements

Implementing a CIMS will affect all levels of personnel in your company. It will change the nature of the skilled labor force. The job of a worker who sets up a gear cutter and

Figure 2: Flow-through design requires fewer moves and smaller in-process queues.

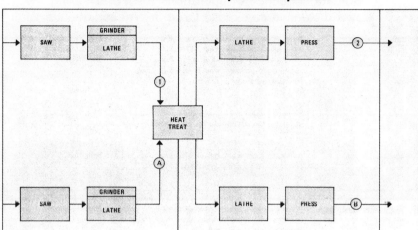

monitors the cutting process, for example, will change significantly.

To accommodate the new CNC machines, the operator must add the setup of the machine controller to his or her list of tasks while eliminating other familiar tasks such as changing tools or dies. In other cases, the operator may be eliminated altogether because the machine cycles automatically.

A CIMS will also require changes in the motivational approaches used by management. In the past, incentives were used to encourage each individual operator to be more productive. With a new emphasis on optimizing the process as a whole, company-wide incentive programs will replace individual performance incentives.

Computer-integrated manufacturing systems will change the jobs of people in the management ranks. An understanding of current manufacturing technologies and philosophies will obviously be one requirement for those in the management ranks.

Another requirement will be the ability to communicate tactics and strategies effectively and to motivate others to accept changes and work cooperatively as a team. Knowing how to apply group technology to the design, plant layout and manufacturing problems they encounter will demand the flexible thinking of multi-function specialists.

Because of the emphasis CIMS places on accurate, timely information, the role of management information services (MIS) will change significantly. Software for a CIMS requires far more complex functions than do the traditional transaction processing and reporting activities. The software will have to be capable of scheduling, analyzing and providing timely notification of problems on the shop floor.

MIS personnel will need the skills to design this complex software and its related data bases. Many companies have data bases that are not suitable to support flow-through manufacturing. These data bases will have to be changed or possibly even completely redesigned.

If the CIMS is to be completed within an acceptable time frame, MIS personnel will have to know how to take advantage of current software development technology such as fourth generation programming languages, application generators and design techniques such as prototyping.

Implementation costs

Costs for implementing a CIM system fall into four basic categories: computer hardware, equipment, software and personnel. Computer hardware costs can be difficult to estimate because so many variables depend on the company's particular situation and environment. These variables might include the volume of transactions, the size of the data bases and the communication requirements.

For example, a host mainframe will range in price from $150,000 to over $2,000,000. The small mainframe or large minicomputer needed for the CAD/CAM functions costs anywhere from $20,000 to $200,000. An additional minicomputer ranging in price from $30,000 to $100,000 also will be required to support the process automation functions.

Additional peripheral equipment such as CRTs and disk drives usually will add on another 50% of the computer's cost, and the graphics equipment needed for the CAD/CAM function will add an additional $20,000 to $50,000 to the cost.

Equipment costs should also be considered when estimating the cost of a CIMS. In many cases, existing equipment can be retrofitted with programmable controllers. This approach can cost several thousand dollars per machine. A new FMS can cost several million dollars. Robots range in price from $15,000 to $200,000. Welding robots, for example, might cost $160,000, while a material handling robot may cost $60,000. Accessories and installation costs can add as much as 100% to the overall cost of a robot.

Equipment used in the process control function requires a microprocessor with appropriate sensing devices. These systems can range in cost from several dollars to several thousand dollars per station or machine.

Software costs

Software costs include the purchase price for commercially available software plus the costs to modify or custom develop software. Appropriate software packages are available for many of the CIMS functions. Purchased software systems can often provide a higher level of data integrity and functional capability and are easier to implement than custom systems.

Purchased packages usually require modification to fit specific needs, however, and the software options for some CIMS functions such as process control and process automation are limited. Consequently, these systems are usually custom developed by the user organization or hardware vendors.

CAD/CAM software ranges in price from $20,000 to $250,000. Comprehensive material requirements planning systems that include cost accounting, purchasing, engineering, requirements planning and inventory control range in price from $200,000 to $500,000 for a mainframe computer and from $20,000 to $50,000 for a minicomputer. Financial planning and reporting, distribution and order processing software ranges in price from $15,000 to $100,000 depending on the features selected and the computer hardware environment.

It is difficult to estimate software costs accurately because the range of prices is so wide and because there are so many environmental variables that can dramatically affect software costs. Whatever software combination you select, you'll have to allocate additional time and money to make sure that the software is properly integrated.

At present, no single software vendor offers all of the necessary CIMS software. Moreover, data communications standards in the industry are almost nonexistent. Ensuring data integrity in a mixed hardware and software environment, for example, can be difficult and may involve developing some relatively complex custom software.

Implementing a CIMS will involve two basic personnel related costs: (1) the skills of present personnel must be upgraded; (2) additional professional personnel may have to be hired.

Teaching the new business philosophy to current personnel is an important part of an effective CIMS strategy. This is usually done using more formal forms of education like workshops and seminars. Although much of the knowledge associated with the newer technologies is not generally available, some professional organizations and colleges can provide competent instruction.

Technicians and skilled labor must also be formally trained to use the new equipment. This is usually provided by equipment manufacturers and is augmented by on-the-job training. If training is left out of the implementation project, the benefits of the CIMS will not be achieved.

Because costs can be significant, it is important to establish guidelines and methods for planning and controlling all CIMS-related costs.

Personnel requirements should be identified at the outset of the project. This will ensure adequate project staffing as well as providing input into the overall cost/benefit analysis of the CIMS project. An organization may also need to hire additional professional personnel to reinforce the weaker technical areas and to ensure that the new philosophy is understood and practiced successfully. Equipment, hardware and software costs should also be controlled with guidelines and authorization procedures.

Management commitment

In recent years, many gimmicks have been used by management to increase productivity among various work groups. Generally, management will review and approve small projects and equipment acquisitions based on recommendations from these groups. This approval process

Figure 3: Eight Key CIMS Functions

FUNCTION	DESCRIPTION	TYPICAL HARDWARE
Design and Drafting (CAD/CAM)	Automated design and drafting of products and computer-aided manufacturing.	Minicomputer Mainframe
Production Scheduling and Control	Master scheduling coupled with material management and requirements planning from procurement to shipment.	Host Mainframe
Process Automation	Direct numerical control of processes, inspection and testing. Utilization of robotics, flexible manufacturing systems and other automated equipment to perform certain processes and direct shop floor activities.	Programmable Controller Minicomputer Mainframe
Process Control	Sensing of equipment activities and reporting of conditions that require operator intervention.	Programmable Controller
Material Handling and Storage	Automated storage and retrieval of finished and purchased parts based on picking schedules and requisitions.	Minicomputer
Maintenance Scheduling and Control	Preventive maintenance scheduling and reporting of equipment downtime by cause. Spare parts inventory management and usage reporting.	Host Mainframe
Distribution Management	Order processing, sales reporting and invoicing coupled with warehousing and transportation.	Host Mainframe
Finance and Accounting	Reporting of operating results, forecasting of future results and analysis of costs.	Host Mainframe

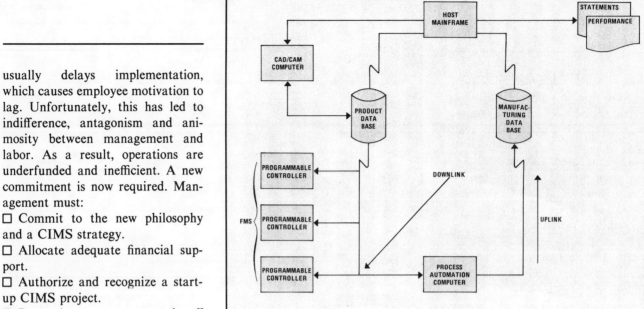

Figure 4: Factory Information Loop Showing Relationships of Computers, Data Bases and Equipment

usually delays implementation, which causes employee motivation to lag. Unfortunately, this has led to indifference, antagonism and animosity between management and labor. As a result, operations are underfunded and inefficient. A new commitment is now required. Management must:
☐ Commit to the new philosophy and a CIMS strategy.
☐ Allocate adequate financial support.
☐ Authorize and recognize a start-up CIMS project.
☐ Devote key management and staff personnel to the CIMS project.
☐ Be willing to risk short-term operational results for the sake of longer term improvements.

Strong commitment at all levels is essential because the changes will affect everyone in the company. If the commitment is not strong, people tend to lose interest, other high priority projects appear and the benefits may not be realized.

Risks and costs can be significantly minimized with judicious planning and use of experienced professionals who know how to manage CIMS projects. A CIMS, like any system, must be carefully planned, designed and installed.

In the planning stage, operations and materials movement are laid out, new equipment is identified, information requirements are defined, costs and benefits are outlined and the project team is organized to design the CIMS. The systems and operations are carefully defined and the implementation timeframe is established in the design stage.

The final stage of a CIMS implementation involves developing the software, installing the hardware, testing the system and training personnel to use the new CIMS.

The implementation of a complete CIMS usually requires several years, significant costs and dedicated effort. Therefore, the plan should include time-phased projects that are prioritized based on benefits, need and costs. In many cases, the plan will include a pilot project selected to allow project team members to design and implement a systems project in a small controllable area before moving on to more complex areas.

Assembly operations or machining operations are usually good pilot projects. As each new area is implemented, extensive prototyping and testing are required to ensure consistent, high-quality systems.

Benefits

The benefits of a successful CIMS are impressive: reduced inventory, less waste and better control over quality, improved productivity, lower personnel costs, more efficient use of machinery and lower occupancy costs for manufacturing facilities.

Because a CIMS uses flow-through manufacturing concepts, work-in-process inventory is reduced and queues are shorter. Raw material inventory is reduced because it is received just in time from vendors. Finished goods inventory can also be reduced because production levels are more flexible with a CIMS. An adequate service level can be maintained with fewer finished goods on hand because manufacturing lead times are shorter.

Adjusting a factory's production level to accommodate seasonal demands, for example, is also easier with a CIMS because it is a more automated process than traditional, labor-intensive manufacturing, and it is far easier to adjust the production rates of equipment than to adjust the number of skilled people you have working, how hard they are working, or how long they work.

The precision equipment and accurate data bases used in a CIMS help reduce waste and control the quality of products. Because computer-integrated manufacturing systems are highly automated, the risk of human error is reduced. And, because smaller lots of materials are moving through much shorter queues, quality-related problems are detected and isolated sooner than they would be in a traditional manufacturing environment.

Over time, a CIMS will also lower personnel costs. As many tasks become more procedural and more tasks are automated, fewer personnel will be required. However, it is important to remember that the skills of the personnel who remain will change dramatically with the implementation of a CIMS. The cost of retraining staff and augmenting the company's work force with new technical personnel must be considered.

As machine setup times are reduced and the flow of materials stabilizes, the amount of time the equipment is up and running and the quantities of products produced in a given timeframe will increase. A CIMS preventive maintenance program should also improve productivity and reduce machinery downtime. Recurrent equipment problems can be more easily identified and corrected when maintenance is routine and carefully reported.

Because a CIMS generally reduces inventory dramatically, it requires much less factory space than a traditional system. Implementing a CIMS may eliminate the need for new plants and warehouses.

To benefit most from a CIMS, it is important that a company set goals for various key operating factors. For most companies, 15 to 30 factors are selected and monitored by the CIMS.

Inventories and sourcing lead times are excellent examples of two important factors. Inventory is a non-performing asset that is susceptible to damage, theft or obsolescence. Therefore, one of the goals of a CIMS is to reduce inventory. Sourcing lead times can be reduced by developing vendor programs. Because quoted lead times for key items can be as long as three to four months, large volumes of these items are kept on hand to accommodate the long lead times.

A CIMS will provide accurate data about the key source items and about vendors who have sufficient volume to provide just-in-time deliveries. At one large equipment manufacturer, lead times of sourced parts were reduced from 11 weeks to 11 days, and they will soon go to five days.

When you consider that the cost of purchased materials usually represents 40 to 50% of the cost of goods sold, 90% reduction in lead time and inventory represents significant reductions in inventory value. In most companies, this alone is enough to justify the cost of a CIMS. Similar inventory reductions are possible in work-in-process and finished goods inventory.

Summary

Because of the pervasive change in philosophy that implementing a CIMS requires, it may seem like a drastic measure. It requires a full commitment from management and a substantial sum of money over a relatively short period of time. It involves changing the physical layout of plants and retraining personnel.

And yet, the payback generally occurs within a short time frame. If properly planned, cost reductions and other benefits can easily exceed the financial commitment by a wide margin. Inventory is reduced and productivity increases. Overhead and payroll costs are reduced and new products can be developed and introduced much faster. Being able to bring new products to the market quickly keeps the company competitive and ensures continued growth.

Computer-integrated manufacturing systems are the future for the manufacturing industry. The companies that succeed will be those that accept the challenge, commit themselves fully to developing and implementing a workable CIMS strategy, and successfully manage the transition from traditional manufacturing to computer-integrated manufacturing.

For further reading:

Bylinsky, Gene, "The Race to the Automated Factory," *Fortune*, February 21, 1983, pp. 51-64.

"CIM: Total Manufacturing Integration," *CAD/CAM Technology*, Spring 1982, pp. 18-20.

Dorf, Richard C., *Robots and Automated Manufacturing*, Reston Publishing Co., 1983.

Gettleman, Ken, "Step by Step to the Automated Factory," *Modern Machine Shop*, September 1982, pp. 53-62.

"Managing MRP and CIM," *CAD/CAM Technology*, Winter 1983, p. 25.

Roger Willis is a partner in the management information consulting division of Arthur Andersen & Co. in Chicago. He has a broad range of experience in working with manufacturing companies in such areas as material and manufacturing management systems, inventory reduction programs, operation productivity improvement and application of advanced manufacturing technology. Willis has BS and MS degrees in industrial engineering from Purdue University. His professional affiliations include APICS and CASA/SME.

Kevin H. Sullivan is a manager in the Chicago office of the management information consulting division of Arthur Andersen & Co. Sullivan has worked in industry for eight years as a consultant and supervisor. His experience includes planning, designing and implementing integrated manufacturing systems in a variety of manufacturing environments. He received a BS in industrial management from Purdue University and an MBA from the University of Wisconsin-Milwaukee.

Reprinted from the 1984 Fall Industrial Engineering Conference Proceedings.

The Economics of Computer Integrated Manufacturing

Jack Meredith
University of Cincinnati

ABSTRACT

In this paper we review the current competitive environment facing manufacturing and identify some of the inherent benefits of computer integrated manufacturing (CIM). Yet, the justification of these expensive advanced manufacturing systems presents a formidable obstacle for most firms, particularly in the U.S. We review the standard justification procedures, noting two significantly different approaches, and then conceptually develop a procedure that recognizes the synergy inherent in CIM systems that is commonly overlooked in single-machine justification methods.

INTRODUCTION

The manufacturing environment is extremely competitive these days compared with even a few years ago. The explosive emergence of Japan as a formidable international competitor has set new standards of quality, reliability, and performance while simultaneously requiring higher productivity, lower delivered cost, and significantly better overall management of the firm. Japan's ability to effectively employ workforce loyalty while proceeding with automation and other forms of new manufacturing technologies in an environment of manufacturing simplicity has confounded U.S. management practitioners and academicians alike. In the U.S. we have typically operated in an environment characterized more by complexity of manufacture and negativism between management and the workforce.

It is clear that the U.S. must quickly improve productivity, quality, market responsiveness, and costs simultaneously or lose even more markets to strong foreign competitors. Given the poor historical record of labor-management cooperation in the U.S., and the extemely positive relationships in many other countries, this approach holds little hope for most firms.

A better alternative for these firms lies in advanced manufacturing systems, particularly automated systems. Many such systems fall in this category such as computer aided process planning (CAPP), computer assisted design (CAD), automatic storage/retrieval systems (AS/RS), group technology (GT), flexible manufacturing systems (FMS), computer aided manufacturing (CAM), robotics, computer aided engineering (CAE), manufacturing resource planning (MRP II), computer assisted testing (CAT), and many others.

Although some of these systems are extremely expensive, not all are, and quite a few are less expensive than other common alternatives. The benefits of these systems are many. They are more productive than manual or simple stand-alone machines, offer more consistent and higher levels of quality, exhibit shorter product lead times and manufacturing cycle times, require less space, minimize inventory levels, need less management attention, are safer, yield more flexibility, reduce scrap, and so on. In terms of the marketplace, these systems offer another advantage against foreign competitors who hold the strong loyalty of their workers-- they minimize the need for extensive workforce-management cooperation.

Yet, in the face of the overwhelming competitive environment, and the numerous, clear benefits these systems can bring to firms, there is a tremendous reluctance on the part of many companies to move in the direction of automation. This is due, in large part, to the extremely sizable investment required by a firm with a minimal, or total, lack of experience with such systems. It literally becomes, for many such companies, a game of "you bet your company".

This lack of experience impacts the firm two ways. First, top management is often totally unacquainted with such systems and must rely on the expertise of their manufacturing executive. However, the manufacturing executive has been often been out of touch with the details of the shop floor for many years, and particularly is not up-to-date on new technology.

Additionally, most of these executives have rarely been consulted in advance about manufacturing strategy, usually being simply handed the task of implementing a preconceived (often impossible) manufacturing strategy. To initiate such a change in manufacturing would normally be done only under rare circumstances, and then only if the manager was perfectly confident in the outcome. Thus, the manufacturing manager is typically no more anxious to try these new technologies, perhaps even less so, than the top executives.

In lieu of guidelines for investing in these manufacturing systems, executives opt for standard financial measures that have been used relatively successfully for single machines and other stand-alone equipment in the past: ROI, payback, IRR, etc. Although it is arguable whether these are appropriate guidelines, the problem encountered when these measures are used for complete manufacturing systems is that they fail to take the natural synergy of the systems into account. This is particularly true for computer integrated manufacturing (CIM) where all the above mentioned subsystems are tied together into one overall system. Further below we will propose a conceptual scheme for handling this difficulty.

COMMON JUSTIFICATION APPROACHES

Historically, the formal justification of new equipment has been conducted as if all costs and benefits were deterministic, in time as well as amount. The return on investment, net present value, internal rate of return, and payback (or payout) period are examples of this approach that have been well documented in the literature and appear commonly in various types of textbooks (e.g., see [3]). Formalizations of these approaches specifically for machinery are the MAPI and related approaches (see [2], [3]).

These methods have come under increasing fire of late (e.g., see [1]) as being shortsighted and emphasing the near term rather than long term benefits when applied to more complex systems of automation.

Other complaints have been that the indirect cost savings are difficult to quantify, the most important noncost benefits are ignored, only local rather than systemwide effects are considered, and that the inherent assumption that the benefits will decrease with time through obsolescence is inappropriate with flexible automation. (We will address some of these criticisms later.) Perhaps more important than any of these arguments however is the warning given by Gold [1] that if conventional (e.g., direct cost) methods of justification are used in acquiring a new manufacturing system, it will tend to be applied in that limited manner and thus not obtain the rich variety of benefits that can accrue to firms with skillful and imaginative users.

All the above applies to the formal justification approach for new equipment. However, another method also is frequently used but as an informal procedure. Many large firms, particularly in the defense and aerospace industry, consider new manufacturing technology as a basic element of their manner of doing business, even their basic strategy. Without the latest technologies these firms would be considered less suitable competitors in the RFP-Proposal Award business. The justification process in these business is not in acquiring the technology as much as in making sure that the technologies are used in as cost-effective a manner as possible. Below we briefly discuss each of these approaches.

Stand-Alone Financial Justification

In this subsection we identify the direct and indirect benefits commonly included in the justification of stand-alone equipment. There are many articles in the literature, as well as vendor documentation, concerning how to go about justifying equipment (e.g., see [4]). The following are drawn from these.

Direct Benefits: labor in production, labor in setup, WIP inventory, raw materials inventory, finished goods inventory, downtime, fatigue, learning rate, capacity.

Indirect Benefits: Inspection, scrap, space, flexibility, safety, quality, materials handling, variable overhead, rework, training, better scheduling/workflow.

Competitive Justification

Firms regularly invest in segments of their business without a formal financial justification. Advertising campaigns, new personnel hires, different grades of raw materials, and the company picnic (or executive retreat) are examples of investments made by firms that are based almost solely on "faith" that their benefits justify their expense and trouble. More to the point, firms invest enormous sums in maintaining their "image" through advertising, community PR, and direct attributes of their product or service such as quality, reliability, friendliness, service, competence, and so on.

In a number of large firms, the acquisition of a new manufacturing system is perceived to be of the same nature as the investments identified above. Special requirements in governmental or military RFPs (request for proposal) often mandate the availability of special hardware, systems, or capabilities to fulfill unique specifications. Firms without these capabilities for defect-free electronics, greatly foreshortened lead times, extensive service requirements under unusual environmental conditions, etc. are deemed unacceptable proposal candidates and barred from competing. Firms with such new technologies or capabilities may thus have a significant competitive advantage over other competitors.

The justification procedure in these cases evaluates the capability of the new technology and the likelihood it will be needed and utilized in upcoming proposal awards. The high profitability of successful proposals subsidizes the cost of acquired but unprofitable technologies.

SYSTEMS JUSTIFICATION

A common criticism of using the formal stand-alone financial justification procedure for advanced manufacturing systems is the inherent tendency to neglect synergy between the systems. For example, the development of a part-family classification and coding system for CAD can be used extensively for CAPP and CAM also, thereby initiating the implementation of the group technology philosophy. Although other criticisms are directed against this justification procedure, such as the fact that it ignores the inherent flexibility of these processes, the failure to account for synergy between the subsystems seems to be the major criticism.

Below we propose a conceptual procedure, illustrated through an example, to account for this oversight. In essence, we propose an exponential function to approximate the benefits accruing from synergy betweeen subsystem elements. Thus, if a new manufacturing subsystem, such as CAPP, is proposed and only appears to offer a 20% ROI when the hurdle rate is 25%, the recognition of the synergy with the CAD subsystem may raise the ROI to over 25%, thus justifying the subsystem. And if a later subsystem such as CAM only offers a 15% ROI when considered alone, by considering the synergy with the CAD and CAPP subsystems together, the new ROI may again exceed 25%.

This, in fact, is exactly what the firms who currently employ advanced manufacturing technologies already do. It is standard advice to enter high tech manufacturing one small step at a time, learning from each subsystem as you go and preparing for the next one. In this way expertise is built up without incurring substantial, and expensive, risk of failure.

The Synergy Function

It has been often said that the majority of benefits from CIM come only as the pieces are tied together. Others have added that the greatest benefits come from tying the data base subsystems together first and then the hardware subsystems.

The implication of these statements is that an exponential-type function is operating rather than a linear function. In Figure 1 some typical exponential functions are plotted which demonstrate this characteristic. The symbol S represents the annual dollar savings or benefits due only to synergy between the elements of the CIM subsystems. The value S_t is the estimated annual dollar savings due to full automation in the firm. The symbol r represents the proportion of full automation that has been achieved, or is being contemplated. It can be represented in cost terms as the expended cost as a proportion of the full cost to implement CIM. At complete automation, $r = 1.0$ and $S = S_t$.

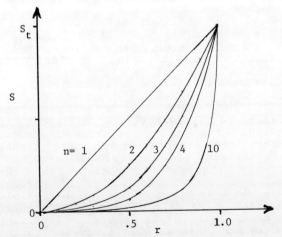

Fig. 1 Graph of Synergy Function $S = S_t r^n$

The graph need not include all aspects of CIM; for example, it could include only the full automation of a portion of a plant. However, care should be taken that multiple elements that can be tied together are included in the value r = 1.0. That is, if only a CAD system for the drafting department is being considered, there is no synergy between having one quarter of the system installed and having two quarters of the system installed. At least four separately functioning subsystems should be considered to make up a full system.

The exponential function relating the synergy benefits to percent completion is

$$S = r^n$$

where n is a characteristic of the CIM technologies. If there is virtually no synergy until almost every element is in place then n will be large, such as 4 or 5. If there is immediate synergy in direct proportion to the added subsystems then n = 1. More commonly, particularly with the technologies we see being employed today (see later example), n would be expected to fall in the range of about 2.

In addition to verifying n, the task of the analyst in the justification process is to determine S_t, the maximum annual synergy benefit due to full CIM. This may be approximated, for rough calculations, as follows. Meyer [4] indicates that the indirect benefits in automation, largely due to synergy between subsystem elements, may be as much as twice the size of the direct benefits, or 2D. Some of these indirect benefits, such as the inventory reductions, will be recognized and included in the normal justification process but many more will be missed.

If we approximate the overlooked indirect benefits as 75% of the total indirect benefits then

.75(2D) = 1.5D

The _recognized_ direct and indirect benefits are

D + .25(2D) = 1.5D,

or in other words, the same amount as unrecognized. Thus, S_t may be approximated as the expected level of annual savings due to full CIM. If even this value is impossible to approximate, it may be estimated by using the cost of full CIM implementation and the expected payback period. For example, if full CIM would cost a million dollars and would generate savings sufficient for a five year payback then the annual CIM benefits would be

$1,000,000/5 years = $200,000

which would then be taken as the value of S_t.

Example

Let us use an example to demonstrate. Suppose a firm has already installed a CAD system and is using it with the benefits that were expected, an ROI of 26% based on a full system cost of $400,000. The hurdle rate for this firm is 25%. A CAPP system, synergistic with the CAD system, is now being considered that costs $200,000 but only exhibits an ROI of 18%, insufficient for funding. It is estimated that full CIM for this plant would include CAM and MRP II, which would bring the total cost of CIM to $2,000,000 (includes the CAD and CAPP systems). The full system is expected to exhibit about the same range of ROI, 23%.

The calculations would be:

S_t: .23 ($2,000,000) = $460,000

r (CAD): 400,000/2,000,000 = .2

r (CAPP + CAD): 600,000/2,000,000 = .3

S (CAD): $.2^2$(460,000) = 18,400

S (CAPP + CAD): $.3^2$(460,000) = 41,400

Therefore the _additional_ return from the CAPP synergy with the CAD system is 41,400 - 18,400 = 23,000. Based on the $200,000 cost of the system this increases the ROI by 23,000/200,000 = 11.5% so the adjusted ROI is now 29.5%, sufficient for full funding.

CONCLUSION

The above example is not intended to be an adequate calculation for justification but rather a method of rough approximation to aid the analyst in considering whether it is worthwhile trying to dig deeper to more accurately determine the benefits of synergy with existing, or proposed, automation systems.

In the case of proposing two or more systems, with nothing currently in place, it may well be that no one system by itself meets the specified hurdle rates. But when the synergy of the multiple systems is considered (and then the full synergy would be added, as opposed to the difference calculated above) the ROI might be acceptable.

A variant of this system is proposed by Gold [1]. He suggests developing a portfolio of projects whose net returns, calculated at 3 to 5 year intervals, justify the set of projects as a whole, rather than trying to justify each project by itself. With this approach, the stronger projects protect the weaker ones through a subsidy effect. But this takes a very strong project to boost the return on the "package" sufficiently to justify the weaker projects as well. With the project synergies considered, the strong project provides the base that lets the weaker projects stand on their own. Also, multiple weak projects, none of which meet the ROI requirement, would fail to meet Gold's criteria whereas if their synergy is considered, they could well meet the ROI hurdle as a package.

REFERENCES

1. Gold, B. "CAM Sets New Rules for Production" Harvard Business Review, Nov-Dec, 1982.

2. Leung, L. C. and J. M. A. Tanchoco "Replacement Decision Based on Productivity: An Alternative to the MAPI Method", SME Technical Paper MM83-926, 1983.

3. Meredith, J. R. and T. E. Gibbs The Management of Operations, 2nd edition, Wiley, 1984.

4. Meyer, R. J. "A Cookbook Approach to Robotics and Automation Justification", SME Technical Paper MS82-192.

BIOGRAPHICAL SKETCH

Dr. Jack Meredith is Associate Professor of Production/Operations Management at the University of Cincinnati and Director of Industrial and Operations Management Programs. He is Editor of Operations Management Review and coauthor of The Management of Operations (Wiley, 1984), Fundamentals of Management Science (Business Publications, Inc., 1985), and Project Management: A Managerial Approach (Wiley, 1985). He is also an Advisory Editor to John Wiley & Sons, Inc. and their Series Editor in Production/Operations Management. His research and consulting interests are in the management aspects of advanced manufacturing technologies such as CIM. He is currently working on a grant to determine the teaching implications for the management of high technology, funded by the Cleveland Foundation. He is also a member of the Manufacturing Management Council of the Society of Manufacturing Engineers and a founding member and Vice-President of the Operations Management Association.

Reprinted from *Industrial Engineering*, April 1984.

Productivity Benefits Of Automation Should Offset Work Force Dislocation Problems

By **Mikell P. Groover**
Lehigh University
John E. Hughes, Jr.
Air Products & Chemicals Inc.
and Nicholas G. Odrey
Lehigh University

In recent years, manufacturing industries in the United States have accelerated the move toward automating their operations. Factors that have motivated this move include the need to increase productivity, the high cost of labor, competition from abroad and the desire for closer management control over production operations.

As more and more automated systems are installed, society will feel their effects, and so will industrial engineers. Some of these effects are impossible to predict. Other effects, however, can be forecasted with reasonable confidence simply by extrapolating current trends into the future and logically assessing their impacts.

This article deals with those effects of automation that can be anticipated. The major areas that will be impacted by continued implementation of factory automation can be broken down into three categories: labor, professional staffs (including industrial engineers) and society in general. We will begin by examining the trends in production automation.

Certain trends are obvious. The use of computers, minicomputers, microcomputers and related software systems in manufacturing has grown tremendously during the last five years; perhaps not as rapidly as the trade press might suggest, but the rate of implementation has been substantial.

There is a very real interest by manufacturing people in computer automation of their production operations. They have seen the impact of computers in other areas, and they sense that similar opportunities for productivity improvement exist in their factories. Indeed, the use of computers has already brought about important improvements in several functions related to manufacturing.

Computer technology has been utilized in design (CAD—computer-aided design of the product), manufacturing planning (CAPP—computer-automated process planning), production and inventory control (MRP—manufacturing resource planning), machine control (numerical control and robotics) and quality control (computer-aided inspection and testing).

Computerization of these functions is usually carried out individually at present, but the technology of CAD/CAM holds the promise of integrating these various activities into one continuous flow of information and work.

Other trends in manufacturing also seem clear. The long-term economic growth rate in the metalworking industries will not be as great as in certain important segments of the electronics industries. Metalworking production increased in 1983 by about 5 to 6% after suffering a decline of 10% in 1982, according to *American Machinist* magazine (See "For further reading").

The semiconductor industry has been growing at a much faster rate and is expected to add another 25% in 1984, according to a recent *Business Week* forecast. Metalworking must be considered a mature industry, while semiconductor and related electronics manufacturing is very much a high technology growth industry.

Metals processing and fabrication is, however, becoming more high tech in its orientation. American industry has found that its stock of machine tools is older and less productive than those of its foreign competitors. According to sources cited in Ayres and Miller (See "For further reading"), 34% of the machine tools in the U.S. are over 20 years old. By contrast, in Japan the corresponding figure is 18% and in West Germany it is 26%.

Manufacturers in this country have embarked on a program of replacing the older machines with more modern versions and more highly automated systems of machines. Uses of industrial robots, computer numerical control machining centers, computer-controlled material handling and storage systems, and computer-aided inspection techniques (such as machine vision) are examples of opportunities for the integration of advanced computer control systems into manufacturing operations.

The implementation of many of these systems also reflects a relatively new philosophy: flexible automation. Flexible manufacturing systems are capable of producing a variety of different parts or assemblies on the same basic setup without significant time losses due to changeovers.

The variety of parts that can be handled by these systems is limited. The parts must have similar geometries (same basic sizes and shapes). Group technology principles are used to design the systems and organize the work that goes through them.

In manufacturing process technology, gradual changes are increasing

47

productivity. Using metalworking as an illustration, the development of new super-hard materials has led to advances in cutting tool technology that are gradually increasing the cutting speeds with which metals can be machined.

These improvements have obvious implications for productivity. Cutting speeds are becoming so high that mechanized conveyor systems must be installed to carry away the large quantities of chips created in the operations. Machine tool technology is adapting to the new tool materials by using spindles capable of much higher speeds.

Modern machines not only operate at higher speeds, they are also more accurate and more intelligent. Feedback control systems are being installed that utilize mechanical probes to inspect the work part at critical steps in the machining cycle. These systems then make adjustments in the machine axis positions to reduce errors in locating the tool relative to the work. Greater precision, time savings in inspection and fewer mistakes are the benefits of using these in-process probes.

Machine tools are also being built with smarter computer controls (computer numerical control—CNC) that can assist the NC part programmer in editing programs and can also help maintenance personnel diagnose the reasons for machine failures.

Process technology is also changing the fundamental methods by which materials are shaped. Non-conventional processes, such as plasma arc cutting and laser machining, are being substituted for conventional machining in order to increase productivity and to deal with new materials that cannot be cut easily with traditional machining methods.

The introduction of these newer processes can have dramatic effects on throughput rates. In one instance, introduction of a CNC plasma arc machine in a metal fabrication shop increased production capability several times. The new machine was so productive that the upstream and downstream operations could not keep up, and it became difficult to keep the machine fully utilized.

These trends in computerization, automation, tooling and processing will make future manufacturing more productive, more flexible and more oriented toward high technology. Fewer workers, who will have to be technically smarter, will be producing more products of greater variety than their counterparts of generations past.

The growth of automation will have a pronounced influence on our work and our way of living. We have identified three separate impact areas for consideration in this article: Labor, professional workers and society in general. For each category, what are the effects of automation going to be?

Automation versus labor

To the labor sector of the economy, automation means substitution of machines for human workers. The implications for employment are a serious concern to labor leaders.

According to the Bureau of Labor Statistics, there are currently 19 million workers employed in all U.S. manufacturing out of a total work force of 103 million. As tasks continue to be automated, the number of persons involved in direct manufacturing will decline.

However, economic forecasts by Data Resources Inc., as reported in a recent issue of *U.S. News & World Report* (See Karmin, "For further reading"), indicate that there will be a net gain of about 2.5 million jobs in manufacturing by 1990. This suggests that automation will result in a net increase in employment over the long term.

The difference is that the production work force will shift from jobs in direct labor (which are generally well defined, methodized, planned and controlled) to indirect labor jobs (which have not been subjected to the same high level of rationalization and control as direct labor tasks). These jobs will have the high technology content required to keep pace with the trend toward computerization and automation described above. The new workers will include computer programmers and operators, maintenance personnel, and technical support specialists.

These forecasts are no cause for optimism among labor unions. What it means to them is a continual reduction in their memberships.

The technically trained knowledge workers who will be needed to run the automated factories do not sense the same need to join the union ranks as direct line workers. The more highly skilled semi-professional workers have a greater confidence about their employment and a closer identification and association with management and professional staffs.

So, although manufacturing activity in the United States might indeed increase because of automation and economic growth, this is not likely to benefit trade unions.

The unions have recognized the impact of automation and other technological advances on their members, and many seem to be dealing with it in a positive way. Rather than resisting the introduction of automation, the unions are trying to participate in it, bargaining for contracts aimed at retraining and reemploying their members in the new skills that are needed to make it work.

The United Auto Workers Union (or more precisely, the United Automobile, Aerospace and Agricultural Implement Workers of America) has stated that it recognizes the introduction of new technologies as essential in promoting economic progress which will ultimately benefit its membership in the form of higher wages and fringe benefits, better working hours and greater safety in the work place (See Weekley, "For further reading"). What the union expects from management is ample notice when the new technologies are introduced, protection of union workers against displacement and retraining of members to perform the jobs brought about by new technology.

> **"If those functions are not automated, the alternative might well be 100% unemployment in specific plants or industries when business is lost to competitors that either enjoy lower labor rates or have themselves automated."**

The International Association of Machinists is a good example of a union which is aggressively bargaining for benefits from automation (See Kuzela, "For further reading"). Its "Technology Bill of Rights" defines not only a bargaining position for union-management negotiations, but also a proposal for federal legislation that would augment existing labor laws.

The union proposes a "tax" on automation and related new technologies that displace factory workers. This tax would represent a sharing of the gains from cost savings and productivity improvements, and would be used to finance the retraining and subsequent reemployment of displaced workers.

There is no doubt that automation has the immediate effect of reducing the number of jobs in those production functions that are being automated. And there is no denying that this is a negative aspect of automation that is distressful to the workers who are displaced.

However, if those functions are not automated, indeed if whole factories are not automated, the alternative might very well be 100% unemployment in specific plants or industries when business is lost to competitors that either enjoy lower labor rates or have themselves automated.

Impact on professional staff

Professional workers who will staff the automated factories of tomorrow must possess a set of technical skills that are different from those in use today. This has important implications for the discipline of industrial engineering, because so much of IE practice is involved with production work.

The trends in manufacturing and automation discussed above suggest that the focus of attention in production is shifting away from the factory production worker and toward the process and equipment. As automated systems gradually replace humans, the functions and requirements for managing the factory will change.

Instead of management systems designed around the activities of production workers, the systems in future automated factories must be designed to manage highly sophisticated machines and integrated combinations of machines. The professional staff must be more knowledgeable and proficient in the technologies that relate to the operation and maintenance of these machines and the manufacturing processes associated with them.

With the gradual elimination of manual labor in the automated factory, work measurement as traditionally defined by industrial engineers will not be needed, at least not for jobs that have been completely automated. New measures of work and new measuring techniques must be devised for indirect labor and other types of workers who will operate the plants.

It is unclear whether work measurement as it is conceptualized today will even exist in the automated factory. Other ways of motivating workers and new forms of incentives will be needed for highly automated production plants. Industrial engineers should take the lead in addressing these issues as labor oriented manufacturing operations give way to new automated processes.

An interesting feature of the technologies associated with production automation is that they are simultaneously highly specialized and highly interdisciplinary in terms of traditional academic content. The field of robotics is a good example of this interdisciplinary character. There is no doubt that robotics will be an important technology in the factory of the future.

The trend in robotics technology is to make these machines more and more like human beings, so it is quite logical that they would be substituted for human labor in production work. Robots are a combination of computer science, machine design, control systems and human factors, and their applications require a blend of engineering economy, plant layout, electrical engineering and robot programming.

The implementation of robotics and other automation technologies requires not only people who know the individual disciplines, but also people capable of integrating the various disciplines into the design of a successful system. Industrial engineers are often called the "systems integrators."

One of the automation technologies that will significantly affect professional and semi-professional staffing in manufacturing is CAD/CAM

> **" The introduction of the photocopying machine made it so easy to make copies that the demand for copies has increased dramatically. Here is a technology of which it can truly be said that 'invention is the mother of necessity.' "**

(computer-aided design and computer-aided manufacturing). The comprehensive application of CAD/CAM will eliminate the need for many of the manual and clerical planning activities normally associated with the introduction of a new product into manufacturing. These planning activities include process planning, estimating, purchasing, and many of the production planning and control tasks.

As firms commit themselves to the full potentials of CAD/CAM, design and manufacturing will no longer be distinguished as separate functions. They will become a highly integrated continuum of design, planning and operational activities.

Once the design data base for a new product has been created, many of the manufacturing planning activities will be accomplished automatically by computer systems and software developed specifically for the company's particular blend of operations. Examples of these automated planning functions include CAPP, MRP, automated NC part programming, and computerized shop floor control.

The automation of these planning functions will have an obvious impact on engineering and supporting clerical jobs in manufacturing. Manual preparation of manufacturing route sheets, manual part programming, dispatching and other similar jobs will gradually disappear as the automatic systems take over.

How can industrial engineers participate most effectively in the development and operation of the future automated factory? The role of the industrial engineer will probably emphasize project work, including such activities as project evaluation, system simulation studies, introduction of new technologies, and the development of new systems for automating functions like process routing and NC part programming.

Industrial engineers will also be involved in technical management, including the management of factory operations, maintenance management and managing automation projects that bring together various technical and engineering disciplines. If the trends we see in CAD/CAM continue, IEs will be performing fewer routine tasks and fewer functions that are largely clerical.

What are the technical skills and educational requirements for the new breed of industrial engineers who will design and manage the future automated factory? We believe that IEs must become more knowledgeable in the following technical areas:

☐ Robotics and other automation technologies.
☐ CAD/CAM and computer graphics.
☐ Microprocessor and microcomputer applications.
☐ Computer and information systems.
☐ Decision support systems and knowledge based systems.
☐ Materials handling/storage technology and facilities planning.
☐ Manufacturing process technology.
☐ Electrical and electronics manufacturing.
☐ Systems analysis, with emphasis on simulation techniques.

IEs also need to cultivate human relations skills in order to implement and manage automated systems that require the integration of various technological and engineering backgrounds. To be "systems integrators," IEs must be able to work effectively with people.

Impact on society

The benefits and disadvantages of automation to society have been argued for many years. Those who argue in favor of automation claim its benefits include lower prices, greater productivity, higher quality products, a shorter work week, safer and better working conditions and in general a higher standard of living for society.

Those who argue against automation claim it will lead to higher unemployment, worker displacement, forced migration of job seekers, more pollution and subjugation of humans by machines.

Let us examine some of these arguments, recognizing that the social impact and the labor impact will often overlap.

One of the arguments offered to promote automation is that its implementation will relieve workers from hazardous, repetitive, dull and demeaning work. Unfortunately, it will also relieve some workers of their jobs. This employment impact has been the object of more attention by scholars, industrialists, labor leaders and legislators than any other social issue related to automation.

The immediate impact of automating a production job is the displacement of the worker who previously performed that job. Deciding what should be done about the worker displaced by automation is a difficult social issue.

The most reasonable solution would seem to be to retrain the worker in some related skill, as recommended by the labor unions. That is easier said than done.

Educational training is expensive, and the question arises: Who should pay the cost of the training? The company introducing automation? The labor union? Or society in general through some federal or local government program?

Furthermore, once a worker has been retrained in some new skill, what are the obligations of the worker regarding reemployment if it means changing employers or even geographical relocation?

There is another aspect to the retraining issue: With manufacturing and automation technology becoming increasingly complex and sophisticated, will the displaced worker be capable of being retrained in areas requiring higher scientific and mathematical aptitudes? How should society deal with the problems posed by the increased complexity of a technological society?

It is difficult to determine how many workers are or will be displaced by automation or what the relationship between automation and employment is. One area where the impact can be measured with relative ease is robotics.

A typical installation of one robot involves the displacement of from one to three workers. There are approximately 7,000 robots installed in U.S. plants, and these have displaced an estimated 12,000 to 15,000 workers.

This is a small proportion of the total work force in manufacturing. Still, the potential future displacement is a matter of concern.

Projections for robot installations in 1990 range from 70,000 to 200,000. Ayres and Miller (See "For further reading") have speculated that the employment displacement will be largest in the northeastern region of the United States (New England and the middle Atlantic states). This region, where the metalworking industries are primarily concentrated, has been characterized by slow employment growth and migration out of the area. By contrast, net migrations into the sunbelt and western states have occurred.

Some argue that the automation industry itself will provide enough jobs to increase rather than decrease employment. High technology industries related to automation include computer electronics, software development, CAD/CAM, machine vision, vapor-phase technology, medical instruments and biogenetics.

Some of the companies involved in these areas are indeed growing at very fast rates. However, the number of jobs involved in these companies is relatively small compared with employments in the more traditional smokestack industries. The high technology industries employ somewhere between 2.5 million and 6 million people, according to estimates in *U.S. News & World Report*.

Even with their high rates of growth, these companies cannot be expected to provide jobs for all the workers who will be displaced by automation. The very purpose of many of these high technologies (e.g., robotics, CAD/CAM, machine vision) is to improve the efficiency of other companies. And, for most manufacturing firms, improving efficiency means reducing the work force.

According to a major survey of U.S. companies conducted by Carnegie-Mellon University with the cooperation of the Robot Institute of America, the biggest single motivation for installing robots in their plants was to reduce labor costs (see Ayres and Miller, "For further reading"). If this motivation is the primary factor which justifies automation, a rational company would not install automation unless its cost (which includes labor) was less than the cost of labor that would be saved as a result of its installation.

In other words, the labor to produce an automated system must be substantially less than the labor displaced by the system. By this reasoning, there can only be a net decrease in employment from increases in automation.

If the introduction of automation is to play an important role in increasing employment in the United States, that role must be in one or both of the following two areas:
☐ Reducing the flow of jobs to overseas competitors.
☐ Increasing the demand for the products made by automation.

Let us examine each of these roles. Reducing the number of jobs lost to overseas workers is a critical economic goal in the United States. One objective of automating is to reduce the manufacturing costs of American manufacturers relative to their foreign competitors.

Indeed, the issue is more serious for many industries than simply improving their competitive advantage. The more mature industries must increase productivity, or they are not likely to survive.

As General Electric's Executive Vice President James Baker put it in a recent interview with *U.S. News & World Report,* "The choice that confronts American industry is to automate, emigrate or evaporate." Through automating and thereby strengthening the manufacturing industries in the United States, the jobs of many American workers will be saved rather than lost.

Increasing the demand for the products made through automation is a second major area where automation can help the employment picture. If automation can be instrumental in increasing product demand, employment will increase in firms that make and market the product. Several examples will help to illustrate this effect.

One example is in data processing. Before the introduction of the digital computer, data processing was largely a clerical function performed by people. When digital computers began to take over the function, the number of clerical workers needed was of course reduced.

However, the capacity of the digital computer to perform data processing at increasing speeds not only reduced the number of clerical workers, it also accelerated the demand by managers and executives for ever-increasing amounts of information. Automating the clerical data processing functions by computerization has actually had the effect of increasing the demand for the product. In this case the product is information.

Today, the number of persons employed in the computer and information systems industries far exceeds the number of workers that have been displaced from routine clerical jobs.

Xerography is another interesting example. Before the advent of this copier technology, making duplicates of paper documents was a relatively inconvenient chore. The introduction of the photocopying machine made it so easy and convenient to make copies that the demand for copies has increased dramatically. Here is a technology of which it can truly be said that "invention is the mother of necessity."

Examples closer to manufacturing can be found in the electronics industry. The combination of new technology, innovative product design and marketing, and creative manufacturing methods has generated markets for color television sets, audio equipment, personal computers and other electronics-based products where there were no markets before.

Summary

The preceding discussion emphasizes the importance of productivity in economic growth. We believe that automation can be an effective means of achieving productivity improvement that will strengthen our economy. We further believe that the net result will be to increase employment rather than decrease it. History has shown that technical advancements in the past have resulted in economic growth and greater employment, despite temporary worker displacements.

According to economists, productivity is determined by a number of factors, including labor, capital investment and technological advances. Statistical analysis has determined that the relative contributions to productivity of these three factors are roughly as follows: Labor, 14%; capital investment, 27%; and technological innovation, 59% (See Froehlich, "For further reading"). Automation, of course, involves both capital investment and technological improvement.

The challenge to American manufacturing industries is not simply to replace their old equipment with updated versions of the same hardware, but to make investments that improve the fundamental technology of their operations. The problem is to install automated systems that substantially improve the basic methods of design and manufacture.

For further reading:

Ayres, R. U. and Miller, S. M., *Robotics—Applications and Social Implications,* Ballinger Publishing Co., Cambridge, MA, 1983.

Froehlich, L., "Robots to the Rescue?" *Datamation,* January 1981, pp. 84-96.

Groover, M. P., *Automation, Production Systems, and Computer-Aided Manufacturing,* Prentice-Hall Inc., Englewood Cliffs, NJ, 1980.

Groover, M. P. and Zimmers, E. W., Jr., *CAD/CAM—Computer-Aided Design and Manufacturing,* Prentice-Hall Inc., Englewood Cliffs, NJ, 1984.

Harrison, H. L. and Miller, R. J., "The Social Consequences of Automation: A Wide-Angle View," *Computers in Mechanical Engineering,* January 1984, pp. 46-47.

Karmin, M. W., "High Tech—Blessing or Curse?" *U.S. News & World Report,* January 16, 1984, pp. 38-44.

Kuzela, L., "IAM Envisions a Tax on Automation," *Industry Week,* May 30, 1983, p. 19.

Naisbitt, J., *Megatrends,* Warner Books, New York, 1982.

Weekley, T. L., "The UAW Speaks Out on Industrial Robots," *Robotics Today,* Winter 1979-80, pp. 25-27.

Mikell P. Groover is professor of industrial engineering and director of the Manufacturing Technology Laboratory at Lehigh University. He is past president of the Lehigh Valley IIE Chapter, past director of the manufacturing systems division of IIE and a past region vice president of the Institute. Groover holds BA, BSME, MSIE, and PhD degrees from Lehigh. He has written numerous articles and technical papers and is the author or co-author of several books on automation and computer-aided design and manufacturing.

John E. Hughes, Jr., is manager of manufacturing systems development for Air Products & Chemicals Inc. in Allentown, PA. He is past president of the Lehigh Valley IIE chapter, past director of the manufacturing systems division of IIE and a current region vice president of the Institute. He received his BIE from the General Motors Institute and an MS in management engineering from the New Jersey Institute of Technology.

Nicholas G. Odrey is associate professor of industrial engineering and co-director of the Robotics Laboratory at Lehigh University. He was previously on the faculty at the University of Rhode Island and the University of West Virginia, and was a research engineer at the National Bureau of Standards. Odrey received his BS and MS in aerospace engineering and a PhD in industrial Engineering, all from the Pennsylvania State University.

II. Manufacturing Components, Structure, and Technologies

Computer-Aided Technologies

Support Systems

Automated Material Handling

Automated Controls

Manufacturing Technologies

EDITORIAL OVERVIEW of Section II

The recent emphasis on manufacturing has provided the focus for the development of the required components and technologies of CIMS. Many of the existing components have been salvaged from the MRP and manufacturing systems of yesterday. In some cases they have even been repackaged to give the appearance of being new concepts. At the same time, new technologies are being introduced at an ever increasing rate. It is not the intent of this section to provide complete coverage of these topics, but to highlight the important elements of each one.

The selection of papers for this section was both enjoyable and frustrating. In some areas there were numerous existing papers from which to select. In other areas it was difficult to find a single paper which would qualify. Some subjects could not be covered, such as the area of computer-aided design (CAD). Although none of the subsections directly discuss the structure of CIMS, the papers as a whole provide insight into the requirements necessary for such a structure.

The first subsection provides three articles covering available computer-aided technologies which are of prime concern to the IE. Lee Hales provides a comprehensive overview of computerized facilities planning and design. In the second paper Bruce Webster discusses the key integration issue involved with the implementation of the Vought Flexible Manufacturing System. The important CAD/CAM link, computer process planning, is addressed by Joe Tulkoff. These last two papers are examples of the progress being made toward CIMS in the aerospace industries.

In the second subsection, Sarin, Dar-El, Biles, and Zohdi present papers on computer support systems. These papers indicate the need to better utilize existing operations research tools for the design, analysis, and control of CIM systems. The continued development of such tools could well fill a critical gap in the existing CIMS.

Physical integration of automated systems through material handling is covered by Neil Glenney in the third subsection. In addition, Robert Smith discusses the versatility of automated guided vehicles and their impact on the production system. With the continued development of automation, this area will play an even more important role in the future as these systems provide the means to transport and control the flow of goods.

The three papers on automated controls cover some of the key component areas required to make the factory of the future a reality. The first two papers provide a glimpse into the future with the potential of computer vision systems. John Artley looks at automating the inspection task through the joint application of computer vision and robotics. David Banks provides an historical perspective and discusses the use of vision systems in future flexible automation. Edmund Andersson covers the more established concepts currently in use.

The last papers in this section cover the concept of cellular manufacturing which is a cornerstone in the development of the newer automated manufacturing systems. J. T. Black presents a comprehensive overview of the entire area, while Dumolien and Santen stress the importance of the approach to CIMS at Deere & Co. The last paper by Dunlap and Hirlinger discusses these concepts as they apply to the manufacturing system.

One of the major problems in putting together this section was how to effectively deal with such a large topic in such a limited space. The problem was eased considerably by the absence of papers on many of the topics which the reader may feel should have been included. However, this should provide guidance for the preparation of future articles for publication in IIE literature. Keep in mind that the question is not <u>should</u> we proceed toward accepting the concept of CIMS and the factory of the future, but <u>how</u> can it be done, and <u>when</u> will it become a reality.

Computerized Facilities Planning And Design: Sorting Out The Options Available Now

By H. Lee Hales
Management Consultant

In the past year, the costs of computing and computer graphics have come down dramatically. At the same time, they have become easier to use. As a result, computer support has become more cost-effective than ever.

This year's facilities planning and design survey includes 88 commercially available aids—software and complete turnkey systems—and more are announced each month. The fact that facilities planning has been "discovered" by a growing number of vendors presents the would-be user with a variety of opportunities—and some pitfalls.

Computer technologies

To make sense of the available products, and to guide internal programming efforts, planners need to understand three computer technologies:
☐ Decision support systems (DSS).
☐ Computer-aided design (CAD).
☐ Management information systems (MIS).
The natures of these technologies and their places in the planning process are summarized in Figure 1.

A DSS includes statistics, modeling, algorithms and calculations. Business or display graphics (charts, graphs, block diagrams) are used to interpret results. DSS are applied to the early phases of a project for sizing, assigning locations, block layouts and providing feasibility.

CAD produces plans, detailed layouts, drawings and visualizations. In the design of structures and mechanical systems, CAD may include interfaces with engineering routines for various load, stress and sizing calculations. Visualizations may include static views or dynamic simulations in two and three dimensions. Nearly all systems provide ancillary outputs such as take-offs and area measurements.

MIS include the routine storage and retrieval of data, with and without graphic displays. In the latter phases of a project, the MIS is used for scheduling and control. MIS is widely used after a facility is occupied to keep track of space and equipment. Sometimes, the data contained in various information systems provide the input to decision support.

As shown in Figure 1, these technologies benefit different phases of a planning project. In general, the capacity, location and conceptual decisions made early in a project have more impact on profit and efficiency than detailed layout, design and construction. As a corollary, computer technologies that aid the early phases of a project have the greatest potential. The message is clear: Exploit DSS first, then CAD and MIS.

Recent developments

The past year has produced a steady stream of useful products, especially for micro and personal computers. In the DSS category, Pritsker & Assoc. Inc., Systems Modeling Corp. and others have extended important simulation capabilities to the IBM-PC and compatible micros. The IIE's own micro software series already includes one offering in this area, with two more packages—Statistical Analysis and Plant Layout—expected to be released later this year. These are useful developments for planners of process facilities and handling, storage and assembly systems.

Group technology is being adopted at a faster pace, giving planners more of the data they need for cellular layout plans. Moore Productivity Software, with its SLPCALC, has provided a clever exploitation of the spreadsheet for layout planning. Computervision recently acquired OIR, and now offers its Group Technology Applications software.

New forms of integrated software have appeared: 1-2-3 from Lotus, VisiON from Visicorp, MBA from Context and many others that combine spreadsheets, file management and business graphics. Using these products, planners with only modest computing experience can develop useful DSS applications on popular microcomputers.

For office space planners, the past year produced ready-to-use decision support for space projections, clus-

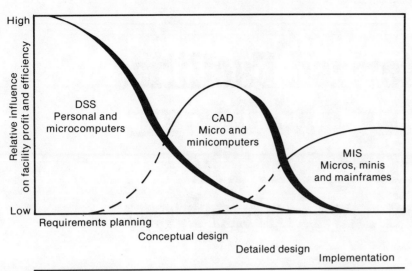

Figure 1: Using Computer Technologies

	DSS	CAD	MIS
Factory & warehouse planning	• Calculating/statistics • Curve-fitting routines • Product-quantity plot • Parts grouping routines • Layout algorithms • Flow diagramming • Process planning • Simulation languages • Linear programming • Business graphics • Financial modeling • Query language to MIS	• 2-D; 3-D drafting • Symbol libraries • Bill of materials • Report generator • Area calculations • Dynamic simulation • Robotic simulation • Interface to civil & structural analyses	• Facilities database • Project management • Capital budgeting • Maintenance management • Interface to engineering database • Interface to production database • Interface to fixed asset management
Office & laboratory planning	• Calculations/statistics • Curve-fitting routines • Proximity/clustering • Layout/stack algorithms • Business graphics • Financial modeling • Query language to MIS	• 2-D; 3-D drafting • Symbol libraries • Bill of materials • Report generator • Area calculations • Color fill & shading • 3-D visualization • Lighting simulation • Acoustical analysis	• Facilities database • Project management • Capital budgeting • Maintenance management • Lease management • Interface to fixed asset management

ter/proximity analysis, vertical stacking and even layout planning. Such programs are now available for micros running the CP/M, MSDOS and UNIX operating systems.

CAD continues to be the most active technology, with most of the larger vendors taking aim at the needs of facilities planners. Once-exotic capabilities such as solids modeling, color fill and shading, dynamic simulation and symbol dragging have become more common at lower prices.

This has given office space planners the ability to produce electronic "renderings" of work stations and interiors. Steelcase Inc. will even simulate interior lighting schemes on a service bureau basis.

For industrial planners, such capabilities are making possible the graphic simulation of robots and work cells. McAuto's PLACE system, perhaps the best current example, includes post-simulation software for programming popular robots.

Clearly, CAD is becoming a tool for better planning in addition to its proven value in drafting. In fact, drafting and symbol placement are becoming widely available software utilities, much like word processing or business graphics. Several drafting packages were recently introduced for the IBM-PC and other microcomputers.

This year's survey includes 30 systems and software packages intended for architectural drafting. Unfortunately, the productivity of CAD systems varies a great deal. One may require a dozen or more commands and cursor moves to insert a door into a previously drawn wall. Another may do this in five or fewer. Some systems cannot measure areas in a repeatable way.

As a rule, products that do not give special attention to floor plan drafting (walls, symbol placement, area measurement, input of existing plans) are unlikely to be productive for facilities planning.

With so many vendors and products, there is a great deal of confusion about prices and performance. Today's popular CAD products fall into three categories, as shown in Figure 2:

1.) *Personal computer systems*—off-the-shelf and enhanced with extra circuit boards. Price: $10,000-$50,000, depending upon capacity, display and plotter. Best used for small, single-line 2-D work using simple, repetitive symbols. Little, if any, reporting capability.

2.) *Microcomputer systems*—and low-end microcomputers. Price: $50,000-$100,000, again depending upon hardware chosen. Useful on most facilities work, including limited 3-D applications. Wide range of drawing-related reports.

3.) *Minicomputer systems*—and high-end micros. Price: $100,000-$200,000 depending upon hardware and software options. Useful in all applications, including piping and process plant design. Full range of reports from associative, non-graphic databases.

Planners interested in CAD should examine systems in each category to acquaint themselves with the very real differences in speed, capability and storage capacity. Several vendors offer decision support and MIS applications running on the same system with CAD. Although this is not always necessary or desirable, it suggests a grasp of facilities planning that sets these vendors apart from those who concentrate on general purpose drafting and mechanical and

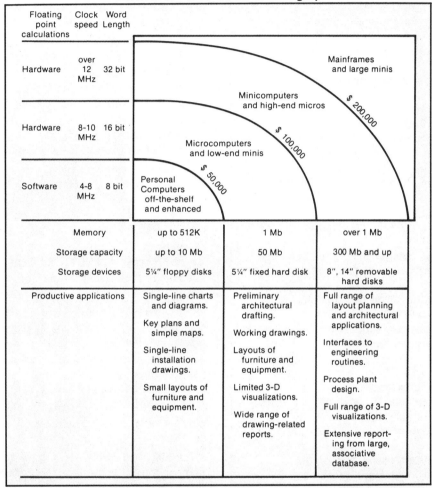

Figure 2: CAD Price-Performance Categories (Price is for a single station, fully equipped with CPU, disk drive and storage, screen copier, printer, large plotter, small digitizer tablet and software—late-1983 offerings.)

circuit design.

MIS have had their share of recent activity. Project management software has been "down-sized" for micro and personal computers. So, too, have programs for lease and property management, preventive maintenance and asset control.

Databases of facilities information are now much easier to create, in-house, with the recent flood of database management systems (DBMS) for micros and main-frames. New products have made it easy to link geographically and functionally dispersed micros to central mainframe databases. Planners at headquarters and in the field, or those in different departments, can share data while still maintaining locally unique systems to address specialized needs.

As companies press for tighter control and better utilization of facilities, the role of MIS will become more important. (See Table 1.) Readily available asset information, and the flexibility to report it in a variety of ways, will be the mark of a good facilities planner.

Integrated approaches

If you plan offices and labs, you can choose from a number of integrated DSS/CAD/MIS systems. These take space projections and relationships as input and give back clusters, multi-story assignments and, in a few cases, block layouts—all for current and future projections.

Once block plans are complete, the same system is used for detailing with full CAD capabilities. The MIS component may be weak, but is at least capable of maintaining space and furniture inventories.

CPMI, Decision Graphics, Resource Dynamics, Autotrol, Calcomp and TRICAD have all offered varying degrees of integrated systems for some time. Intergraph and Computervision are expected to demonstrate theirs shortly. *(Editor's note: Computervision's new Facilities Management Package is announced in the survey that follows this article.)*

If you plan factories and warehouses, you cannot yet find commercial systems with the same degree of integration or functional standardization. Although the office-oriented systems can be used for plant layout, they are missing critical functions for group technology and material handling analysis. At present, these must be performed with separate systems or software.

There is nothing inherently wrong with using discrete systems for DSS, CAD and MIS. In fact, a good case can be made for separate, specialized aids that do not compete for a single computing resource, or a single color display, or degrade response time when used simultaneously.

Even in office space planning, not every facilities group needs an integrated solution. The two chief conditions that favor this solution are:

☐ A single facilities group has comprehensive internal responsibility, from concept through design to installation and ongoing management. All tasks are performed in-house, at one location.

☐ The output of one planning step is *directly* usable as the input to the next. The same original data are used repeatedly for different planning

Table 1: Uses of INSITE Facilities Database

1. Maintain up-to-date inventory of facilities.
2. Provide accurate and timely inventory lists by user, by building, by floor, etc.
3. Provide automatic response for federal, state or local report requirements.
4. Hold information about space operations, allocation and planning.
5. Experiment with a variety of space allocations.
6. Review space allocation policy.
7. Match activities to appropriate spaces.
8. Track door key assignments and room access.
9. Inventory capital equipment by location.
10. Locate major pieces of machinery with an eye towards consolidating their use.
11. Analyze space change costs by square foot, room type and space user.
12. Analyze telecommunications usage.
13. Analyze building use by occupant, room function, organizational use and program use.
14. Project future space needs based upon existing use.
15. Provide planning data to assist in "build" or "renovate" decisions.
16. Assist in long-term and immediate planning decision-making.
17. Prorate building costs directly to building users.
18. Provide building maintenance budgeting/scheduling.
19. Inventory parking facilities and analyze parking assignments/users.
20. Analyze existing building maintenance by trade.
21. Project future building maintenance by trade.
22. Track exterior spaces by type, function and work requirements.
23. Track and analyze crime occurrences.
24. Relate space inventory data to personnel and financial data for analysis.
25. Study and produce overhead recovery/patient care cost allocation reports for submittal to government agencies.

Courtesy of Office of Facilities Management Systems, Massachusetts Institute of Technology.

tasks, and are maintained during the ongoing management of the facility.

Short of a single system that does it all, planners have a large array of possible configurations, as shown in Figure 3. These include local service bureaus for CAD and time-sharing for DSS and MIS.

PLANPAK, from Moore Productivity Software, is an example. Through the GE Information Services network, users can access popular layout algorithms such as CRAFT, CORELAP and COFAD, as well as other routines for material handling and layout evaluation.

Planners in large corporations usually have internal, time-shared access to a variety of software for statistics, display graphics, financial modeling and, often, project management.

For those who believe an integrated system is realistic and desirable, careful planning is required. With today's networking and telecommunications products, it is becoming easier and less expensive to link various pieces of hardware. But accessing or sharing data and applications software once all the hardware has been connected continues to be difficult.

Industrial facilities planning needs to draw from a variety of other systems. It also needs its own dedicated capabilities. As suggested in Figure 4, the desired functional and software integration will determine the most desirable physical or hardware configuration.

The systems in use at Deere & Co. are a good example. Facilities planning is supported by five partially integrated systems:

1.) JD/GTS—John Deere Group Technology System—is a collection of group technology databases, one for each product unit. These hold non-graphic part descriptions and related data on production rates and routings.

2.) FROM-TO ANALYSIS—actually part of the JD/GTS—extracts a variety of material flow measures from the routing files of the parts database. Output can be in tons, pieces, number of parts or even number of containers per day, week or month. Output can be entered directly as the volume array for CRAFT.

3.) CRAFT—the original layout improvement routine—accepts volume manually or automatically from JD/GTS (via FROM-TO). The planner enters the distance array and move costs manually.

4.) GPSS—one of the first and still the most powerful of simulation languages—is used to model conveyors and other automated systems. Once constructed, the models of newly installed systems are kept intact for subsequent analysis of changes and improvements.

5.) GMS—geometric modeling system—uses the CADDS 4 system from Computervision. This system contains graphic part descriptions. It is used for 2-D and 3-D graphic simulations of clearances and movement of parts and equipment. It is also used for detailed plant layout once CRAFT has verified the best block plan.

According to David Scott, Deere division engineer, these systems have been evolving since 1976. The JD/GTS was developed in-house by engineer-programmers after initial experimentation with a commercial group technology system.

The CRAFT code was installed without modification, as was GPSS. Roughly 70% of the GMS commands and menus had to be developed by Deere's engineers. Only 30% came from the basic, vendor-supplied software.

All but GMS reside on IBM mainframes in the corporate data center. They are accessible from any manufacturing plant. Each plant with a CADDS 4 system can use the GMS menus and commands. Files are shared between plants by exchanging magnetic tapes. Future goals include

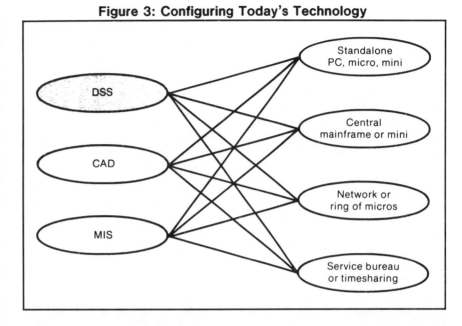

Figure 3: Configuring Today's Technology

Figure 4: Computer-Integrated Facilities Planning

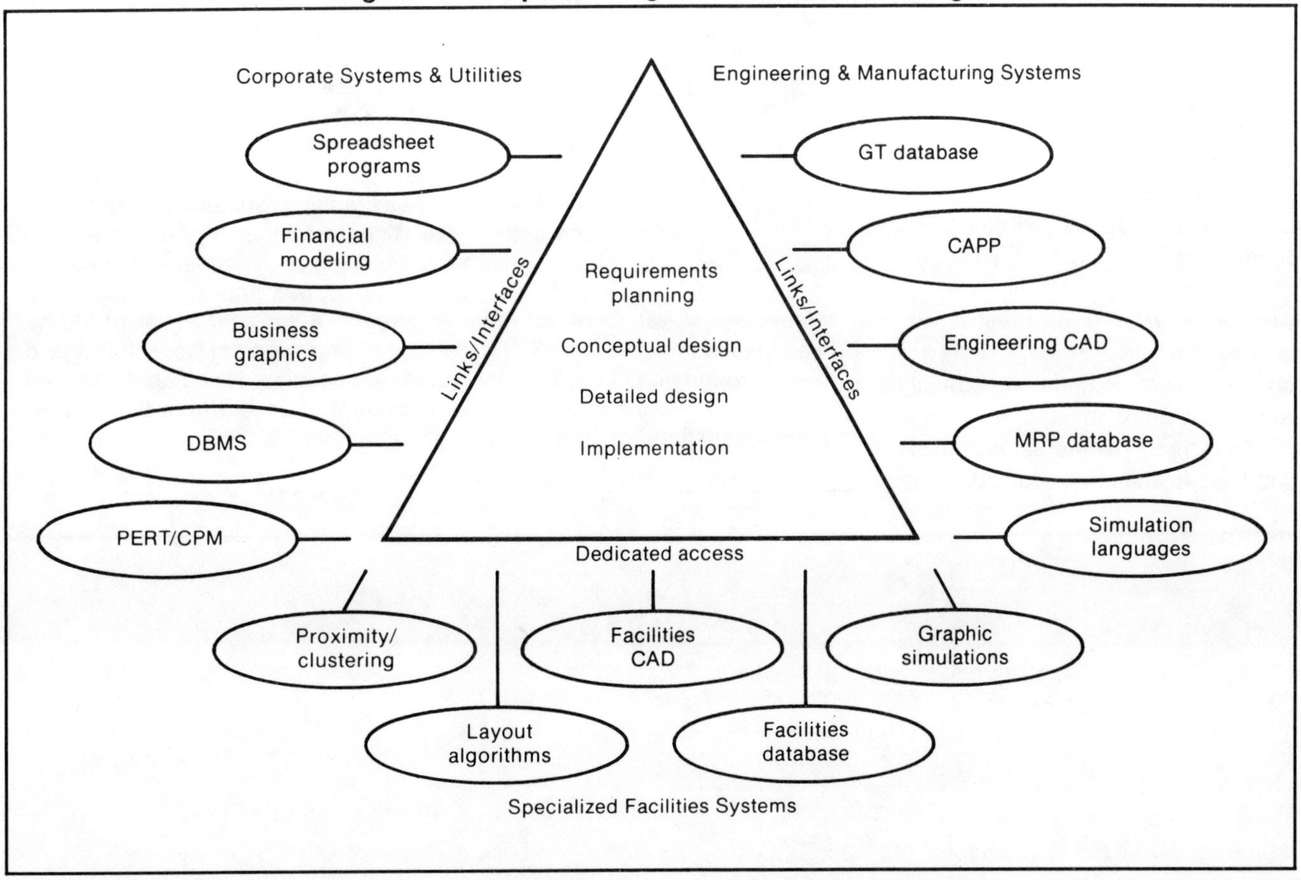

an interface between the JD/GTS and GMS and a system for process planning.

Clearly, these accomplishments required foresight and many man-years of sustained commitment. Commercial products have played an important role, but major developments and all of the system integration have been accomplished in-house.

Agenda for action

All too often, the facilities planner's options are limited by the decisions of others. Choices of hardware, database software, programming languages and networking are frequently dictated by data processing Industrial planners are often captives of product engineering and manufacturing when it comes to group technology, CAD and production (MRP) database design.

As a result, planners may be left with less than desirable computer and data access. How do we avoid this outcome and the restricted opportunities that will follow? Here is a five-point agenda:

1.) *Systematize first; computerize second.* Little coherent use can be made of computers without first committing to standard procedures and techniques that can be shown to work *manually*. If planning is hit-or-miss, without method, strictly ad-hoc, the use of computers is unlikely to make it more effective.

2.) *Investigate available aids* once you have installed a good system for planning. Review the survey that follows. Attend relevant shows and seminars. Do not rely solely on the vendors who happen your way. Seek out others already using computer aids.

3.) *Develop a long-range plan,* based on your system of planning and your new-found knowledge of what is available. Identify the aids you must develop in-house and any overlaps with other computer systems, planned or in place. If in doubt, seek help.

4.) *Make your needs known* by sharing your plan with data processing, engineering and manufacturing. Don't be an afterthought on engineering's CAD request. Work together to ensure a coordinated approach that satisfies your long-term needs.

5.) *Take the lead.* Be a catalyst for better computer systems throughout your organization. If there is a vacuum in engineering or manufacturing, help fill it with informed decisions that will benefit all.

Well planned facilities are central to productivity and successful automation. Better utilization of facilities provides significant economic returns, and computer aids can help companies achieve this. David Scott of Deere & Co. reports that Deere's systems helped achieve annual savings of more than $500,000 on a large rearrangement. The potential is there; but the payoff requires a plan.

Note: Portions of this article are reprinted from the forthcoming book: Computer-Aided Facilities Planning, *by H. Lee Hales, copyright 1984, Marcel Dekker Inc., New York, NY.*

H. Lee Hales is a Houston-based consultant who specializes in facilities planning and factory automation. He works with engineers and data processing on a variety of computer and automated systems and advises software and systems suppliers on the needs of industrial managers and planners. Hales is the author of a forthcoming book, *Computer-Aided Facilities Planning* (September 1984), and co-author with Richard Muther of the two-volume set *Systematic Planning of Industrial Facilities*. He is a contributor to Auerbach's *CAD/CAM Series* and to McGraw-Hill's *Management Handbook for Plant Engineers*. A senior member of IIE, he also belongs to CASA-SME, the Institute of Management Consultants and the Independent Computer Consultants Association. Hales holds BA and MA degrees from the University of Kansas and an MS from the Massachusetts Institute of Technology.

Production and Order Planning in the FMS Environment

W. Bruce Webster
Vought Aero Products Division
LTV Aerospace and Defense Company

ABSTRACT

Implementation of advanced manufacturing technologies to meet cost and schedule requirements for current and future aerospace programs presents a number of challenges to the industry contractor seeking to apply new automated methods to specific program needs, while expanding the potential for cost-effective production of future programs. The Vought Aero Products Division (VAPD) of the LTV Aerospace and Defense Company conducted extensive analyses in determining that a Flexible Machining System (FMS) offered the greatest benefit to the current B-1B aft and aft intermediate fuselage (AIF) assembly subcontract effort.

However, installation of the hardware and software elements of the FMS by no means completed the task of implementing the automated system into production operations. Considerations for integration of the FMS into VAPD planning and order release systems had to be addressed. Acceptable procedures had to be developed that would coincide with both program schedule requirements and the system's advanced production capabilities. This discussion presents the issues and observations concerning FMS production and planning integration with a large population of part numbers in a low rate environment, as experienced at VAPD.

INTRODUCTION

To meet the need for machined parts in the B-1B Program's aft and AIF assembly efforts, VAPD was faced with the decision to either purchase large quantities of machined parts or to add to inhouse capacity. Extensive trade studies and analyses indicated that the most cost-effective and advisable solution was the purchase of a Flexible Machining System.

Part Population

The population of B-1B machined parts that were considered in the FMS selection process and are now being produced by the system consists of more than 500 part numbers that can be machined in a work envelope measuring 32 by 32 by 36 inches. A typical part is machined from a 15- by 5- by 2-inch or smaller piece of plate stock. The largest part is machined from a piece of plate stock measuring 33 by 22 by 5 inches. The machining of the parts within the identified part population involves a total of just under 1,100 operations, or part orientations.

System Configuration

The VAPD FMS, as illustrated in Figure 1, is designed to fulfill the fabrication needs of the defined part population, and is comprised of the following elements:

o Eight 4-axis computer numerically controlled (CNC) machining centers equipped with 90-cutter tool magazines and pallet delivery/discharge systems

o Two coordinate measuring machines (CMM)

o Automatic wash station

o Four wire-guided vehicles

o Forty machine work pallets

o Two carrousel storage devices each with two load/unload stations

o Cutter automated storage and retrieval system (AS/RS)

o Two fixture buildup stations

o Dual-flume chip removal/coolant distribution system to support the machining of ferrous and nonferrous materials

o System control computer network including direct numerical control (DNC) capability.

The system functions without human intervention except for the part loading/unloading operation and support areas such as fixture building, and cutter loading and delivery.

Tooling Philosophy

Each of the 40 machine work pallets in the system is mounted with one of eight different styles of risers, as shown in Figure 2. Some riser styles require a part-specific fixture, while others do not. Typically, a part-specific fixture

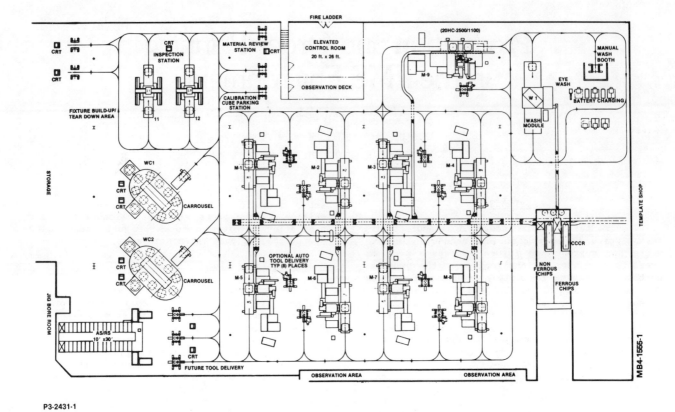

Figure 1   Flexible Machining System Layout

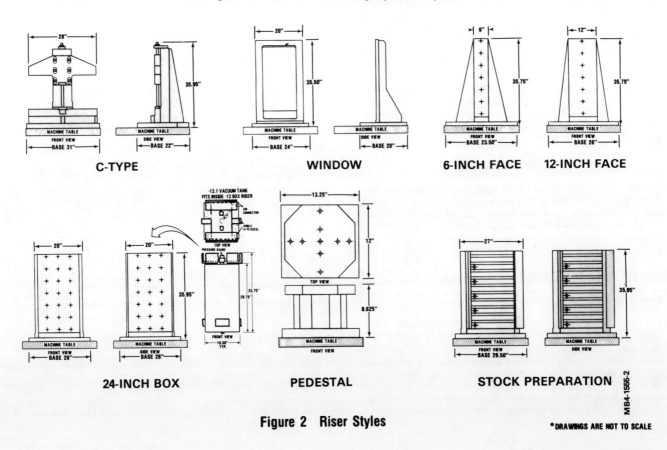

Figure 2   Riser Styles

*DRAWINGS ARE NOT TO SCALE

is a sub-plate that bolts to the riser and contains special locating pins and back screws for mounting a piece of plate stock. Only one copy of a part-specific fixture is generally available in the FMS.

Pallet Loading Philosophy

Depending upon which riser style is used, up to eight different part numbers can be loaded onto one riser. The FMS risers were designed to allow the mounting of multiple parts on each one in order to maximize the utilization of the wire-guided vehicles and to reduce the loss of spindle time that occurs when shuttling parts in and out of the center section of the machine tool. The mounting of different part numbers was driven by the requirement of the system to support "serial" production.

Serial production is a philosophy wherein parts are produced in lot quantities containing only the numbers required to support the fabrication of only a single ship set. Since only one part of a given part number is required per ship set in a majority of cases, a number of different part numbers had to be mounted to meet the multiple-parts-per-riser requirement.

PRODUCTION PLANNING AND CONTROL

Successful operation of the FMS within the VAPD production environment was, to a great extent, contingent upon the integration of the automated machining function into the production planning and control operations required to meet B-1B assembly schedule indentures. At issue in the developmental level was the actual work release philosophy to be followed in controlling the production of parts within the automated environment.

The determination of a viable, adaptable release philosophy for the dramatically different FMS operation required the consideration of a number of data requirements that were critical to the definition and actual use of a work order scoping and control approach for the system. These data requirements included:

o Manufacturing routing

o Order requirements

o Numerical control part programs

o Inspection programs.

In addressing these data requirements that are critical production control factors, the component resource and functional elements of each were defined and analyzed in terms of FMS operations.

Data Requirements

To support an automated system such as the VAPD FMS, extensive data is required in machine-readable form. This data must be of a very detailed nature since the system runs with a minimum of manual intervention. This data, to the maximum extent possible, must come from existing company data systems. The origination of data from existing systems eliminates redundant data input, enhances data accuracy, and ensure consistency between system activity and overall company plans.

Manufacturing Routing. These data are part-number specific and call out each location (load/unload, machining center, wash, CMM, etc.) through which a part is required to be processed. Location data address the following factors:

o Riser style

o Part-specific fixture identification (ID)

o Numerical control program ID

o Inspection program ID

o Operation "run-with" ID to support the multiple-part-per-pallet loading concept

o Required standard hours.

Order Requirements. These data are part-number specific and include:

o Work order ID

o Number of pieces

o Completion date for each location in the manufacturing routing, which is derived from the assembly demand of the part number.

Numerical Control Part Programs. These data consist of the APT post-processed output that is stored in the DNC library. In addition to the machine motion and part geometry, a list of cutter assemblies and the aggregate time in cut for each assembly is included.

Inspection Programs. These data comprise the control programs for the CMMs. The programs are loaded in a library and maintained in a manner analogous to the DNC library for the machining programs.

Pallet Set Grouping

To implement a workable release philosophy, a the pallet loading approach had to be determined as well. Once the overall approach to be followed was identified, the software had to be developed to organize the part attributes that govern the grouping of parts. The primary objective of the pallet set grouping software is to identify a group of part machining operations that are made from the same type of material, require the same style of riser, have compatible geometry, are due within the same time frame and have high commonality among the cutters required.

After the manufacturing planning was modified to reflect the FMS routing, a file of the involved parts was extracted, and the Pallet Set Grouping report, as shown in Figure 3, was prepared. This report sorts operations by material type, riser style, and setback day (the number of days in advance of product shipment that the part is required to be complete) time bands. Within each time band, the cutter lists for each job are compared to determine the highest cutter commonality.

PALLET SET GROUPING REPORT

PROD - B-1B MATLTYPE - ALM RISER TYPE - 13 STYLE - BOX AVAILABLE FACES 4 MAX. PARTS/FACE 2
FROM SET BACK DAY 136 TO SET BACK DAY 126

TRIAL GRP	PS ID	PRODUCT	SBK	QTY	PART NUMBER	OPNO	ITEM	IN	PGM S/N	SFX	FIXT S/N	SFX	PPF	LGTH	THICK
1	130455-02	B-1BAFT	125	1	L5414151-003	0116	448	02	R821782		R704375		1	30.8	5.650
	130455-04	B-1BAFT	125	1	L5414151-004	0116	448	02	R821785		R704891		1	30.8	5.650
	130452-01	B-1BAFT	125	1	L5414187-003	0290	448	02	R821817		R704380		1	20.6	3.150
	130452-03	B-1BAFT	125	1	L5414187-004	0290	448	02	R821819		R704381		1	20.6	3.150

COMBINED CUTTERS	L5414151-003	L5414151-004	L5414187-003	L5414187-004
12000005	12000005	12000005	20000171	20000171
20000171	20000171	20000171	50000049	50000049
50000036	50000036	50000036	50000075	50000075
50000094	50000094	50000094	50000128	50000128
50000102	50000102	50000102		
50000115	50000115	50000115		
50000169	50000169	50000169		
50000247	50000247	50000247		
50000248	50000248	50000248		
52000001	52000001	52000001		
55000003	55000003	55000003		
50000049				
50000075				
50000128				

TRIAL GRP	PS ID	PRODUCT	SBK	QTY	PART NUMBER	OPNO	ITEM	IN	PGM S/N	SFX	FIXT S/N	SFX	PPF	LGTH	THICK
2	130463-04	B-1BAFT	136	2	L5226212-003	0085	468	01	R822252		R704596		1	4.5	2.150

COMBINED CUTTERS	L5226212-003
50000033	50000033
50000035	50000035
50000223	50000223

TRIAL GRP	PS ID	PRODUCT	SBK	QTY	PART NUMBER	OPNO	ITEM	IN	PGM S/N	SFX	FIXT S/N	SFX	PPF	LGTH	THICK
3	130454-02	B-1BAFT	125	1	L5414151-003	0122	448	02	R822352		R704375		1	30.8	5.650
	130454-04	B-1BAFT	125	1	L5414151-004	0122	448	02	R822353		R704891		1	30.8	5.650

COMBINED CUTTERS	L5414151-003	L5414151-004
12000005	12000005	12000005
20000171	20000171	20000171
50000036	50000036	50000036
50000073	50000073	50000073
50000089	50000089	50000089
50000094	50000094	50000094
50000102	50000102	50000102
50000115	50000115	50000115
50000169	50000169	50000169
50000247	50000247	50000247
50000248	50000248	50000248
52000001	52000001	52000001
55000003	55000003	55000003

TRIAL GRP	PS ID	PRODUCT	SBK	QTY	PART NUMBER	OPNO	ITEM	IN	PGM S/N	SFX	FIXT S/N	SFX	PPF	LGTH	THICK
4	130453-01	B-1BAFT	125	1	L5414151-003	0082	448	02	R821781		R703463	1	1	30.8	6.500
	130453-03	B-1BAFT	125	1	L5414151-004	0082	448	02	R821784		R704890		1	30.8	6.500

COMBINED CUTTERS	L5414151-003	L5414151-004
54000001	54000001	12000005
55000003	55000003	20000077
80000006	80000006	20000122
12000005	12000005	20000171
20000077	20000077	20000174
20000122	20000122	30000007
20000171	20000171	30000016
20000174	20000174	30000042
30000007	30000007	50000075
30000016	30000016	50000094
30000042	30000042	50000106
50000075	50000075	54000001
50000094	50000094	55000003
50000106	50000106	80000006

MB4-1555-3

Figure 3 PSG Report

The jobs are further evaluated based on their size since this determines whether multiple parts can fit on a particular riser face.

The pallet set grouping data are further analyzed to group jobs on a particular pallet. This analysis is conducted to assure that:

- o Multiple jobs requiring the same fixture are not grouped on the same pallet
- o Different operations on the same part number are not grouped on the same pallet
- o Thickness differences between two parts on a face do not exceed 1 inch.

The result is a recommended grouping of operations for a pallet. These recommended or "trial" groupings are then evaluated by numerical control (NC) programming for collision potential. After evaluation and adjustment, pallet set IDs are assigned and input into manufacturing planning, as required.

Release Philosophy

To optimize the operational advantages offered by the automated, flexible machining system, a release philosophy that is responsive to the system's functional characteristics is necessary. In the FMS design and implementation process, investigations into how work is controlled through the conventional, manually paced shop and the requirements for control through the computer-driven shop forced a rethinking of the manner in which orders are released.

Conventional Method. In the traditional, manually paced shop, jobs are released to the factory floor in accordance with an assembly demand schedule. These work orders are formatted in large lot sizes in order to reduce the setup cost per piece. This is necessary because the machine tool, the most expensive component of the shop overhead rate, is nonproductive when an individual workpiece is set up for machining. By machining a quantity of the same part number, this initial setup effort is effectively "spread" across all the parts in the designated lot.

In this conventional environment, the resulting shop workload is most commonly characterized by a work content profile with pronounced peaks and valleys. To smooth out this "roller coaster" load and manage the machining operation, the traditional machine shop is staffed to a level-load manpower requirement and jobs are worked ahead of schedule. This mode of operation requires a significant amount of informal manual coordination. It may also force the splitting of orders into smaller quantities to meet assembly demands thus losing the savings in setup time costs originally intended by the large lot sizes.

The problem can be even further aggravated if the remaining orders of the split are left with the original schedule. This situation results in overstating the behind-schedule load, since the remaining parts are, in fact, required to support a later article in the assembly line.

FMS Method. Since the selection of jobs for the FMS is computer driven, it is essential that the work order release plan prepared for the FMS is an achievable one. In an environment where large machined part lot sizes feed small assembly lot sizes, machining resources are thus committed to the fabrication of parts substantially ahead of assembly need.

The upper portion of Figure 4 illustrates this condition. The area under the large curve represents the work content required to machine a lot comprised of the number of parts needed for the assembly of 15 ship sets. The pieces produced in this lot will be used in assembly Lots A, B, and C. The spread of the distribution through time is driven primarily by the indented assembly structure of the end product. This distribution is positioned in time to support the first assembly point in time of Lot A; thus the parts used for building assembly Lots B and C are produced substantially ahead of need. It should also be noted that the distribution peaks <u>above</u> the capacity of the machining equipment.

The lower portion of Figure 4 represents the same work content (the area under the three small curves equals the area under the one large curve in the top of the illustration), but it has been

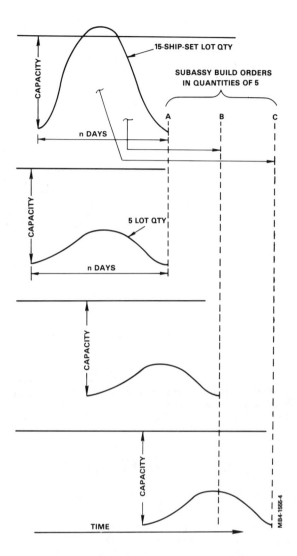

Figure 4 Assembly Use of Detail Parts

distributed to support the individual assembly need points. The loads in the three small curves peak well under the capacity line. It is this small lot philosophy that is recommended for the FMS environment.

In the above discussion, only the machining content is included, that is, the setup and load/unload content have been excluded. It is true that the fixture setup effort for three small lots is greater than for a single large lot. However, the fixture setup in the FMS environment is done away from the machining center at a lower overhead rate. The off-machine setup operation also makes the spindle available more hours per day for cutting chips. The effect of this release philosophy through time is illustrated in Figure 5. This diagram includes a series of seven releases of five ship sets each that are spread through time to support assembly lot quantities of five parts. These small lots overlap each other significantly. The addition

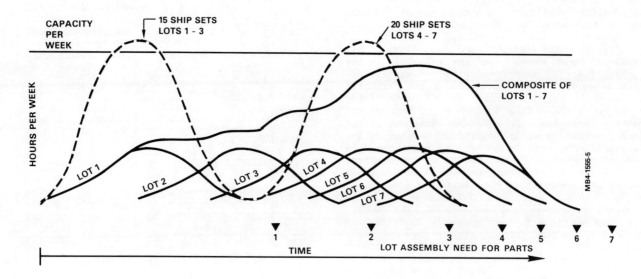

Figure 5 Small vs Large Lot Release to the FMS - Cumulative Effect of Lot Releases

of this overlap results in a composite load curve that is smooth and does not peak above the capacity line. The figure also references the two comparative conventional lot releases that include one lot for 15 ship sets of parts and another for 20 ship sets of parts.

INTEGRATION OF FMS INTO THE CONVENTIONAL FACTORY

The installation of an FMS at VAPD facilities in Dallas, Texas, was the Division's first major step into unmanned factory automation for discrete parts manufacture. VAPD includes personnel with a host of experience in the technological elements that comprise the system, such as NC parts programming, and order release and shop floor control. However, the integration of these into an FMS presented new and significant challenges.

Existing Factory Environment

The environment into which the FMS was installed consisted of the following functional elements both prior to and after the actual part machining operation:

o Material cut from plate stock is a batch-oriented operation that is performed by operators using circular saws (Larger lot sizes are used because such quantities tend to reduce the material waste caused by the overrun of the arc of the saw blade)

o Chemical surface treatment and paint operations include some automated activities

o Subassembly is oriented toward small, lot-size batch production (usually one-third to one-fourth of a machined part lot size)

o Part movement and location readings are manually regulated.

Existing Information Systems Environment

The following information systems were well established at the time of the FMS installation.

Master Schedule. This system provides dates at the item and indenture level.

Manufacturing Planning. This system provides routing, fixture requirements, instructions, material requirements, and item and indenture for individual part numbers. Data in this system is regulated by effectivity; therefore different methods can exist in the file for the same part.

Order Requirements and Work Order Release. Part requirements are accumulated in this system with their respective schedules. Miscellaneous requirements, such as spares and refabrication requirements, are entered directly with their schedules. Basic requirements are generated upon receipt of a work authorization for a designated quantity of ship sets. The system selects effective parts from manufacturing planning and schedules them based on their item and indenture. A daily survey of the requirements combines them for a given part number, as required, and then releases a work order to the shop in accordance with the appropriate schedule.

Open Order. This system is used to track a released work order through the factory. Records in this file are established by the order requirements and work order release system.

Considerations for FMS Adaptation to Conventional Factory Environment

Although the FMS was designed to support serial manufacturing (one ship set of parts at a time), it was recognized that at the time of implementation it was really a "serial" island in a "batch" ocean. It was also recognized that the FMS, though constituting a rather significant capital investment, was only one element in the overall factory. Therefore, implementation of a single "island of automation" could not unnecessarily force major changes to existing business systems controlling such functions as manufacturing planning and order release.

Certain considerations had to be assessed and weighed to determine the overall affects of the FMS on existing factory operations, policies and procedures. Those critical to the cost-effective and efficient operation of the new automated system were identified and scoped in order to provide the greatest benefit to the FMS with the least negative effect on the existing factory areas. These critical factors are presented in the paragraphs that follow.

Lot Size. A lot-size policy of producing in subassembly lot sizes was adopted. As discussed earlier, this method allows for the formulation of a smooth achievable load. It also provides parts in a manner that is consistent with the way in which the subassembly areas are planned and budgeted. Only minor adjustments to the existing order release system were required to allow machined part orders to be released in subassembly lot quantities.

Planning Modifications. Existing planning procedures had to be modified to establish and input inspection program serial numbers, riser IDs, and pallet set IDs for each FMS-produced part number. This modification required only minor changes to the planning system.

Standardization of Cutter Assemblies. To minimize cutter changes and to provide a tool for analyzing cutter requirements, cutter assemblies were standardized and cataloged by means of cutter assembly identification numbers. These cutter assembly ID numbers are also used to identify cutters in the NC programs.

Considerations for Information Flow between Conventional Factory and FMS

In any automated manufacturing system, the key to successful utilization of unmanned capabilities is the continuous and directed flow of information. Information is the ultimate controlling mechanism in the FMS environment. However, in VAPD's existing conventional factory, this information-centered environment had to interface with the human-centered environment of traditional manufacturing. Figure 6 illustrates the clearly defined, yet complex relationship between VAPD's existing information systems and those of the FMS.

Passing NC and CMM Programs to the FMS. After a job is post-processed on the existing factory mainframe computer, the NC data is transmitted to the FMS computer and stored on a library within it. The NC program library also contains a composite list of cutter assemblies for each respective program along with detailed data that gives the total time in cut for each assembly. When an NC program is altered, this processing and filing activity is repeated.

Passing Order Data to the FMS. Transferring order data from existing business systems to the FMS computer required the development of a computer file that would allow such a data exchange without negatively affecting conventional or FMS systems. To minimize the impact to VAPD's existing business systems, a new FMS load file was created. This file reads the open order and manufacturing planning files, and creates a separate file of data to support the scheduling activity performed by the FMS computer.

A record in this file supports the processing of one part orientation for one work order. The record contains information such as part number, work order serial number, pallet set ID, NC and CMM program IDs, scheduled due date, material availability and stocking location, and loading instruction ID. This file is run once for each FMS planning cycle.

Scheduling FMS Work Orders by Pallet Set Groups. The FMS scheduler file reads the order data passed to the FMS and groups work orders by pallet set ID. Based on schedule and material availability, the file selects pallet sets that will adequately load the system for the designated planning period. This pallet set load data is augmented with information from the NC program library regarding the required cutter assemblies and their times in cut.

These identified pallet sets are then distributively assigned to the machine tools in order to minimize cutter changes and to level machine tool loads. The schedule plan that results from these calculations is stored in a data base for use by the FMS processor file. Reports are prepared showing the support activity required for the planning period. These reports include cutter assembly load/unload kits, and pallet fixturing requirements.

Although the relationship of a particular operation to a pallet set is fixed, every occurrence of a pallet set through the system will not necessarily include all operations on it. A partial pallet set may be loaded into the system when a spare of a given part number is needed or when a scrapped part is refabricated. In this partial pallet set arrangement, a single part number may be processed through the FMS with none or only some of its pallet mates in the load. A partial pallet set load is accommodated by a test that assigns an operation to a pallet load when its due date becomes imminent. Any pallet mates of the required part number, for which material

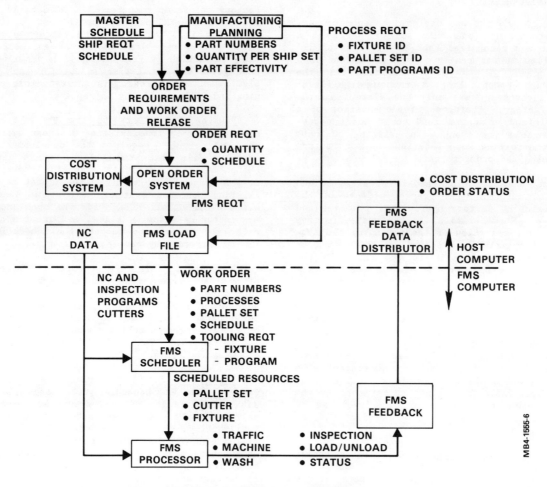

Figure 6 FMS Information Flow

is available and that have due dates falling within a designated future need time period, are also included in the partial pallet load.

The FMS scheduler file plays an additional critical role in the automated machining operation. Since machine tools are totally unmanned, the wear of cutters are modeled the FMS computer in order to eliminate the production of out-of-tolerance parts and the occurrence of cutter failure in the machining process, which adversely affects productivity and schedule compliance. When preparing cutter kit lists, the FMS scheduler file accumulates the times in cut for each assembly required for the designated pallet load. These times are compared to the projected remaining life of cutters already on the machine tools. If the total required cutter life is greater than what is available, additional cutters are listed for loading. Cutter removals are specified when required cutter additions would cause the capacity of the tool magazine to be exceeded. Cutter removals are prioritized in the following descending order:

o   Assemblies with cutter life exceeded

o   Assemblies not required in the next kit, in sequence by lowest frequency of usage.

Statusing and Monitoring FMS Operations. Statusing and monitoring of FMS operations is performed by the FMS processor file. This FMS processor is an on-line monitor that oversees the status of the system elements and directs the system in order to achieve the plan developed by the scheduler. It captures and stores order completion and cost distribution information in a data base.

The FMS processor also monitors and controls cutter use. Time in cut for a cutter assembly is decremented against its remaining life. Cutters with exceeded cutting lives are so noted in the file and are not called for again. This cutter status data is used by the scheduler in the preparation of cutter kits for the next planning period.

The FMS processor directs the load/unload of workloads into and out of the system by means of a terminal at the load/unload station. This terminal instructs the operator as to which positions on the riser are to be loaded or unloaded. The processor keeps track of completed pieces for each work order in the workload. If quantities are unequal among the work orders comprising a workload, and the order quantity for one of the part number positions has been exhausted, only those part number positions that remain to be loaded will be so noted by the processor file. When one of these less-than-full workloads is processed across a machine tool or CMM, NC programs will be executed only for those part number positions that are loaded.

Passing Data from the FMS. Once each planning period, the FMS feedback file collects order completion and cost distribution data that is accumulated in the data base and transmits it to the VAPD business systems host computer.

Distributing Data to the Business Systems. The FMS feedback data distributor receives the data sent from the FMS computer and reformats it in a form compatible for input to the existing conventional order status and cost distribution systems.

FUTURE EVOLUTION OF FACTORY AUTOMATION

The installed FMS, as described in the preceding discussion, is just a beginning for factory automation at VAPD. The Division's plans for program-oriented nondisruptive technology implementation include automation of the manufacturing operations that occur before and after the currently automated machining task. Planned automated operations that occur prior to machining include an AS/RS for plate stock, NC-driven plate cutting using either plasma arc or a band saw, and AS/RS of cut blanks. An integral part of the plate cutting operation will be the use of software that nests part blanks to allow plate stock to be cut in lot sizes to support a single ship set, while maximizing utilization of plate stock.

Planned automated operations that occur after machining in the FMS include chemical surface treatment, painting and subassembly. Automation in these operations will allow for integrated serial production in a computer-controlled manufacturing environment. The experience of development, design and implementation of the FMS and the continued pursuit of plans for establishment of a VAPD Multiproduct Factory of the Future demonstrate to us, as participants, and to our colleagues throughout industry, who are pursuing similar efforts, that the future in computer-integrated manufacturing has just begun.

BIOGRAPHICAL SKETCH

W. Bruce Webster is Manager of Industrial Control Systems within the Industrial Modernization organization of the Vought Aero Products Division of the LTV Aerospace and Defense Company, which is based in Dallas, Texas. Mr. Webster is responsible for the identification, design, development and implementation of information systems for operations departments. His areas of interest include production planning and routing, order requirements and release, and order status tracking. Mr. Webster holds the bachelor of science degree in industrial engineering from Purdue University. He is a member of IIE and is a past treasurer of the Pittsburgh, Pennsylvania, chapter.

Implementing a Computer Process Planning System Based on a Group Technology Classification and Coding Scheme

Joseph Tulkoff
Lockheed-Georgia Company

Using group technology concepts in concert with a unique coding and classification scheme, Lockheed-Georgia was able to automate the process planning function to a state-of-the-art generative system that yielded vast benefits to the manufacturing process. The Genplan computer system has eliminated inaccuracies in manufacturing planning and offered enhanced productivity and quality throughout the factory environment.

The whole may be equal to the sum of the parts in mathematics, but in manufacturing today the whole easily exceeds its components. Manufacturing converts raw materials into products: its very essence and success depend upon the right linkages of materials, technologies, and people being integrated systematically to create a new entity. Because manufacturing activities are interactive, if not kinetic, the computer is now seen as the foremost organizational tool to coalesce complex manufacturing processes. Computer-integrated manufacturing (CIM), in fact, promises to reshape the nature of the industry by investing in more automation, redesigning entire plants and processes as integrated flow systems, and integrating mini- and micro-computers into factory tools.

A host of benefits can be expected by using computers to integrate diverse functions into a single, continuously flowing manufacturing system. These range from such overall benefits as improved product quality, reduced manufacturing costs, and more manufacturable designs to more bounded benefits, like shorter lead times, reduced throughput and inventory, and less burdensome start-up and learning costs. Not only can computers streamline the so-called hard components of manufacturing, like equipment and machinery, but they also can simplify critical soft areas such as information flow and data bases.

An area in which computers particularly promise vast potential for productive change is that of integrating the manufacturing and design process. Two manufacturing disciplines that largely are the focus of such attempts are CAD and CAM. Thus, computer-aided design uses computers to create, modify, and evaluate product design whereas computer-aided manufacturing uses computers to plan, control, and produce the actual product. The crossroads between design and finished product is process planning, an area in which computers have made tremendous inroads.

Process planning involves the transfer of information or knowledge in a form that allows workers to physically transform raw materials into the final product. In effect, process planning is the development of work instructions that comprehensively detail all aspects of the ways in which a product is to be manufactured, from machining and tooling to assembling and inspecting. At Lockheed-Georgia, this process once was entirely manual. Using group technology concepts in concert with a unique coding and classification scheme, however, Lockheed-Georgia was able to automate the process planning function to a generative system that yielded vast benefits to the manufacturing process.

By taking advantage of the similarities and likenesses of component parts, group technology divides them into categories or families based upon their commonalities. This allows batch manufacturers to achieve economies of production scale usually attained only in mass production industries. In the application of group technology, there are two primary types of classifications: those related to product engineering and those concerned with manufacturing processes. Product engineering similarities include factors such as shape, size, material, or function; manufacturing similarities relate almost exclusively to those processes and machines used to produce a specific family.

Group technology concepts were applied to divide parts into coded families based upon such similarities as size, shape, materials, and manufacturing processes. This, in fact, was critical to establishing an interactive computer-assisted process planning, or CAPP, system. Eventually, it led to reduced machine down time for set-ups and lengthened production runs by manufacturing similar parts together as a single family.

Process planning is a cornerstone to factory operations. Because it systematically determines the methods by which a product will be manufactured, process plans must stipulate and synthesize functional requirements, desired production volumes, equipment, tools, manufacturing operations, and costs into sheets that detail necessary tools and facilities and sequence of operations.

In effect, process plans represent an outline to allocate scarce organizational resources, such as labor, machines, and materials. The process plan itself reflects a human planner's expertise, in-depth knowledge of shop capabilities and limitations, and experience in manufacturing operations. Because no two planners will produce the same part in the exact same way, however, numerous plans can result, all detailing different ways to make the same basic part.

The computer, however, takes the best of this manufacturing knowledge. It optimizes the plans for a specific manufacturer's needs and standardizes procedures and processes for manufacture. Thus, a computer-aided system, therefore, can organize and store completed plans and manufacturing knowledge from which specific process plans can be accessed readily and synthesized. Authomated systems also allow plan updates and revisions, reduce the variety of methods employed, and allow new planners to get up to speed rapidly by standardizing production processes and structuring much of the decision-making capability within the computer.

Lockheed-Georgia's generative process planning system, or Genplan, captured the logic and rules of manufacturing by interrogating a comprehensive data base with part particular requirements identified by a classification code. Hence, the Genplan system is able to determine the sequence of operations, select the appropriate machines and tools, and calculate production times based on manufacturing logic resident in the computer. In order to accomplish this, however, the capacities and capabilities of all shop equipment had to be inventoried and stored in the computer.

A unique classification scheme also had to be created with a baseline software system that would support the construction of process plans. With this development, the Genplan system could determine automatically the sequence of operations, select the proper machines and tools, and calculate the proper times based on manufacturing logic.

In using Genplan, a process planner applies an appropriate classification code to an engineering drawing. This code describes the geometry and manufacturing properties of the part. The code, once synthesized by the computerized manufacturing logic network, formulates the most desirable manufacturing plan. Thus, the plan reflects the optimum manufacturing sequence, tool selection, standard language, tool codes, and other process parameters. Once the planner reviews and edits the screen displays, the plan is ready to be stored and, if desired, printed.

As the planner assigns the code based upon part description, the computer summarizes the data. It then evaluates the alternatives and makes the best planning decision. When entered into the system, the code produces a detailed manufacturing plan that is consistent not only in methodology but also in sequence, format, and technology.

Many computer-aided process planning systems rely upon the use and modification of extant standard process plans. These variant systems rely on standard process plans for similar parts developed according to the decision logic inherent in the system. There is a major difference between these variant systems and Lockheed-Georgia's generative system. Genplan actually can synthesize and create a complete process plan without relying on a standard process plan for a similar part. Genplan synthesizes process plans using the manufacturing logic and rules that were programmed into the system. Its technological data base consists of process decision logic, machine data, factory rules, tooling data, and labor formulas.

The Genplan system includes an extensive man-machine communication system to provide planners with a means to edit and supplement the decision-making process. Genplan consists of a keyboard, cathode ray tube, printers, and a host IBM 3083 computer. When entered into Genplan by the cathode ray tube, the code produces a detailed manufacturing plan requiring only minor fill-in. The process plans, therefore, are consistent in methodology, sequence, format, and technology.

Careful planning for implementation was a key to Genplan's success. The system for machined parts was put into production after production flow analysis, part classification, data base development, and systems software programming. Other applications of sheet metal, extrusions, and tubing were added in 1979; in 1980, the remainder of fabrication planning, subassembly planning, and wiring was added. The most recent expansion to the system involved major assembly planning capability.

Prior to the development of the assembly portion of Genplan, planners would analyze assembly drawings and manually copy parts requirements in order to make parts assignments. The planner, relying on research or personal knowledge, would then originate the work instruction language necessary to install and assemble parts and described related functions such as paint and inspection. Once written, the documents would be loaded via CRT into Lockheed-Georgia's Universal Manufacturing Requirements Master system. The parts data stored in this system would be reconciled periodically to the Engineering Requirements Master system. Of course, this method could not support standardized assembly steps as two planners do not proceed in the same way. The periodic reconciliation indicated errors in parts listings that were detected ex post facto.

The assembly segment of Genplan as used now is employed by tool planners to create production job sheets (PJS), lists of assembly work instructions, and parts lists for assemblies. Tool planners are provided with an assembly engineering drawing and the parts list of this drawing. Using a preplan created by a manufacturing planner as a guide and an assembly group technology code book, planners analyze the engineering drawing to assign the appropriate generative codes which generate assembly work instructions. These codes describe the assembly conditions and methods necessary for manufacturing to meet the engineering and manufacturing requirements of the assembly of aircraft.

Although the creation of new assembly group technology codes was necessary, the Genplan system retained the same architecture and general matrix system used in machining and other manufacturing operations. The expanded technological data base required the creation of new codes to describe such factors as structures, holes, hole making techniques, fillers, fasteners, fixtures, and so forth. Sealants, for example, are essential for aircraft corrosion protection and numerous other uses. Seven major types of sealants are used at Lockheed-Georgia for aircraft applications. These may include sealants used for aerodynamic smoothness, wiring and plumbing brackets, pressurization, fuel tank integrity, and so on. For each category, there are specific types and formulations intended for specific applications, each of these being characterized in the assembly coding system.

A code dealing with assembly can be broken down into as many as 19 fields or classifications of features. These are displayed horizontally and typically provide information including assembly characteristics, tooling configuration, method of hole preparation, component part and assembly configuration, next assembly requirements, blueprint and process specifications, paint and finish requirements, identification requirements, and so on. Descending vertically from each of these fields in matrix form, there may be as many as three dozen values that further describe the assembly and the type of work invoked when a specific code is applied. In the assembly characteristics field, for instance, the code may indicate the type of aircraft is used on, the type of assembly, the requirements for painting or washing, and so on.

Although the codes are vital to the system, the planner does not have to code every characteristic of an assembly into the system. The data handling aspects of the software are the same as the original version of the system designed primarily for machining operations. Hence, when a planner invokes a code, the software triggers a cascade of related information as the code interacts with the technological data base. Only a few codes, therefore, can invoke a great deal of manufacturing information and logic owing to the cascading architecture of the software. By invoking the code for a particular hole structure, for example, the system may automatically trigger the proper fastener, sealant, installation tools, drill size, and other related assembly information. The assembly logic files residing within the technological data base contain both the sequence logic and the procedural command statements or instructions for effecting assembly. The computer program analyzes the generative codes and selects a list of appropriate work instructions via the logic files. The generative codes, in turn, indicate an array of information including aircraft part position locations, dimensions, quantity of mechanical fasteners, blueprint numbers, tool numbers, item numbers, and production job sheet numbers. The parts in the parts list also are assigned to the appropriate work instructions. The final production job sheet then is stored in the computer for time standards to be added, retrieval for future revisions, and for issuance to the production shop.

Automating the process planning function at Lockheed Georgia offered distinct benefits. With logic files being maintained by manufacturing engineers, as new equipment, technologies, and manufacturing procedures arise, they become instantly available and uniformly to all planning engineering users. Liaison planners, in fact, now prepare updates and revisions directly on the shop floor through CRTs and other printers. Improved process planning document quality and consistency also are major benefits. Standardized language for work instructions, optimum manufacturing and processing sequences, and selection of the best tools result in top quality plans. Automated technological know-how virtually eliminates inaccuracies in manufacturing planning. Finally, of course, the time and cost of moving from design to manufacturing were drastically reduced.

The manufacturing process is in a constant state of flux, both in rate and direction. The rate of change continues to accelerate and the direction of change promises to streamline the intricate manufacturing process through increased efficiencies in data access and management. The implementation key is cogent integration: of design, tooling, manufacturing, and engineering, all working in unison from a common data base. CAD/CAM is only one step, albeit a giant one, toward attaining the factory of the future. The succeeding generation of automation--CIM--builds on CAD/CAM to merge it with all related functional areas. By automating process planning, based on a group technology classification and coding scheme, Lockheed-Georgia stepped into the CIM world and leaped the intellectual distance to come that much closer to the automated, paperless factory of the future.

BIOGRAPHICAL SKETCH

Mr. Joseph Tulkoff is Director of Manufacturing Technology at the Lockheed-Georgia Company, Marietta, Georgia. He graduated from the Georgia Institute of Technology, receiving a BSIE in Industrial Engineering, and received graduate management instruction at the Lockheed-Emory University Institute.

Mr. Tulkoff is currently directing state of the art manufacturing systems and technology programs including an Air Force/Lockheed Technology Modernization program and Lockheed's Factory of the Future Program. He also serves as Lockheed's Coordinator of the Air Force Integrated Computer-Aided Manufacturing (ICAM) program and also is Lockheed's Corporate Coordinator of Generative Process Planning Technology.

Mr. Tulkoff has 30 years' experience in both technical and managerial roles in many phases of industrial engineering and manufacturing engineering, covering such areas as production engineering, systems engineering, management information systems, manufacturing research, and computer-integrated manufacturing. He is the principal architect of several major computer-aided manufacturing systems including Genplan, an advanced computerized process planning system.

Director of the Aerospace Division of the Institute of Industrial Engineers (IIE) for 1984-85, Mr. Tulkoff also is Chairman of the Advanced Technical Planning Committee of Computer Aided Manufacturing-International (CAM-I); a senior member of the Society of Manufacturing Engineers (SME); the past President and Director of the Atlanta Chapter of IIE; and Instructor of the computer automated process planning course sponsored by SME/CAM-I; a certified Manufacturing Engineer of the Computer and Systems Association (CASA) of the Society of Manufacturing Engineers; and a member of the National Management Association.

Mr. Tulkoff is a Fellow of the Institute of Industrial Engineers. He received the International IIE Excellence in Productivity Award for 1982 and the IIE Aerospace Division Award in 1982. He received the Lockheed-Georgia cost reduction Member of the Year in 1967 and was named National Management Association Lockheed-Georgia Chapter Member of the Year in 1961.

Mr. Tulkoff is an author and speaker in the fields of computer-aided planning, group technology, automated manufacturing systems, industrial modernization, and factory of the future planning.

Approaches to the Scheduling Problem in FMS

S.C. Sarin
Department of Industrial Engineering
and Operations Research
Virginia Polytechnic Institute
and State University
Blacksburg, Virginia 24061

and

E.M. Dar-El
Faculty of Industrial Engineering
and Management
Technion — IIT
Haifa
Israel

ABSTRACT

In this paper we briefly review the literature regarding scheduling in FMS. A variety of approaches are described. A new approach is introduced for real time scheduling in FMS which gives fairly high levels of machine utilization. This approach incorporates four levels of flexibilities. An inplementation of this approach is presented on its performance is studied under various operating conditions.

INTRODUCTION:

A flexible manufacturing system is a production system consisting of programmable machine tools which are capable of performing multiple operations and are inter-connected by computer-controlled automated material handling equipment. The parts once loaded on to this system are processed automatically. The term "flexible" is coined to these systems because of two reasons: (i) the ability to perform a given set of operations on a part by using alternate routings through the machines and (ii) the ability to simultaneously work on several types of parts. Both (i) and (ii) are achieved due to the use of multiple tool carrying numerically controlled machines. The impact of both of these flexibilities is in reducing work-in-process inventory and setup time while changing from the production of one type of parts to another. Consequently, FMSs are suitable for medium volume production which encounters significant amounts of work-in-process inventory and setup times. Medium volume production also comprises a major proportion of manufacturing activity.

Several new problems arise for the design and operation of a FMS. Among others, these include (i) selection of the type and number of machine tools, (ii) selection of the material handling equipment (iii) selection of the type of parts and the part mix to be run on a FMS (iv) size and location of storage for work-in-process inventory (v) number of pallets in the system (vi) selecting parts for loading in the system (vii) loading of tools on the machines (viii) scheduling parts to the machines and (ix) specification of sequences in which parts are to be processed on the machines. Several studies have been reported in the literature to address these problems. An overview of the proposed models is given in Buzacott and Yao [5], Sarin and Wilhelm [24] and Wilhelm and Sarin [29]. For a thorough background on FMS the reader is referred to the five folume report: Flexible Manufacturing System Handbook [12]. Here we briefly review the approaches developed in the literature to solve the scheduling and sequencing problem of FMS mentioned in (vii) above and present a new approach.

APPROACHES FOR SCHEDULING IN FMS

The scheduling problem of FMS is like that of a dynamic job shop. A variety of parts are simultaneously processed through the machines. The operations of a part can be accomplished by routing the part through alternate machine sequences. Several operations can be performed on a machine with negligible tool change-over time. The decision to schedule next operations of a part depends upon the status of the machines and the mix of parts available for processing at that time. The objective is generally to finish parts in a minimum amount of time, minimize part tardiness or maximize machine utilization. As the machines involved in a FMS are very expensive, maximizing machine utilization is one of the more important objectives. In practice, the dual objective of maximizing utilization while minimizing a function of job tardiness is very common. Various approaches have been proposed to schedule parts in FMS. We review these briefly next. But first a few words about the selection of part mix to be run on a FMS. Generally, the type of parts to be simultaneously processed on a FMS is selected based on their geometry, weight, material etc. First a list of candidate parts is constructed and the unsuitable parts (those technologically infeasible or impractical) are screened out. Then those parts that are economically less attractive are deleted. It is better to have several parts in process than one but too many part types can be counter productive due to excessive tool requirements, overburdening of the material handling system and production control overhead. In a study by Nof et. al. [19] it is shown that there exists some optimum set of part types to be run concurrently although the performance is dependent on particular part types nd process used. The optimum set of parts is generally believed to consist of part types whose processing requirements are complementary to one another such that their processing requirements balance the work load on all machines. A two step procedure to achieve such a part mix is proposed by Vaithianathan [27]. In Step 1 all the feasible candidates are clustered into subgroups in accordance with the similarity in visitation sequence. Work is then loaded on to the system according to their dissimilarity.

Use of Dispatching Rules:

This is the most practical approach of scheduling parts to machines because of the simplicity of its implementation. Since the scheduling problem in FMS is like that of a dynamic job shop the dispatching rules of relevance here are those used for dynamic job shops. A review of literature on this subject is given in Moore and Wilson [18] and Day and Hottenstein [10]. Panwalker and Iskander [21] summarize various dispatching rules used in scheduling. Two studies have been reported in the literature on the use of dispatching rules in FMS. In a study by Nof et.al. [19] the ratio rule which schedules a part next on a machine if the ratio of its remaining production requirement to original requirement is larger than the ratio of the remaining production time to day's production time is found to be better than the first come first served (FCFS), shortest remaining processing time (SPT) or largest remaining processing time (LPT) rules. An experimental study of a real system is reported by Stecke and Solberg [25]. A priority rule which is based on dispatching next a part with the smallest ratio obtained by dividing the shortest processing time for the operation by the total processing for the part is reported to give better results than the SPT, LPT or related priority rules.

Network Analysis

A FMS can be viewed as a network in which the multiple tool carrying machines are represented by nodes, and arcs between nodes represent routes for performing operations on the parts. Since there are queues of parts building up in front of nodes, the network is more precisely termed as the network of queues. The number of parts in the network are usually kept the same by replacing completed parts by an equal number of new parts. The system can therefore be modelled as a closed network of queues. A methematical programming approach based on the network flow model is given by Kimemia and Gershwin [17]. The objective is to determine the flow of different parts on arcs so as to maximize the production rate of parts which flow through the machines in a predetermined part mix. The network of queues model is used to account for the congestion effects. The network of queues analysis is based on the first come first served discipline for the sequencing of parts to machines. The mathematical model is as follows:

Let S_n = strategies to produce part type n
t_{nmk} = time for the kth operation for part type n on machine m
α_n = fraction of total production for part type n
X_{nmk} = flow rate of part type n to machine m for operation k
Y_{nj} = flow rate of part type n on arc j
$I(X,Y) = \sum_m q_m(X) + \sum \sum \tau_j Y_{nj}$
= average number of parts in the system in which
$q_m(X)$ = average queue at machine m
τ_j = time to traverse arc j

The objective for part routing is to maximize output, i.e.

$$\text{maximize} \sum_{n=1}^{N} X_{n11} \qquad (1)$$

st.

$$\sum_m X_{nm1} / \sum_{nm} X_{nm1} = \alpha_n \qquad n = 1, 2, \ldots, N \qquad (2)$$

$$\sum_n Y_{nj} = \sum_{j'} Y_{nj'} \qquad j = 1, 2, \ldots, J \qquad (3)$$

$$\sum_{nk} t_{nmk} X_{nmk} \leq 1 \qquad m = 1, 2, \ldots, M \qquad (4)$$

$$I(X,Y) \leq C \qquad (5)$$

$$\sum_n Y_{nj} \leq d_j \qquad j = 1, 2, \ldots, J \qquad (6)$$

$$X_{nmk}, Y_{nk} \leq 0 \qquad \forall_{nmkj} \qquad (7)$$

Constraint (2) pertains to specified production ratio, Constraints (3) represent conservation of flow at each node, Constraint (4) represents limited capacity of machines, Constraints (5) correspond to the average level of work in process limitation and Constraints (6) represent ARC capacities.

Due to the complexity of $q_m(X)$ (which appears in the function $I(X,Y)$), this is a nonlinear problem which can be solved using an augmented Lagrangian method in combination with Dantzig-Wolfe decomposition. The solution gives optimal plans for routing parts to maximize production outputs. If arrivals and processing times are deterministic, the model reduces to a linear program consisting of (1), (2), and (4). This approach considers aggregate flow of parts between nodes and not the individual movements of parts through the machines and is therefore not useful for the real time operation by operation scheduling of parts.

Hierarchical Approach:

A multilevel approach for real-time control of FMS is proposed by Hildebrandt and Suri [16] for application to large-scale systems. The problem is to schedule parts to failure-prone machines to minimize total completion time. The problem is divided into three stages of decision making. The Stage 1 problem considers the machines and the pallets as limited resources and determines the aggregate flow of parts through machines during the machine failure conditions. The objectives of lower stage problems are based on the modelling assumptions made at higher level. The Stage 2 problem is based on the assumption regarding the distribution of parts in the system to have constant profile of characteristics. The Stage 3 problem resolves short-time conflicts for resources to minimize the average delay of tasks waiting to use the resources. The Stage 1 problem can be formulated as follows:

Let I = Set of system failure states
N = Set of different parts
R_{ni} = Set of routes parts can take through the system during the system state i, $n \varepsilon N$, $i \varepsilon I$.
$\bar{N}_n$ = Total number of part n to be produced
P_i = Expected proportion of the time the system spends on state i
T = Time of completion for all parts
f_{nri} = Average number of pallets devoted to route r of part n during condition i
b_{nrm} = Operation time of part n, using route r, on machine m
τ_{nrm} = Average time of part n, using route r, on machine m. This includes operation time and queueing time
$\bar{F}$ = Maximum number of pallets to be in the system
F_n = Maximum number of pallets available for part n

Then the Stage 1 problem is to determine f_{nri} and T so as to

Min T
s.t.

$$\bar{N}_n / \sum_{i \varepsilon I} P_i \sum_{r \varepsilon R_{ni}} [f_{nri}/\Sigma_m \tau_{nrm}(\underline{f},\underline{t})] \leq T \quad (8)$$

$$\sum_r f_{nri} \leq F_n \quad \forall n,i \quad (9)$$

$$\sum_n \sum_r f_{nri} \leq \bar{F} \quad \forall i \quad (10)$$

$$f_{nri} \geq 0, \ T \geq 0 \quad (11)$$

where $\tau_{nrm}(\underline{f},\underline{t})$ are nonlinear functions of $\underline{f}$ and $\underline{t}$. The determination of $\tau_{nrm}(\underline{f},\underline{t})$ is not straight forward and requires the analysis of the network of queues. Mean value analysis [2,23], a recently developed technique for the analysis of closed network of queues, can be used for the determinaiton of $\tau_{nrm}(\underline{f},\underline{t})$. An iterative procedure for the determination of $\tau_{nrm}(\underline{f},\underline{t})$ using mean value analysis is given by Hildebrandt [15]. The left side of constraint (8) represents the total time required to process the given set of parts and constraints (9) and (10) represent the availability of pallets for part type n and the total number of pallets in the system respectively. Solution of the Stage 1 problem determines allocation of work to resources during each failure period but does not determine when that work is to be performed. This is done in Stage 2 by using a dynamic programming based procedure.

Integer Programs:

The above approaches assume that appropriate tools are available at various machines for processing of the parts scheduled on them. Stecke [26] considers the problem of loading tools and assigning operations to machines simultaneously. This gives rise to the consideration of constraints pertaining to tool magazine capacity. Duplicate assignment of operations to machines is permitted up to a limit. Various objectives can be considered for the loading problem: (i) balance processing times on machines, (ii) minimize movements of parts among machines, (iii) balance or unbalance workload per machine for a system of machines which are pooled together in equal sizes, (vi) fill the tool magazine as densely as possible with not more than one tool assigned to the same machine, or (v) duplicate operations to multiple machines only for critical operations by maximizing a sum of priorities assigned to each operation. The nonlinear integer formulations of the loading problem under these objectives is presented in [26] and computational procedures based on the linearization of the nonlinear constraints are used for their solution.

An integer programming formulation for the real time scheduling of parts in FMS is proposed by Chang and Sullivan [6].

Let t_{nmk} = processing time of operation k of part n on machine m
$K(n)$ = last operation of part n
$d_{m_1 m_2}$ = travel time from machine m_1 to m_2
S_{nmk} = start time of operation k of part n on machine m
T_{nmk} = finish time of operation k of part n on machine m i.e., $T_{nmk} = S_{nmk} + t_{nmk}$
R_{nmk} = 1, if operation k of part n requires machine m
     = 0, otherwise
$X_{nm_1 m_2 k}$ = 1 if operation k of part n is assigned to m, and operation k + 1 is assigned to m_2
     = 0, otherwise
$Y_{n_1 k_1 m n_2 k_2}$ = 1 if operation k_1 of part n, precedes k_2 of n_2 on machine m
     = 0, otherwise

Then the problem is to determine X's, S's, and Y's to

$$\text{Min } \sum_m \sum_n T_{nmk_n} \quad (12)$$

s.t.

$$\sum_{m_1} X_{nm_1 m_2 k} - \sum_{m_1} X_{nm_2 m_1 k+1} = 0 \text{ for all } m_2, n \text{ and } k \quad (13)$$
$$\text{except for } k = K(n)$$

$$\sum_m R_{nmk} S_{nmk} \geq \sum_{m_1} R_{nmk} (S_{nm_1 k} + t_{nm_1 k} + \sum_{m_2} d_{m_1 m_2} X_{nm_1 m_2 k}) \quad (14)$$
$$\text{for all } n \text{ and } k$$

$$S_{n_2mk_2} + Q(1 - Y_{n_1k_1mn_2k_2}) \geq T_{nmk} \text{ for all} \quad (15)$$
$$m, n_1, n_2, k_1, k_2$$

where Q is a very large positive number. The objective depicted by expression (12) is to minimize the sum of the completion times of all parts. Constraint (13) maintains the consistency concerning the start of operation k+1 as represented by variable X. Constraint (14) links the operation of each part. The fact that only one operation is performed by a machine at a given time is represented by Constraint (15). The formulation as such is very large in number of variables and the number of constraints. A two phase method which involves the use of a (0,1) integer programming code is proposed to obtain the optimal solution.

Artificial Intelligence:

Artificial intelligence is a branch of computer science which tries to get machines (computers) to make decisions like human beings. Human intelligence is defined as the ability to adjust when encountered with new situations and to interrelate presented facts in order to take an action toward a desired goal. The field of artificial intelligence develops programs which tend to capture these aspects of human intelligence. The operation of complicated systems like the flexible manufacturing system is controlled by computers because of their speed in data processing, however, for greater effectiveness several decisions need to be made by humans. The artificial intelligence methods can be used instead to automate the entire decision process. Scheduling in FMS is one such area. The dynamic behavior of the system is difficult to capture mathematically and may be better handled by the artificial intelligence methods. A discussion to this effect is given by Bullers et. al. [4] and simple artificial intelligence based procedures are introduced. The artificial intelligence method discussed is based on predicate logic and mechanical theorem proving using resolution principle. For more detailed discussion on the artificial intelligence methods and their applications the reader is referred to the three volume series of the handbook on artificial intelligence [3]. Some references pertaining to the applications in scheduling are Barber [1], Fikes [11] and Goldstein and Roberts [13].

NEW APPROACH

A generalized concept of flexibility in the context of FMS is depicted in Figure 1 and described in [7]. Four levels of flexibility are defined. The level one flexibility pertains to part mix. In most of the studies part mix is generally prespecified. The second level of flexibility pertains to operations set. Each part can be completed quite often using alternative operation sets OS_1, OS_2 . . . OS_q. For example, an operation can be executed using a heavy cut single pass, or else in two sequential passes using lighter cuts on the same, or two different machines. This gives rise to two alternate ways of completing that operation and consequently to two ways of completing the part. The third level of flexibility is the sequencing of operation (SO). Each operation set can generally be executed in alternative sequences. For instance,

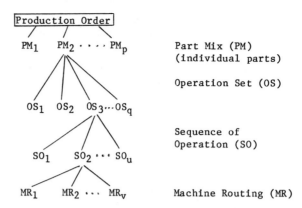

Figure 1: Levels of Flexibilities in FMS

drilling could be done before or after turning thereby resulting in alternate ways of accomplishing an operation set consisting of drilling and turning. This is illustrated in Figure 1 by SO_1, SO_2 . . . SO_t. Both the flexibilities of defining operation set and sequence of operations should be left as a part of the operational control process rather than to prespecify it. This would result in better utilization of the equipment and a better design. Finally, the fourth level of flexibility pertains to performing an operation on alternative machines giving rise to machine routings represented by MR_1, MR_2 . . . MR_v. The literature is reviewed from this viewpoint of flexibility [7]. Most of the researchers have considered only the fourth level of flexibility. Our new approach attempts to cover all four levels.

We view the problem of scheduling of FMS as that of allocating resources. Limited resources in the form of machines are to be allocated to the parts to be processed on them. Scheduling in FMS essentially involves determination of (i) how much work should be performed by each machine and (ii) when should that work be performed. The latter point pertains to the determination of when a part ought to be diverted to a machine for performing an operation and how should the parts at a station be sequenced. The scheduling models presented in the current literature pertain to (ii) and result in (i). These models consider the aggregate flow of parts between machines and use heuristic rules like FCFS, SPT and others to sequence next part on a machine. An improved approach for tackling this problem is to first consider aspect (i) so that machines are well utilized and then to determine the routing policies to achieve that utilization by effectively making use of the available flexibilities of performing an operation at several machines.

The option of performing an operation at alternate machines leads to different resource consumptions by a part at different machines. Such an approach was developed by Dar-El and Tur [8] for tackling the multi-resource, single project scheduling problem. By suitably modifying this approach for FMS scheduling each of these alternate ways of performing operations on a part would lead to an alternate routing combination (ARC).

The proposed procedure is as follows. First we fix a planning period, designated by T, for which the schedule is to be determined. This may be an hour, a day, a shift or other. Usually a set of parts dedicated for FMS are available at the start of the planning period. A subset of this set is selected for determining the schedule for the planning period based on the criterion of minimizing a function of part tardiness. Denote this subset by N. Next, we determine a set of ARCs among those of N parts for maximum machine utilization. Several such sets of ARCs may exist and each one, in turn, would be treated in the manner described. Let V_{ij} be a vector representing ARC type i of part j. The dimension of V_{ij} is K and each element of V_{ij} represents the amount of resource type r (r = 1, 2, ... K) consumed by (ij). Let R be a K dimensional vector representing the availability of resource type r. If we define

X_{ij} = 1 if (ij) is selected for execution

= 0 otherwise

then the problem of selecting optimal ARCs to perform operations on a set of parts from among N parts to maximize utilization is the following:

$$\text{Max} \sum_{r=1}^{K} \sum_{j=1}^{N} \sum_{i=1}^{n_j} X_{ij} V_{ij}(r)$$

s.t.
$$\sum_{j=1}^{N} \sum_{i=1}^{n_j} X_{ij} V_{ij} \leq R \qquad (P)$$

$$\sum_{i=1}^{n_j} X_{ij} = 1$$

$$X_{ij} = 0,1$$

where n_j is the number of ARCs of part type j. The machine availability vector R is defined for the planning period, T, for which the above problem is solved. The solution of (P) gives machine utilization in the absence of congestion or blocking. The ARCs determined in (P) are considered for scheduling and minor variations are made during this step to avoid congestion, blocking or machine idleness. The set N is then determined for subsequent planning period and the whole process is repeated.

A formulation for the real time scheduling in FMS using ARCs in order to maximize machine utilization is as follows:

Let $X_{jakmm''t}$ = 1, if the kth operation of the ath ARC of job j (jak) starts at time t on machine m" after coming from machine m

= 0, otherwise

Z_{ja} = 1, if the ath ARC of job j is selected for processing

= 0, otherwise

T = target finish time of jobs

n_j = number of operations of Job j

A_j = number of ARCs Corresponding to job j

$g_{mm''}$ = travel time from machine m to m"

P_{jak} = processing time of operation k of job j for ARC a

Q = a large positive number

Then the problem is to determine $X_{jakmm''t}$ and Z_{aj} so at to

minimize NT −

$$\sum_{j=1}^{N} \sum_{a \in A_j} \sum_{t=0}^{T} t \cdot X_{jan_jmm''t} \cdot Z_{ja} \qquad (15)$$

Subject to

$$\sum_{a \in A_j} Z_{aj} = 1 \qquad j=1,\ldots,N \qquad (16)$$

$$\sum_{a \in A_j} \sum_{t=0}^{T} t\, X_{jak+1mm''t} Z_{ja} \geq \sum_{a \in A_j} \sum_{t=0}^{T} t \cdot X_{jakm''mt} Z_{ja}$$

$$+ \sum_{a \in A_j} Z_{ja} P_{jak}$$

$$+ \sum_{a \in A_j} \sum_{t=0}^{T} g_{mm''} X_{jak+1mm''t} Z_{ja} \qquad (17)$$

$j = 1,\ldots,N;\ k = 1,\ldots n_j$

$$\sum_{j=1}^{N} \sum_{a \in A_j} \sum_{k=1}^{n_j} X_{jakmm''t} Z_{ja} \leq 1 \qquad m'' = 1,2,\ldots,m$$
$$t = 0,\ldots,T \qquad (18)$$

$$\sum_{a \in A_j} t\, X_{jakmm''t} Z_{ja} + (1 - \sum_{a \in A_j} X_{jakmm''t} Z_{ja})Q$$

$$\geq \sum_{j' \neq j} \sum_{a \in A_j} \sum_{k=1}^{n_{j'}} \sum_{\ell < t} (\ell + P_{j'ak''})X_{jak'mm''\ell} \cdot Z_{j'a}$$

$m'' = 1,\ldots m \quad k = 1,2,\ldots n_j \qquad (19)$
$t = 0,\ldots T \quad j = 1,\ldots,N$

Note that for a given a and j mm" is known therefore the summation over m does not appear in the formulation.

The objective as represented by (15) is to minimize total machine idle time. Constraint (16) represents selection of only one ARC for a job j. The fact that a subsequent operation of job can start only after its previous operation is finished, plus the travel time between machines is captured in Constraint (17). Constraint (18) limits the assignment of at most one operation to a machine at a given time t while machine availability to start an operation at a machine is represented by constraint (18). This formulation explictly considers the travel time of a job between machines. To account for the congestion effects an expression for the expected waiting time, if known, can be added to the travel time. Also, this formulation can be easily modified to account for different starting times for machines. If S_m represents the available starting time of machine m then the required modification is simply to change the lower limit of the summation of t from 0 to S_m.

Instead of a frontal attack on the problem using either of the above two formulations which would in all likelihood result in a branch and bound method we develop a procedure which tries to obtain a schedule whose utilization falls in a prespecified range if such a solution exists. The procedure is easy to use and generates solutions fast. It is currently applied to the FMS system of Ingersol Rand in Roanoke, Virginia plant. This is explained next.

Algorithm:

The proposed algotithm is a heuristic algorithm to obtain a solution within the prespecified range of desired machine utilization. A flowchart of this algorithm is shown in Figure 2. The main steps are as follows:

(i) Selection of Jobs:

The jobs to be processed are first selected from among those available. Any user supplied selection procedure can be used to select jobs. The procedure used in the program essentially consisted of preparing a priority list based on the sum of the ranking from job tardiness and the minimum cost. The cost is determined as follows. Machines are differentiated based on their operating costs. An expensive machine will have higher operating cost than a less expensive machine. For a preferred ARC of each job cost is thus computed based on the processing times of that job on respective machines. The number of jobs selected (NOJ) depends upon the number of scheduling periods.

(ii) Specification of Upper and Lower Utilization Limits:

The upper and lower utilization limits are user specified. Using a preferred ARC of each job, the program considers several combinations of jobs from among NOJ and compute their projected machine utilizations without actually scheduling jobs. Those combinations falling within the upper and lower utilization limits become candidates for subsequent scheduling considerations. Excessively high utilization limits may not lead to a feasible schedule. The program has the capability of lowering utilization limits if no feasible schedule can be obtained within the specified limits. However for more effective implementation of procedure it is essential to specify realistic utilization limits. The upper and lower utilization limits are designated by UUL and LUL respectively.

(iii) Binary Specification of Combinations

A binary method described in [22] and summarized in [7], is used for creating all combinations of NOJ jobs. Resource demands for each combination are aggregated and compared with the available machine capacities for the period in question. The 50 combinations giving the best fits are chosen as candidates for the schedule.

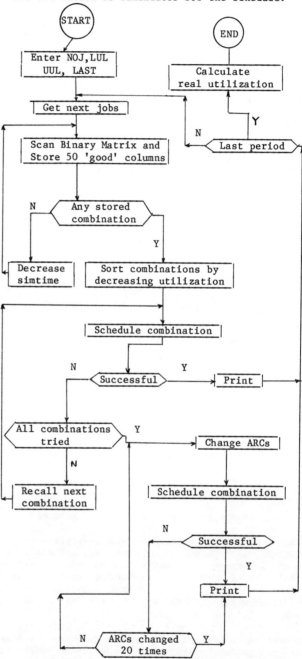

Figure 2: Flow Chart of the Algorithm

Attempts at scheduling always begin with the candidate set having the best potential utilization, then the second best, then the third, and so on. No attempt is made at altering ARCs, until the last of the 50 sets are tried with no success at obtaining a schedule. Then follows a routine that changes the ARCs of the last set that was tried in an attempt at finding a schedule. If after 20 ARC changes a solution is not found, the program simply utilizes the best fit obtained from a shortened list (from NOJ) & proceeds to the next period.

Machine capacities for each period are determined as indicated in figure 3 where each horizontal row represents the time scale for the machine indicated. A schedule for period k is "complete" when the scheduled period for at least one machine is close to, or, at the period boundry and no other parts can be scheduled on any machine without "violating" the boundry. The incompleted portion in period k can then be added to the block time for the next period to give the total machine availability for period (k+1). In this manner, higher machine utilizations can be achieved. Traditional minimum makespan conditions for job shop scheduling are obtained when the incompleted portion in period k is reduced by selecting higher values for UUL and LUL.

The scheduling itself is achieved through the WINQ rule which was shown to have good anti-tardiness properties[10].

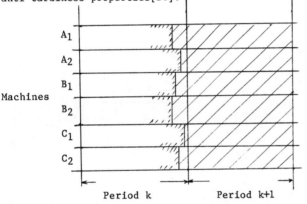

Figure 3: Machine capacities for each period

The Implementaiton:

A computer program was written for implementation of the procedure on the Ingersol Rand's FMS installation in Roanoke, Virginia. The program was written in PLI and was extensive in nature incorporating about 20 subroutines. Ingersol Rand's FMS installation in Roanoke consists of 3 pairs of machines - two 6 degree of freedom (dof) Universal milling machines (A); two 5(dof) boring machines (B) and two 4 (dof) drilling machines (C). The six machines have a fixed conveyor system connecting each machine to the others. Each machine has one pallet storage space associated with it. This storage space is used for in-process storage of the job waiting for being processed on that machine. This helps load a job on the machine quickly after the job already on the machine is finished. Since the machines are different the ratio of machine costs of A:B:C was taken as 4:3:2, thus, machine A is twice as expensive as machine C but only a third more costly than machine B. Whereas machine B is 50% more costly than machine C.

There are 25 parts that need to be processed. Each part may have one, two or three ARCs and each ARC has three, four or five operations. A total of 30 operations are defined each having its own machine processing time so that each ARC has a different total processing time. It is given that machine A can also do the work performed by machines B and C, and that machine B can also perform the work done on machine C. However, this flexible condition can be constrained with each machine working on mutually exclusive operations. The performance under this condition will also be studied. Other assumptions are as follows:

i) The number of ARCs for each job and the number of operations in each ARC are randomly generated from uniform distributions.

ii) Operation times are randomly generated from a normal distribution with $\mu=50$ and $\sigma=10$.

iii) Each operation is executed in the same processing time on any machine which can perform the operation.

iv) Each machine type (i.e. A, B and C) has a single queue from which work can be drawn.

v) Each job is an order for a specific part and the file of orders are sorted by decreasing lateness penalty.

vi) The number of pallets in the system is equal to the number of jobs.

vii) Machine utilizations are weighted in proportion to their cost so we are comparing the resource requirements in each combination with the weighted resource availability based on 100/18 {MA(A).4 + MA(B).3 + MA(C).2} where MA(i) represents machine availability of machine type i.

The objective was to obtain the maximum machine utilization and minimum job tardiness for the condition that different ARCs are available, and that a storage space is available in front of each machine, and that different machines either work on mutually exclusive operations, or are able to execute operations performed on the simpler machines.

The Results

A number of preliminary simulation studies for varying conditions were carried out and results are shown in Table 1. Overall weighted utilization was based on minimizing the idle time within the machine completion time boundry e.g., the unshaded area within period k in figure 3. In effect we are measuring the effectiveness of packing in the work in order to minimize the idle time blocks that may occur. Our motivation for doing this is that a "period" may only be a half hour or an hour in duration and that work over one shift (as a minimum continuous work span) is a continuing process. Hence only the idle time as described applies during all but the last period which must also include the idle time to the end of the shift.

Table 1: Results of the simulation studies

Factor	Study 1	Study 2	Study 3
NOJ	6	15	15
Total periods simulated	2	2	2
UUL	95%	100%	95%
LUL	80%	98%	80%
Utilization			
M/Cs A	95.5%	100%	100%
M/Cs B	96.6%	98.2%	95.0%
M/Cs C	77.5%	94.7%	96.3%
Overall Weighted Utilization	91.9%	98.2%	97.5%

The utilization figures are seen to be exceptionally high especially for the two cases of NOJ = 15. Consequently it was decided to calculate the overall weighted machine utilizaitons as if these were the <u>last</u> period in a given shift. This meant that utilization efficiencies are based on the <u>entire time block</u> of the period –in much the same way that machine utilizations are calculated in traditional flow shop and job shop problems.

Tables 2 and 3 contain the results for the condition that the more sophisticated machines can also do the work performed on the simpler ones, whereas Table 4 gives the results for the three machines types executing mutually exclusive operations.

Several conclusions are indicated from the results in Table 2. These are as follows:

1) For fixed values of UUL = 85% and LUL = 70% the average machine utilization are increased as the NOJ value is increased form 8 to 16 jobs. This is also clearly seen in Figure 4 with "dots" for the data points.

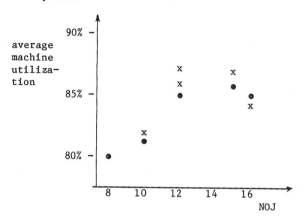

Figure 4: Average machine utilization verson NOJ for varying conditions

2) By raising the UUL (with or without altering the LUL), we are able to achieve higher machine utilization values. These are seen as the "crossed" data points in Figure 4.

3) When schedules are performed over two periods as in Table 3, we see that machine utilizations are lower than when the two periods are considered as one in Table 2. We can only compare this for NOJ = 8 in Table 3, since over 2 periods we complete 16 jobs and NOJ = 16 data is available in Table 2. Thus, we are comparing column 2 in Table 3, having an average machine utilization of 76.4%, with column 10 in Table 2 which yields an average machine utilization of 85%.

Table 2: Smarter machines able to also do the work of the simpler machines (Period = 1)

Item	Experiment No:										
	1	2	3	4	5	6	7	8	9	10	11
NOJ	8	10	10	12	12	12	15	15	15	16	16
LUL	70	70	75	70	75	80	70	75	80	70	80
UUL	85	85	90	85	90	95	85	90	95	85	95
Period	1	1	1	1	1	1	1	1	1	1	1
Machine Utilization											
A	.830	.898	.820	.827	.830	.827	.872	.872	.882	.832	.825
B	.883	.706	.845	.924	.875	.924	.866	.895	.867	.862	.843
C	.673	.877	.836	.800	.897	.900	.821	.821	.824	.824	.875
Overall	.813	.829	.832	.850	.860	.876	.869	.869	.864	.864	.843

Table 3: Smarter machine able to also do the work of the simpler machines (Period = 2)

Item	Experiment No:									
	1	2	3	4	5	6	7	8	9	10
NOJ	8	8	10	12	15	15	15	16	16	16
LUL	60	70	70	70	70	75	80	70	75	80
UUL	85	85	85	85	85	90	95	85	90	95
Period	2	2	2	2	2	2	2	2	2	2
Machine Utiliaztion										
A	.783	.824	.758	.831	.881	.855	.867	.854	.872	.868
B	.749	.754	.780	.811	.813	.859	.857	.833	.852	.848
C	.673	.661	.703	.802	.788	.843	.849	.851	.880	.876
Overall	.747	.764	.753	.818	.837	.854	.860	.846	.867	.863

4) Table 3 exhibits the same characteristics as conclusion (1) given for the data on Table 2, i.e., the average machine utilization increase with NOJ.

Table 4: All machine types do exclusive work

Item	Experiment No:					
	1	2	3	4	5	6
NOJ	14	15	20	24	25	30
LUL	60	60	75	75	70	75
UUL	90	90	90	90	90	90
Period	1	1	1	1	1	1
Machine Utiliaztion						
A	.567	.642	.688	.68	.693	.717
B	.577	.574	.650	.737	.768	.734
C	.884	.821	.854	.87	.895	.871
Overall	.634	.659	.712	.741	.756	.757

5) Table 5 shows results when the problems of Table 4 are run under the condition that smarter machines can do the work of simpler machines. Comparing Table 4 with Table 5 one can conclude, as expected, that the inclusion of the constraint that machine types work on mutually exclusive operations, reduces the overall average machine utilization.

6) Once again we see that increasing NOJ from 14 to 20 increases the average machine utilization from 68.6% to 82% which could indicate that some trade-off in the number of pallets and expected utilization could be beneficial for the systems design.

Table 5: Problems of Table 4 run under the condition that smarter machine are able to also do the work of the simpler machines (Period = 1)

Item	Experiment No:					
	1	2	3	4	5	6
NOJ	14	15	20	24	25	30
LUL	60	60	75	75	70	75
UUL	90	90	90	90	90	90
Period	1	1	1	1	1	1
Machine Utiliaztion						
A	.603	.550	.755	.769	.657	.761
B	.700	.867	.891	.807	.850	.828
C	.830	.807	.803	.871	.891	.958
Overall	.686	.713	.820	.804	.773	.827

7) At of the time writing, it appears that the algorithm does not utilize the ARC characteristics very efficiently and more work is needed for strengthening this aspect. The authors are convinced that with a greater utilization of the ARCs concept, that much higher utilizations can be obtained.

8) We need to stress, that the bulk of our analyses are directed towards considering the utilizations of only the final period of a shift i.e. when all machines are shutdown e.g. for preventive maintenance. Consequently the overall utilization taken over a continuously working period (1 or 2 shifts), even with our

unsophisticated scheduling model, is likely to be a lot higher i.e., most likely over 90%. This is far in excess or the utilization value obtained by Nof et. al. [19] and Stecke & Solberg [25]. Certainly, they are far superior to the levels actually obtained in the field and their adoption would make a significant improvement in their system performance. CPU times were all sufficiently short to enable the algorithm to operate on an on-line mode, but this will have to be validated in a simulation study under real conditions.

CONCLUSIONS

In this paper the literature on scheduling in FMS is reviewed. The term "flexibility is shown to involve flexibility factors at four levels - the selection of: part-mix, the operation set, the operation sequence and the machine routing. Most FMS researchers have covered only one of these four factors.

A new approach is presented here for incorporating all four flexibility levels found in FMS. Its performance is studied under various operating conditions. Preliminary results of the implementation are very encouraging with fairly high weighted machine utilization values being obtained. Further modifications can be incorporated in the procedure for use under general operating conditions of an FMS. Efforts in this regard are currently underway.

ACKNOWLEDGEMENT

The authors wish to acknowledge the help of Mr. Chin-Sheng Chen Graduate Student in Department of IEOR at VPI & SU for making the computer runs of the algorithm.

REFERENCES

[1] Barber, G., "Supporting Organizational Problem Solving with a Work Station", ACM Transactions on Office Information Systems, 1, (1983)

[2] Bard, Y., "Some Extensions to Multiclass Queueing", G320-2124, IBM Cambridge Scientific Center, Cambridge, Mass., (1978)

[3] Barr, A., and E. A. Feigenbaum, (Editors), The Handbook of Artificial Intelligence, Vol. 1 through 3, HeurisTech Press, Stanford, California, (1983)

[4] Bullers, W. I., S. Y. Nof, and A. B. Whinston, "Artificial Intelligence in Manufacturing and Control", AIIE Transactions, 12, (1980)

[5] Buzacott J. A. and D. W. Yao, "Flexible Manufacturing Systems: A Review of Models", Working Paper # 82-007, (1982)

[6] Chang, Y. L., and R. S. Sullivan, "Real-Time Scheduling of Flexible Manufacturing Systems - A Conceptual and Mathematical Formulation", Research Report, Graduate School of Business, The University of Texas at Austin, (1984)

[7] Dar-El E. M. and S. C. Sarin, "Scheduling Parts in FMS, to Achieve Maximum Machine Utilization" paper presented at the special ORSA/TIMS Conference on FMS, Ann ARbor, Michigan, (1984)

[8] Dar-El, E. M. and Y. Tur, "Resource Allocation of a Multi-Resource Project for Variable Resource Availabilities," AIIE Transactions, 10, 3, (1978)

[9] Dar-El, E. M. and R. Wysk, "Job Shop Scheduling - A Systematic Approach," J. of Manufacturing Science, 1, 1, (1982)

[10] Day, J. E. and M. P. Hottenstein, "Review of Simulation Research in Job Shop Scheduling", Production Inventory Management, 8, (1967)

[11] Fikes, R. E., "Odyssey: A Knowledge Based Assistant", Artificial Intelligence, 16, (1981)

[12] Flexible Manufacturing System Handbook, Volume 1 through 5, prepared by C. S. Draper Labs, Cambridge, Mass., (available from NTIS, U. S. Department of Commerce), (1983)

[13] Goldstein, I. P. and B. Roberts, "Using Frames in Scheduling", Artificial Intelligence: An MIT Perspective, Ed. bu P. H. Whinston and R. H. Brown, MIT Press, 1, (1982)

[14] Halevi, G., The Role of Computers in Manufacturing Processes. John Wiley and Sons, Inc., (1980)

[15] Hildebrant, R. R., "Scheduling Flexible Machining Systems using Mean Value Analysis," IEEE Conference on Decision and Control, (1980)

[16] Hildebrant R. R. and R. Suri, "Methodology and Multi-level Algorithm Structure for Scheduling and Real-Time Control of Flexible manufacturing Systems," Proceedings, 3rd International Symposium on Large Engineering Systems, Memorial Univ. of Newfoundland, Canada, (July 1980.)

[17] Kimemia J. and S. B. Gershwin, "Network Flow Optimization in Flexible Manufacturing Systems," Proceedings of the IEEE Conference on Decision and Control, (1979)

[18] Moore, J. M. and R. C. Wilson, "A Reveiw of Simulation Research in Job Shop Scheduling", Production Inventory Management, 8, (1967)

[19] Nof, S. Y., M. M. Barash and J. J. Solberg, "Operational Control of Item Flow in Versatile Manufacturing Systems," Int. J. of Prod. Res, 17., 5, (1979)

[20] Olsder G. J. and R. Suri, "Time Optimal Control of Parts – Routing in a Manufacturing System with Failure Prone Machines," Proceedings of 19th IEEE Conference on Decision and Control, Albuqurque, N.M. (December 1980)

[21] Panwalker, S. S. and W. Iskander, "A Survey of Scheduling Rules", Operations Research, 25, (1977)

[22] Project Report, "A Computerized Model for Increasing Utilization," E. M. Dar-El, Department of Industrial Engineering and Management, Technion, Haifa, Israel (1983)

[23] Reiser M. and S. S. Lavenberg, "Mean Value Analysis of Closed Multichain Queueing Networks", IBM Research Report, RC-7023, Yorktown Heights, New York, (1978)

[24] Sarin, S. C. and Wilhelm, W. E., "Flexible Manufacturing Systems: A Review of Modeling Approaches for Design, Justification and Operation," Working Paper, The Ohio State Univ., (1983)

[25] Stecke K. E. and J. J. Solberg, "Loading and Control Policies for a Flexible Manufacturing System," Int. J. Prod. Res., 19, 5, (1981)

[26] Stecke, K. E., "Formulation and Solution of Nonlinear Integer Production Planning Problems for Flexible Manufacturing Systems", Management Science, 29, (1983)

[27] Vaithianathan, R. "Scheduling in Flexible Manufacturing Systems", IIE Proceedings, (1982)

[28] Wilhelm, W. E. and S. C. Sarin, "Models for the Design of Flexible Manufacturing Systems," IIE Conference Proceedings, (May 1983.)

BIOGRAPHICAL SKETCH

Subhash C. Sarin is Assistant Professor in the Department of Industrial Engineering and Operations Research at Virginia Polytechnic Institute and State University, Blacksburg, Virginia. Previously, he has held a faculty position at the Ohio State University. He received his bachelor's degree in Mechanical Engineering from Delhi University, India, his master's degree in Industrial Engineering from Kansas State University and his doctorate in Operations Research and Industrial Engineering from North Carolina State University. His research interests are in the areas of production scheduling and applied mathematical programming and currently its application in manufacturing systems design.

Exey M. Dar-El is the Harry Lebensfeld Professor of Industrial Engineering at the Technion, Israel and is currently a Visiting Professor in the IEOR Department of Virginia Polytechnic Institute and State University, Blacksburg, Virginia 24061. His areas of specialization include the design of production systems, safety research, and on productivity development programs. He has consulted in over 50 companies in Australia, U.K. and Israel. He is a member of the Israel National Council of Safety and Occupational Health and a Council member of the Israel Productivity Center.

Operations Research and Computer-Integrated Manufacturing Systems

William E. Biles
Professor of Industrial Engineering
Louisiana State University
Baton Rouge, Louisiana 70803

and

Magd E. Zohdi
Professor of Industrial Engineering
Louisiana State University
Baton Rouge, Louisiana 70803

ABSTRACT

This paper examines the application of the well-known techniques of Operations Research to the design, installation, analysis and control of computer-integrated manufacturing systems (CIMS). CIMS is shown as a fertile area for the application of classical OR techniques.

INTRODUCTION

A CIMS is commonly thought of as a truly integrated CAD/CAM system, encompassing all the activities from the planning and design of a product to its manufacture and shipping. As shown in Figure 1, a CIMS includes such elements as production planning and control, computer-aided design (CAD), computer-aided engineering (CAE), computer-aided manufacturing (CAM) in parts fabrication, CAM in assembly, automated material flow, automated material storage and handling systems, and a hierarchical computer control system linking and integrating these elements. The essential feature of a CIMS is that computers replace the human function in every way possible, including design (computer graphics in CAD), processing (robots and CNC machines), assembly (robots), testing and quality control (computer-aided testing or CAT), material flow (computer-controlled conveyors and monorails, automatic guided vehicle systems), material storage and warehousing (high-rise automatic storage and retrieval systems), process planning (CAPP), production planning and inventory control (MPS, MRP), and database management systems. In the limit, a CIMS could conceivably be operated without human assistance or intervention, relying completely on artifical intelligence. More pragmatically, the advent of the "ultimate" CIMS is a long way off, and systems involving human activity in design, maintenance, and computer systems is a necessity in the near term.

Where does Operations Research fit in? In a word - everywhere. Linear programming in optimal production scheduling, branch-and-bound zero-one programming in tool and tool path selection in CNC part programming, zero-one programming in computer-aided process planning (CAPP), queueing analysis in material flow optimization, and computer simulation of manufacturing cells and flexible manufacturing systems (FMS) are just a few applications of OR techniques. Again, the essential feature of these OR techniques in CIMS is that they are usually applied in an automatic manner in order to obtain "good", but not necessarily classically optimal, solutions that are compatible with the total manufacturing environment.

The following sections describe the application of specific OR techniques to elements of a CIMS. The techniques described represent up-to-date research results that might not have yet found their way into actual CIMS applications but which typify the applicability of OR research efforts to manufacturing systems.

SIMULATION OF FLEXIBLE MANUFACTURING SYSTEMS

Clearly, the general-purpose computer simulation languages such as SIMSCRIPT 2.5, SLAM-II [11], and GPSS-H have general applicability to a CIMS. The SIMAN language [10] has been specifically tailored to be easily applied to manufacturing systems, incorporating as it does modeling blocks for such manufacturing features as conveyors, transporters, sequences, etc. Fernandes [5] employed SIMAN to compare automobile manufacturing systems in Japan and the U.S., focussing on the effect of fast set-ups and frequency of parts delivery on in-process inventories. The systems studied were much too complex to permit clasical OR analysis. Schroer, Black and Zhang [13] also used SIMAN to model a "Just-in-Time" system for small batch manufacturing.

PASAMS [2], a Pascal-based simulation language for flexible manufacturing systems, requires only that the modeler be able to input data describing each unit of equipment in a manufacturing system. Figure 2 illustrates a typical system modeled by PASAMS. Obviously, PASAMS is meant to be applied to very specific subsystems within a CIMS, and is more applicable to modeling material flows than information flows. A PASAMS model of a flexible manufacturing system, for instance, would yield such results as production rate for each machine in the system, total production rate for the system, up-time and down-time for each item of equipment, distributions of inventory levels for workparts at various queues in the system, etc. It would enable the modeler to evaluate the effects on these performance measures of such factors as the number of maintenance man-hours allocated, batch sizes, setup times, etc. The availability of PASAMS for 16-bit microcomputer systems enables such evaluations to be undertaken on the shop floor as production schedules and resource allocations are being prepared.

QUEUEING

Queueing theory involves the mathematical study of waiting lines or "queues." The basic process in queueing involves the arrival of items to a service process and the waiting in queue for service. Thus, the basic elements of a queueing process are the arrival process, the service process, and the queue discipline.

In CIMS, queues involve the arrival of workparts or subassemblies at a workstation and the processing of the item at the workstation. "Processing" might involve a metal fabrication operation or an assembly operation. In another sense, the failure and repair of manufacturing or material handling equipment involves a queueing process, since failed items wait in "queue" for repair resources (facilities or personnel). In yet another application of queueing theory, items arriving at a material storage or handling facility might wait for service by automated material handling equipment. For example, a pallet of purchased units of raw material brought from the receiving dock to a high-rise ASRS on a pallet conveyor might wait at a transfer station for a stacker crane to pick up the pallet for insertion into storage.

The analysis of queues in CIMS can be managed through various means, including classical queueing analysis (see Hillier and Lieberman [7]), a computerized procedure for the analysis of network of queues such as CAN-Q by Solberg [14], or simulation as previously described. More often, a specialized technique is employed in CIMS. Typical of such techniques is MVAQ, an acronym for "mean-value analysis of queues", which was developed by Suri and Hildebrant [16]. MVAQ computes throughputs, utilizations, and mean queue lengths in closed networks of queues. It extends the basic functionality of CAN-Q by incorporating multiple parts classes.

INVENTORY MODELING

The optimal operation of a CIMS would involve Just-in-Time/Total Quality Control technology. In-process inventories would be virtually eliminated, stocks of raw materials and purchases items would be kept very low, and finished goods would be shipped to order just as soon as an order is completed. Thus, classical inventory modeling, such as the EOQ model, would be inapplicable in a CIMS. Instead, master production scheduling (MPS) would be accomplished automatically using the highest level computer in a hierarchical computer control structure, and a computer-aided MRP system would be utilized to procure needed stocks of purchased items. Those items manufactured "in-house" would be for automatically scheduled for production. Inventory management would be an almost completely computerized function. The entire scheme of things in inventory management would be different from that studied in classical inventory modeling, whether for deterministic or stochastic systems.

Computer simulation modeling of Just-in-Time/Total Quality Control, as described by Schroer, Black and Zhang [13], would become the dominant "analytical" tool employed in a CIMS. The importance of classical inventory modeling would be left to comparing the actual performance of the CIMS with the ideal of an inventoryless system.

LINEAR PROGRAMMING

Linear programming is one of the most frequently applied tools in the Operations Research venue. CIMS are no exception. LP would be used in CIMS in the following ways:

1. Optimum assignment of jobs or batches to manufacturing cells.

2. Optimum production scheduling.

3. Optimum placement of inventory into automated high-rise ASRS.

4. Optimum assignment of operators to equipment.

5. Optimal assignment of maintenance personnel to production areas.

Bradley, Hax and Magnanti [4] describe strategic and tactical production planning models which, with variations, would be readily employable in the "production planning and control" module in a CIMS. The strategic model utilized LP to consider the effect of such inputs as capacity, levels of plant operation, and price information on production schedules and transportation requirements. The strategic model was essentially a logistics system model. The tactical model considered such factors as on-hand orders, reservations, transportation and inventory costs, and corporate policy considerations. It used LP to generate optimal production schedules and transportation requirements.

These are just a few of the likely opportunities for the application of LP to CIMS, and include specialized tools such as the assignment algorithm and the transportation algorithm.

INTEGER PROGRAMMING

Integer programming is a special case of the linear programming problem, or the nonlinear programming problem, in which some or all of the variables x_i, $i=1,\ldots,n$ are required to assume only nonnegative integer values. Branch-and-bound techniques are particularly useful in solving integer programming problems.

A special case of the integer programming problem is the zero-one programming problem in which some or all of the variables x_i, $i=1,\ldots,n$ are required to be either 0 or 1. Typically, a 1 indicates that a particular variable is "in" solution, whereas a 0 means it is "not in" solution. For example, the problem of selecting the optimal sequence of tool paths in a metal cutting operation can be formulated as a "traveling salesman problem," which is a special class of the zero-one programming problem, and solved using a branch-out-bound technique. Jiron [9] demonstrated the applicability of branch-and-bound techniques to the development of CNC programs for milling operations.

DYNAMIC PROGRAMMING

Dynamic programming involves the stagewise optimization of multistage systems. The term "dynamic" refers to systems varying over time, but more generally refers to any system that has a multistage structure. The essential character of such a system is that there is a dependency of a stage's performance on that of the preceding stage. In general, the output from stage i becomes the input to stage i+1. For example, a multiperiod production planning problem could be formulated and solved as a dynamic programming problem. Allocations of labor, equipment, materials, and funds are made in each time period in a manner that optimizes the total performance of the system over the planning horizon. The scheduling of batches of workparts over a sequence of cells, or machines in a cell, also has a dynamic structure, and could thus be approached using dynamic programming.

Optimal control theory, employing discrete and/or continuous forms of the Pontryagin maximum principle, are generally useful in production and inventory control, as well as maintenance and replacement, in CIMS.

GEOMETRIC PROGRAMMING

Geometric programming is concerned with constrained optimization problems involving a special class of plynomial functions a the objective function and constraints in the problem. Asadzadehfard [1] and Zohdi [18] used geometric programming in the optimization of milling processes and turning operations, respectively. Thus geometric programming can be used as a component of a computer-aided process planning tool for metal cutting operations.

GOAL PROGRAMMING

Goal programming is an optimization technique which enables one to optimize a system in the face of multiple, conflicting objectives. Ignizio [8] has developed a wide range of techniques to apply to such problems. Takakuwa [17] applied goal programming to the solution of manufacturing systems under multiple objectives.

NONLINEAR PROGRAMMING

Nonlinear programming is useful in CIMS in several of the applications cited previously, but it is also important in the area of online optimization of machining operations. An experimental variant of nonlinear programming, response surface methodology, has been shown by Groover [6] to be useful as an online search strategy in supervisory control as well as in establishing optimal metal cutting conditions.

SUMMARY AND CONCLUSIONS

Operations Research techniques are indispensable to the orderly, systematic operation of a CIMS. The computer software that underlies each of the major subsystems shown in Figure 1 must of necessity possess a strongly OR flavor. The instructions that are routed along the data flow channels shown in Figure 1 will very likely have passed through an OR "black box" to establish some optimal configuration, allocation or assignment. CIMS represent perhaps the dominant application of OR for the Industrial Engineer in the 1980's.

REFERENCES

1. Asadzadehfard, A., "Optimization of Milling Processes Using Geometric Programming," unpublished MS thesis, Industrial Engineering Department, Louisiana State University, Baton Rouge, LA, December 1979.

2. Bathina, V. R., "PASAMS: Pascal Simulation and Analysis of Manufacturing Systems," unpublished MS thesis, Industrial Engineering Department, Louisiana State University, Baton Rouge, LA, May 1984.

3. Beightler, C. S., and D. T. Phillips, Applied Geometric Programming, John Wiley and Sons, New York, 1976.

4. Bradley, S. P., A. C. Hax, and T. L. Magnanti, Applied Mathematical Programming, Addison-Wesley Publishing Company, Reading, MA, 1977.

5. Fernandes, C., "A Computer Simulation Model to Analyze Production Processes in the American and Japanese Automobile Industries," unpublished MS thesis, Industrial Engineering Department, Louisiana State University, Baton Rouge, LA, August 1984.

6. Groover, M. P., Automation, Production Systems, and Computer-Aided Manufacturing, Prentice-Hall, Englewood Cliffs, NJ, 1980.

7. Hillier, F. S., and G. J. Lieberman, Introduction to Operations Research, Third Edition, Holden-Day, San Francisco, 1980.

8. Ignizio, J. P., Linear Programming in Single and Multiple Objective Systems, Prentice-Hall, Englewood Cliffs, NJ, 1982.

9. Jiron, D., "Optimization of Tool Path in Numerically Controlled Milling Using Branch-and-bound Techniques," unpublished MS thesis, Industrial Engineering Department, Louisiana State University, Baton Rouge, LA, December 1983.

10. Pegden, C. D., Introduction to SIMAN, Systems Modeling Corp., State College, PA, 1982.

11. Pritsker, A. A. B., Introduction to Simulation and SLAM-II, Systems Publishing Corp., W. Lafayette, IN, 1984.

12. Reklaitis, G. V., A. Ravindran, and K. M. Ragsdell, Engineering Optimization: Methods and Applications, Wiley-Interscience, New York, 1983.

13. Schroer, B. J., J. T. Black, and S. X. Zhang, "Microcomputer Analyzes 2-Card Kanban System for 'Just-in-Time' Small Batch Production," *Industrial Engineering*, Vol. 16, No. 6, June 1984, pp. 54-67.

14. Solberg, J. J., CAN-Q User's Guide, Report No. 9, School of Industrial Engineering, Purdue University, W. Lafayette, IN, 1980.

15. Stecke, K. E., "Design, Planning, Scheduling and Control Problems in Flexible Manufacturing Systems," in *Flexible Manufacturing Systems: Operations Research Models and Applications*, University of Michigan, Ann Arbor, MI, August 1984.

16. Suri, R., and R. R. Hildebrant, "Modeling Flexible Manufacturing Systems Using Mean-Value Analysis," *Journal of Manufacturing Systems*, Vol. 3, No. 1, 1984.

17. Takakuwa, S., "Multiobjective Optimization of Manufacturing Systems," unpublished PhD Dissertation, Department of Industrial and Management Systems Engineering, Pennsylvania State University, University Park, PA, March 1982.

18. Zohdi, M. E., "Application of Geometric Programming in Optimization of Turning Operations," Proceedings of the 1st International Conference on Production Engineering, Design and Control, Alexandria, Egypt, 1980.

BIOGRAPHIES

Dr. William E. Biles, Professor and Chairman of Industrial Engineering at Louisiana State University, received the BSChE degree from Auburn University in 1960, the MSE (Industrial Engineering) from the University of Alabama in Huntsville in 1969, and the PhD in Industrial Engineering and Operations Research from Virginia Polytechnic Institute and State University in 1971. He was a Lieutenant in the U.S. Army in 1961 and 1962, a Development Engineer in the Advanced Materials Laboratory of Union Carbide Corporation from 1962 to 1966, and a Process Engineer with Thiokol Chemical Corporation from 1966 to 1969. He was Assistant Professor, Associate Professor, and Professor of Aerospace and Mechanical Engineering at the University of Notre Dame from 1971 to 1979, and Professor and Head of Industrial Engineering at Pennsylvania State University from 1979 to 1981. Dr. Biles is a member of IIE, ORSA, TIMS, NSPE and SME. He is Director of Research for IIE and a member of the Engineering Accreditation Commission of ABET. He was Chairman of the TIMS College on Simulation and Gaming from 1982 to 1984. Dr. Biles has authored or co-authored two books and almost fifty journal and conference papers. His areas of specialization include simulation, industrial experimentation, and manufacturing systems.

Dr. Magd E. Zohdi, Professor of Industrial Engineering at Louisiana State University in Baton Rouge is responsible for directing and supervising the manufacturing engineering and metrology laboratory at the University. Dr. Zohdi received a BS in production engineering from Cairo University; an MS in Mechanical Engineering from the University of Kansas; and a PhD in Industrial Engineering and Management from Oklahoma State University. He is a member of the Society of Manufacturing Engineers, Alpha Pi Mu, American Institute of Industrial Engineers, Sigma Xi, Tau Beta Pi and American Military Engineers. Dr. Zohdi is listed in the International Who's Who of Intellectuals, England 1981; Directory of World Researchers, 1980s Subjects; Personalities of the South, 1980; Who's Who in America, 1980; and Outstanding Educators of America, 1972.

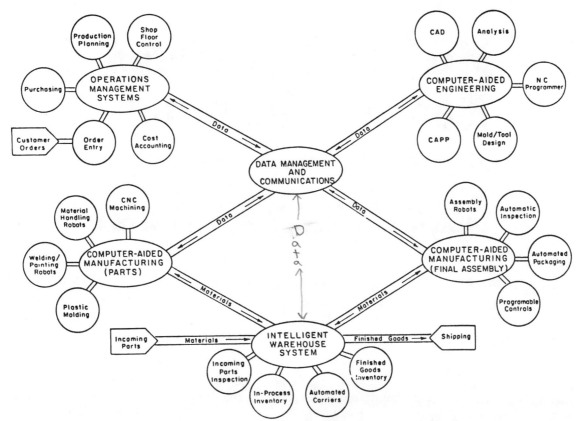

Figure 1. Overview of a Computer-Integrated Manufacturing System

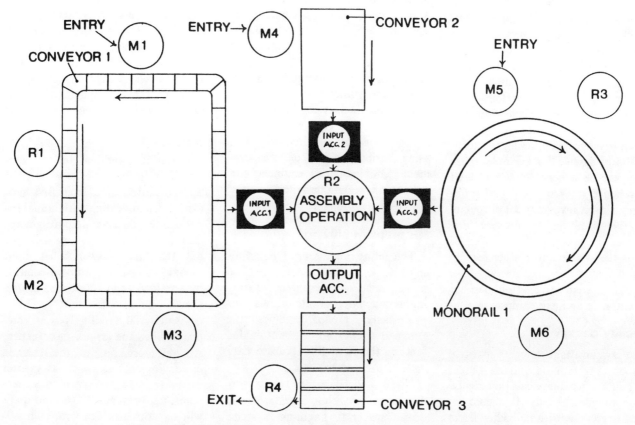

Figure 2. A Flexible Manufacturing System Modeled by PASAMS[2]

Modular Integrated Material Handling System Facilitates Automation Process

By Neil Glenney
The Confacs Group Inc.

Material handling is a critical consideration with respect to the automated factory because in it, great amounts of in-process material are or will be moved at a high velocity from place to place, stored for future use or fed into operations that add value. Material handling provides the movement, machine tools the precision assembly and computers the direction. It is the balancing of these, however, that is the most difficult to accomplish.

The key role of the systems integrator will be recognizing the best practices available for minimizing inventory and maximizing the utilization of assets. For example, there may not be a need for a unit-load warehouse stacker if the real problem is simply inventory out of control or the situation a new machine creates by producing a number of parts the system can neither feed to the machine nor take away. Make sure the system has the right number of entrance and exit ramps.

Facility design

A system that triggers positive responses from those who use it must be built from components that meet a wide range of needs. It should be easy to change, with individual parts that can be replaced or expanded with a minimum of disturbance to the system. When each component or subsystem has a spectrum of uses, a modular approach to problem solving is utilized and flexibility in decision making is achieved. The structural changes made possible by the modular approach will allow the industrial environment to keep pace with increasing demands for change at reasonable cost.

The material flow pattern is the basis for the entire facility design as well as the success of the enterprise. Too much emphasis cannot be placed on the importance of determining the most efficient plan for the flow of materials, information and people.

The importance of the flow pattern in any facility as it influences production can be analyzed in advance for cost, quality and time using various means of measurement. The industrial engineer can evaluate these data based on how the plant is functioning at present.

MH problem solving

The industrial/material handling engineer will soon become one of management's key resources for solving productivity problems. As this professional group becomes the "change agent" in the automated factory, the need for accurate data collection becomes critical.

The skilled IE can see trends, predict their impact on operations and devise material handling systems with flexibility for present and anticipated future requirements. Top management will demand justifications for automated material handling systems. A simulation model will be necessary to show how the proposed system will operate in all situations.

In addition, a tactical material handling plan can indicate the most efficient routing of material through the facility using routing analysis. This technique will allow the material handling engineer to increase his effectiveness by using computer-aided plant layout. Systems will be designed using computer-aided plant layout routines, integration with simulation models, equipment design routings and interactive computer graphics. This equipment and data analysis will soon become a widely accepted planning tool in the automated facility.

Material handling is to some extent an art however, and creative solutions often come from first-hand experience in dealing with actual situations. Production simulations and sophisticated MRP systems will help, but the final answer must come though analysis and implementation by a trained and experienced professional.

As material handling system components become larger, more expensive and more complex, the selection of equipment assumes a greater importance in management's economic performance. Experienced IEs will select and design with compatible components that meet the long-range facility plan. The automated factory design must adhere to the

Figure 1: Material Handling Systems in the Automated Factory

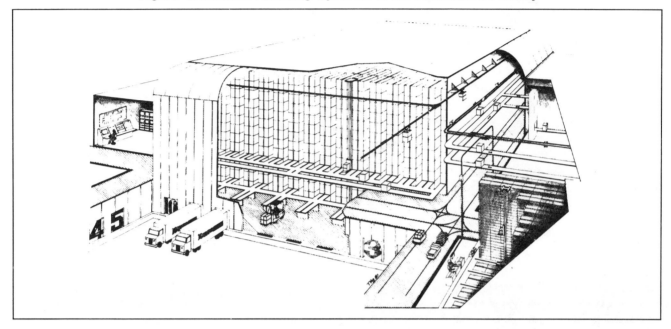

principles of flexibility so that it can meet the future needs of its users.

An example of an integrated material handling system that presents some possible alternatives to conventional flow patterns—the "facility 'U' design"—is described below. The primary objectives of this system are as follows:

☐ Reduce initial building cost.
☐ Reduce construction time.
☐ Reduce contingency investment.
☐ Increase efficient utilization of space over life of building.
☐ Reduce change and relocation costs.
☐ Reduce energy consumption.
☐ Increase comfort, efficiency and satisfaction of users.
☐ Utilize independent production units for use separately in any sequence.
☐ Centralize storage of inventory/work-in-process.
☐ Increase speed of the distribution system that provides material to independent production units.
☐ Modular containerization for storage and handling.
☐ Total "real-time" distributed process control.

The dock location is the focal point of planning and offers the opportunity for expansion. Increases in production demands often require more deliveries and additional shipping capacity. The scenario for the facility plan presented here utilizes a dock door location one level below grade. This is to allow for a lower profile warehouse that meets possibly tough zoning restrictions. The below grade location is useful also in maintaining control of incoming and outgoing shipments with no interference with in-process movement on the second level above. Considerable energy savings also can be achieved when outside air is not able to enter or escape from surrounding operations.

The area of the truck entrance is also critical. Sufficient room to maneuver larger trucks, possibly 60 to 70 ft long, will demand turnaround areas up to 200 to 300 ft wide. Security of the materials being handled in the dock area must be under control at all times.

The dock equipment will also have to be variable enough to negotiate the difference between truck bed heights. A good dock layout can accommodate many options, including new types of material handling equipment that will be introduced in the automated factory.

Special attention must also be given to understanding trends in packaging. The supplier will have to become aware of how his shipments are going to be handled in an automated system. Standardization of pack and load techniques will be an important issue to be dealt with throughout the process.

Since peak loads for truck docks are not easily determined, there should be a truck waiting area within the grounds of the facility. Trucks awaiting assignment from the control station should be held in an area in which they will not interfere with other trucks.

Security is a key element of warehouse and dock design. A single entry and exit point at the front section of the physical plant best

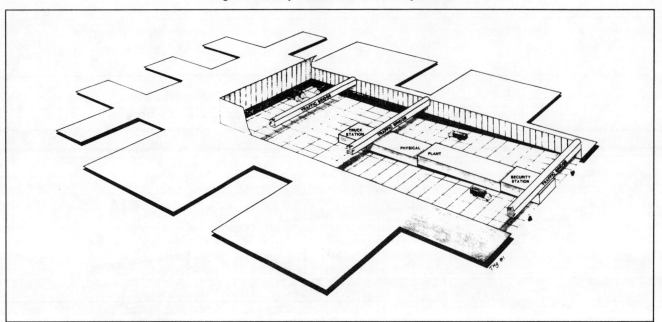

Figure 2: Expansion for Dock Spaces

provides this. The security area is also the preliminary point from which the main computer is notified that a delivery has been made and what dock it is to be delivered to.

Inside the warehouse, total control of inventory can be achieved with an on-line, real-time minicomputer. Hand-held bar code scanners will aid in the identification process required to accurately establish routings.

Storage requirements

Material storage and the smooth flow of materials throughout the automated factory will require complete control of inventory. The optimum system would be one in which material never stood still, but stayed in constant movement from receiving to shipping. But it is doubtful such a system could ever be realized. A critical trade-off is necessary between how to store material waiting to be moved and how long to keep it in storage. The materials and products that are warehoused represent a tremendous investment and one that's crucial to justification of a systems project.

The modeled warehouse requirements are shown to demonstrate that no one system, high rise or conventional, is by itself a practical approach for all levels of materials storage and retrieval. In the automated factory we will find that new and modified technologies in material handling systems for warehouse design will have generated improved utilization of other operational functions.

The model utilizes computer-controlled storage machines that will reduce labor and floor space requirements and increase the accuracy of records. Driverless tractor trains can eliminate long hauls for fork trucks, reduce labor and increase productivity. Laser scanning devices can boost accuracy and reduce cost and process time. Automatic order picking in unit and miniload applications can mean greater accuracy, reduced costs and better control. Computerized carousel systems are also effective and offer added flexibility in terms of access to the smaller components required for kitting operations.

These are but a few of the technological improvements that will be utilized in the automated factory, but at the heart of all of them will be the computer. Management will be able to assume total control with on-line real-time distributed processing set-ups that all interact with the host system.

The proposed model also utilizes a dedicated storage building structure. The rack-supported building concept offers a very attractive tax depreciation write-off. The same structure will support the in-process delivery system.

The model facility now has in place a flexible dock and warehouse layout that has utilized many of the automated sub-systems that exist in the marketplace today. The real challenge in the '80s will be to integrate the movement of materials from storage to manufacturing through various modes of transport utilizing the theory of "on-time delivery" techniques. This will reduce the initial building cost of manufacturing areas. The centralized storage concept will replace the safety stock concept.

The results of locating all material handling equipment in a common aisle will include economized engineering and installation, maintainability, safety, ease of change, use of pre-engineered modular components, energy efficiency and keeping the noise of running equipment out of manufacturing operations. This design stresses the need to have the flow of materials separate from the flow of people. The firewall separating warehousing structures from the material handling aisle will support the conveying equipment and physical plant utilities.

With the materials now flowing at higher velocities throughout the workplace modules, advances in information handling will also be

Figure 3: Various Warehouse Storage Methods

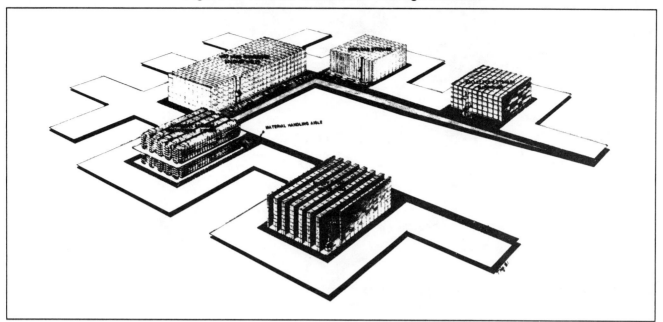

accelerated. The increased demand for material handling equipment will require communication and control systems to be easier to design and install and to be readily available at a relatively low cost.

With the increased demand for handling systems will come a possible shortage of skilled professionals, including IEs, to design and install equipment. To meet such a shortfall, the specified equipment must have built-in diagnostics, not only to make it easy to check out the system as it's being installed, but also to make it possible to monitor operations throughout the system.

Information on every item moving through the automated facility will be fed into the computer at strategic locations in the handling system through various types of fixed or moving beam laser scanners. Scanners now available will read very

Storage and Retrieval Systems for In-Process Storage

By Michael E. Heisley, President
Conco-Tellus Inc.

Automated storage and retrieval systems can be separated into two groups: those in which the system is located remotely from the process and those in which it is adjacent to the process. The remotely located AS/RS requires a transportation system that will link it with the in-process area. The basic idea is to have the AS/RS interfaced with the production control system so that the right parts will be delivered to the correct work station exactly when they are needed.

Many different types of transportation systems are available to take the product from the storage area to the processing area. Some of the more basic and commonly used systems are roller conveyors, power and free conveyors, towlines, shuttle cars and automatic guided vehicle systems (AGVS).

Roller conveyors can be mounted on the floor or overhead. The advantage of mounting the delivery conveyors overhead is that the plant floor is clear. Elevators can be used to lower the loads to the floor level at the production work stations.

Power and free conveyors with automatic transfers from the SR machines also leave the plant floor clear. With accumulation lanes and sortation systems, a very close coordination between the parts and their usage can be accomplished. In addition, the hoisting mechanism can be used to lower the material at the proper time to the production work station.

Towlines can be utilized, with loads automatically placed on a tow cart from the AS/R system, or the tow carts themselves can be stored in the AS/RS and retrieved directly to the tow line conveyor.

The following types of systems are relatively inexpensive and provide high volume throughput; however, they do result in floor congestion: *AGVS systems* can pick up loads directly from the AS/RS conveyor. There are several types of AGVS. The most current development is to have computer controlled vehicles handle individual loads and be constantly in touch with the control system so that they can be routed throughout a plant and even have a destination changed en route.

In addition, driverless tractor trains—where many loads are put behind a driverless vehicle—can provide heavy volume capability and flexibility. With this concept, system changes and expansion can be accomplished easily by re-routing the wire in the floor.

Shuttle cars are often used in combination with the in-aisle transfer car in multiple-aisle systems. Here the shuttle car receives loads from the aisle transfer car and delivers them to the takeaway conveyors, which bring the

Figure 4: Systems for Control of Materials, Information and Employees

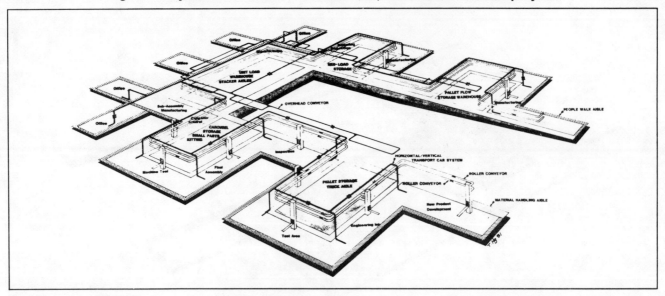

accurately such codes as the interleaved 2 of 5 code or the code 3 of 9 (Code 39), but advanced systems may see the introduction of new techniques. An example of one such new application for the factory could be the use of the optical character recognition or OCR code. This code might be labeled the "human factored code" because of its readable characters. New developments are making it capable of reading accurately at higher rates of speed.

Making all materials and piping accessible from a common aisle and making use of color coding are other techniques that facilitate the flow of products.

Physical plant utilities would also be housed in the area around the fire wall of the warehouse structures and, like the handling systems, be capable of extension with very little engineering design required. Such "intangible" provisions as good natural lighting and ventilation from glazed adjustable louvers above, the bright colors of the equipment, wall graphics, acoustics and worker convenience and comfort are other desira-

loads to the production center.

With all these forms of transportation, a good control system is required to keep track of loads during transit. Besides, the timing of the material's arrival at the point of usage is very critical.

Materials waiting for the next process will be returned to storage by the same type of system. It is obvious that a remote system can only be used where the waiting time between processes is sufficient to route the loads back through the AS/RS.

In most cases heavy volume and a large number of intermediate processes require that a system or systems be installed adjacent to the various product areas. The volume of traffic the transportation system must handle increases dramatically every time a process requires that material be stored while waiting for some future process. However, an AS/RS installed adjacent to the process area requires especially thorough study, since future requirements for manufacturing expansion of that particular area of the plant can be very critical. Once an in-process system is located in the center of a manufacturing facility, flexibility can be severely limited.

In machining centers, investment in expensive computer-controlled machines requires nearly continuous usage to generate an acceptable return on investment. An AS/RS centrally located between machines can feed not only parts, but also the proper tooling to a machine as needed. Such a machining center gives the operator the ability to run his own shop as a closed unit with maximum efficiency.

One type of AS/R system frequently used for in-process storage is the small item or mini-load AS/RS machine. This machine typically handles steel trays with small parts. Normally, the weight of the tray is about 500 lb. In these systems the operator is located at the end of the aisle, where he has a control console and can pick the parts needed for the assembly. This type of system may be completely enclosed except for the front access, and usually fits very well into a small parts manufacturing facility. With larger unit load systems, the position of the AS/RS in the middle of the processing area gives some very unique possibilities for efficient use of the storage system.

Frequently, the bottom tiers of the rack openings are converted to output stations. An operator can pick up material here and bring it to his work station. This type of system can be further automated by installing conveyors back in the bottom of the rack openings and using a shuttle car to pick up the loads and deliver them directly to the conveyor section servicing individual work stations.

In multi-level buildings or in areas with sub-assemblies on mezzanine floors, the AS/RS can be not only the buffer storage between each operation, but also the elevator that brings loads back and forth between different floors. This use of AS/RS with output on different floors has given new life to old multi-floor manufacturing facilities, where proper assembly flow has been hampered by the lack of space available on individual floors.

Finally, building a free-standing AS/RS outside the present building and knocking holes in the adjacent outside wall for conveyor connections at each floor level, can provide for good material flow and superior inventory controls.

Figure 5: Human Factored Work Environment Utilizing Flexible Design

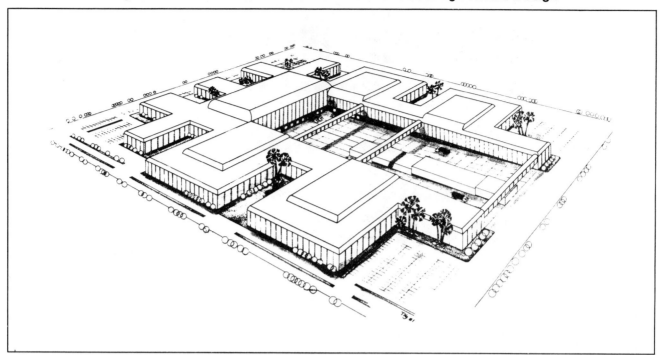

ble elements of the proposed system.

The computer-integrated assembly process will bring forth a new breed of worker in the automated factory. The programmed machines may require cleaner environments to maintain their reliability. But more importantly, a human factored work environment must be provided.

The automated factory will begin to materialize only when the conventional "caste system" that differentiates between "factory workers" and "office workers" is abolished. When this becomes reality new methods of material handling will be presented. Some innovative techniques developed for the office will be used in manufacturing areas (e.g., pneumatic tubes) fluorescent strip guidance mail carts and independent electric cars that travel horizontally as well as vertically on aluminum tracks). The office will utilize techniques now seen more or less only in the factory (e.g., vertical and horizontal carousels for storage of documents, moving-track shelving units and people movers that will convey the slowest moving component of the automated factory—its employees).

In the modular manufacturing environment, open structural members support various work station components, including work surfaces, shelves, cabinets, drawer fixtures, dispensing rails and display panels. Modular components will make optimum use of vertical as opposed to horizontal space.

It is extremely important for management to understand that the success or failure of the automated factory will depend on the flow of materials and not on what equipment is used. In general, equipment is designed to meet the unique needs of its users. Misapplication of equipment to a system design causes the most problems. The success of an automated handling system in this factory setting will depend more on the planning, control and management of these systems.

With the model design described here, a 100,000 sq ft building can be expanded tenfold and still provide interactive utility and convenience between subsystems. The critical task related to introducing automated handling systems will be designing the worker into the system. Long-range plans must include developing the people who will operate the system.

The real challenge in implementing automated facilities handling is achieving the proper balance between time, service, cost, people and flexibility to meet future changes. Achieving this balance will move us toward increasing productivity and controlling inflation. **IE**

Neil Glenney is president of the Confacs Group, a Phoenix-based consulting firm that specializes in concept design for facility systems. The company is presently involved with designing material handling systems for the electronics industry. Previously, Glenney was manager of material handling and facilities systems for ITT Courier Terminal Systems in Tempe, AZ. His paper on evolution of a systems project was awarded first place in the Material Handling Institute's "Concepts for the '80s" competition in 1979. At this year's show he won the Grand Prize Reed-Apple award. Glenney is a member of the College-Industry Council on Material Handling Education of IIE.

Reprinted from Industrial Engineering, September 1983.

Robotic Vehicles Will Perform Tasks Ranging From Product Retrieval To Sub-Assembly Work In Factory Of Future

By Robert E. Smith
Integrated Factory Controls

At present, most factory robotic vehicles (automated guided vehicles, or AGVs, and wire guided industrial trucks) are used in mobile material handling. That is, they move large sub-assemblies from one work station to another, or operate strictly in warehouse functions.

Future automation of vehicles will allow facilities to use flexible unmanned fleets to perform a multitude of functions, such as:
☐ Product retrieval and storage at remote locations.
☐ Identification and selection of random articles.
☐ Performance of multiple axial control with robotic limbs and of sub-assemblies while en route (mobile work stations).
☐ Security surveillance utilizing advanced on-board image processing.
☐ Developing their own routing free of wire guidepaths.

This may sound farfetched, but all the necessary technology exists in various forms today. It's just waiting to be packaged and applied.

Robotic vehicles are not new. Wire-guided tow vehicles were operational in 1965. In 1975, the military had several operational prototype unmanned vehicles. By combining advancements in vehicle control concepts with affordable electronics, it is possible to robotize many existing vehicles.

Practical implementation

In order to get a realistic perspective on this subject, we need to review the applied technology and the associated state-of-the-art. Present applied technology involves AGVS and automated guidance via wire guide control kits for several industrial vehicles. State-of-the-art technology includes on-board microprocessors, intelligent controls, vision systems and communications. Related systems level software involves vehicle tracking, traffic control and fleet supervision.

Today's AGVS

Today's AGVS units are guided using wire guide (signal on a buried wire) in combination with control signal logic. Such control signals are picked up by inductive sensors placed near the bottom of the vehicle. These sensors receive signals from the floor wire that define control functions such as stop/start, forward/reverse, acceleration/deceleration, turns and other logic for axial control.

The AGVS vehicles may receive command signals from a central controller unit. The central controller provides the capability of automatic dispatching, tracking and monitoring of all the vehicles in the system. Microprocessors on the vehicle provide control for routing, automatic load/unload, data communication,

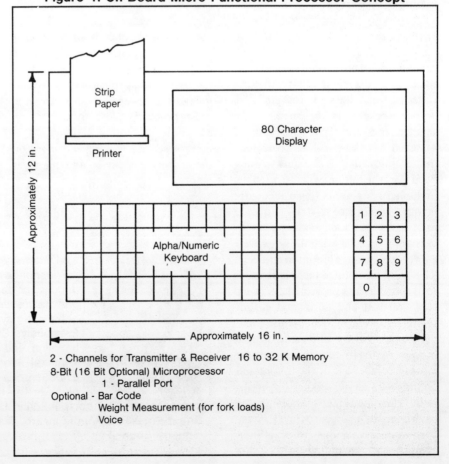

Figure 1: On-Board Micro Functional Processor Concept

2 - Channels for Transmitter & Receiver   16 to 32 K Memory
8-Bit (16 Bit Optional) Microprocessor
        1 - Parallel Port
Optional - Bar Code
        Weight Measurement (for fork loads)
        Voice

This Portec Navigator ULC 310 is used to transfer pallets automatically from a conveyor to an inspection station.

diagnostic information, guidance and safety monitoring.

Kit guidance systems

Vehicle guidance kits are available for a number of vehicles. Some of these kits actually replace entire steering mechanisms or provide for hydraulic retrofits with electronic controls. Table 1 lists makers of industrial trucks supported by these guidance systems.

The kit form of control assures that assorted vehicles operate on the same guidepath systems. These kits provide a sound starting point for retrofitting of existing vehicles into facility plans. The most popular applications today are kit vehicle controls in narrow aisle and drive-through storage racks for warehouse modernization.

Due to the evolutionary state of these products and vendor marketing practices, little standardization has occurred across the industry. Vendor compatibility can become a major problem to users. In other words, equipment supplied by different vendors may not work together on the same traffic guide system.

The user does not have the flexibility necessary to select functional vehicles from more than one vendor. One way to get around this problem is by using kit control systems.

Kit control systems are packaged electronic elements common to all vehicles. These common elements are sensors, microprocessors, communications, and actuator controllers. The electromechanical interfaces may vary slightly from vehicle to vehicle, as may the mounting brackets and PROM (programmed read only memory containing preprogrammed instructions) control programs.

New guidance control kits provide optional sensors that require no wire guidepath. This allows the vehicle to

Table 1: Brands of Industrial Vehicles Supported by Guidance Control Kits

Allis Chalmers
Baker
*Barlow
Barrett
Big Joe
Clark
Crown
Drexel
*Geveke
Hyster
*Lansing Bagnall
Prime Mover
Raymond
Schreck
*Space Master
*Steinbock
*Wagner
Yale

*European vehicles

Table 2: Systems Oriented Product Developments for Automated Vehicles Technology*

Product on-Board Vehicle	Date Applicable
Wire guidance kits	Existing
Microprocessor wire guidance kits	Existing
Infrared microprocessor guidance kits (for areas where wire guidance cannot be installed)	Late 1983
Infrared communications coupling	Existing
Radio dispatch with functional processing	Existing
Tiller steered vehicle (AGVS) kit	Late 1983
Robotic vision	Existing
Functional processing (under $5,000) Printer Bar code Display Keyboard	Late 1983
Free ranging vehicles with vision	1985
Vehicle fleet tracking systems (Non-wire)	1985
On-board robotics (6 axes of control)	Existing

*Vendor names not given due to R&D schedule status.

be guided over aisle floors not suitable for wire guidance. Since the guidance kit is microprocessor based, it will operate with either or both types of sensors.

Functional processing

Functional processing may be required aboard a vehicle to provide a sequence of operator instructions (manned or unmanned vehicles). Functional processing for order picking is the most popular of today's applications.

With traditional order-picking systems, the operator inserts a cassette tape into the on-board computer. The tape has been preprogrammed by a warehouse computer with instructions to the operator indicating picking sequence and orders to be filled. A computer display alerts the operator of operational sequences.

Sometimes added peripherals such as bar code readers, strip printers and operators' keypads are included to provide additional flexibility to the operator. Today, these functions

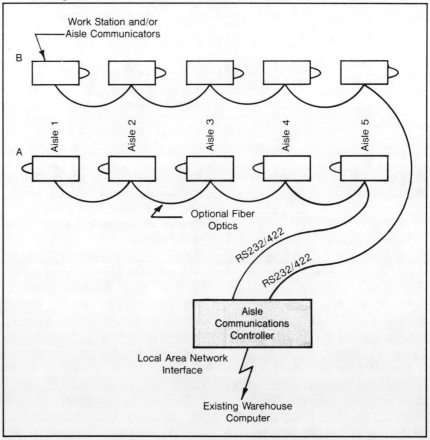

Figure 2: Distributed Vehicle Communications Links

Figure 3: Techniques for Installing IR Systems

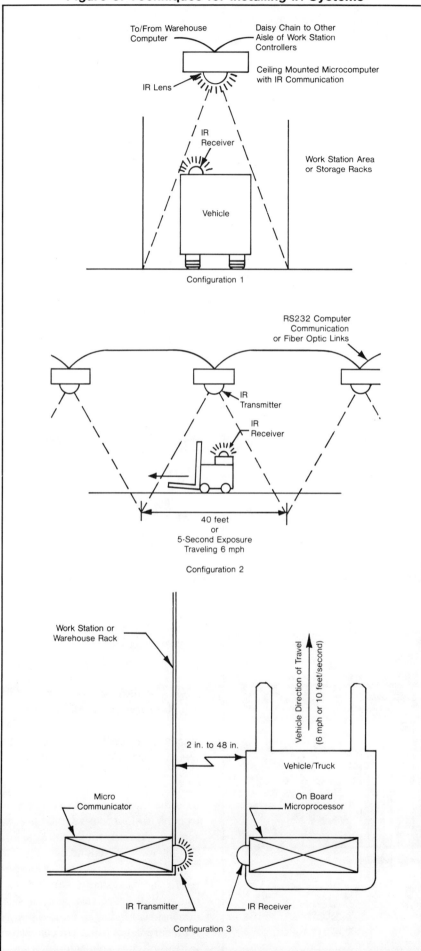

have a higher degree of reliability and can be purchased for under $5,000. Figure 1 illustrates an onboard computer for functional processing.

Optical links

In addition, new techniques may provide direct high speed communications via optical links as the vehicle enters a storage rack area. An aisle controller is mounted on or above storage rack aisles (and could be applied to work stations as well), which provides bidirectional communications with passing vehicles. The communication speed is such that the vehicle need not slow down when passing. Figure 2 illustrates this technique.

Aisle or work station controllers necessary for this function are priced below $2,500. Designs include total CMOS (complementary metal oxide semiconductor) technology packaged for industrial grade use and to operate during power outages.

Communications

Communication with mobile vehicles in manufacturing facilities can be provided for using several types of technologies. These are infrared, radio, low energy microwave and ultrasonics. Infrared (IR) is the most cost effective, most easily maintainable and easiest to install.

Figure 3 illustrates several techniques for installing IR systems. Normal vehicle travel is about six miles per hour. Proper communications exposure to IR requires one second to transmit 1,000 data words (or about 10 ft of exposure; most low cost transmission occurs at 19.5K Baud).

The aisle or work station controllers may be linked into a local area network or factory computer via normal RS-232 links. See Figure 3 for a typical daisy-chain network concept. With free wander vehicles, reliability might require a combination of tech-

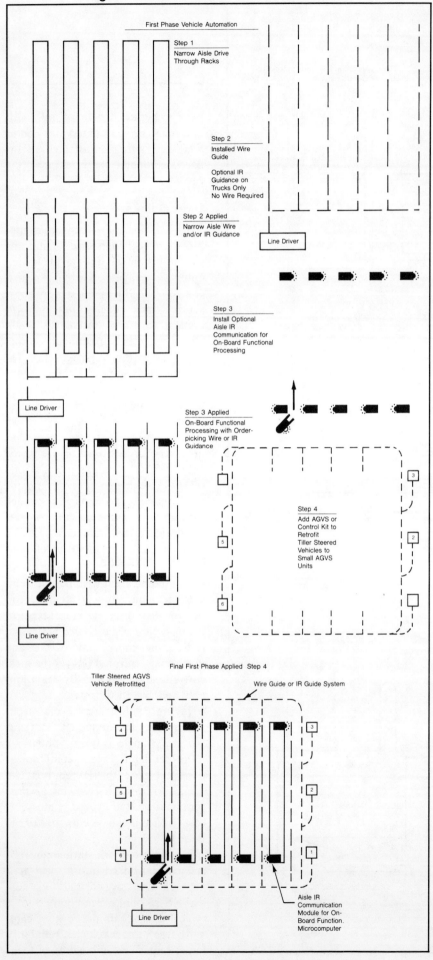

Figure 4: First Phase Vehicle Automation

nologies, such as IR and radio.

Future developments

Robot mobility without wire guide is nearing reality for facility production operations. Present restrictions are in noncontact sensing, location identification and free wandering fleet control. The last two items require significant software development to give existing hardware the functional capability. Non-contact sensing closely parallels robotic vision systems.

Table 2 lists systems oriented product developments in automated vehicle technology and indicates when they are expected to be in application.

Existing technology in vision systems is quite adequate for robotic mobile intelligence. It will allow free movement around objects and down aisleways. Development in object recognition, with depth-of-field, has been achieved by more expensive systems ($20,000 to $30,000).

The prime measurement from vision sensors is distance to objects, with the ability to calculate perspective depths and widths. These measurements are used in control programs that allow intelligent free movement.

On-board microprocessors

On-board microprocessors require a high degree of reliability. Robotic vehicles should incorporate high standards in redundancy, vibration, temperature tolerance, moisture and electronic noise immunity.

A few points can help a user evaluate the quality of such vehicle controls. With inexpensive microprocessors, it is reasonable to expect a certain level of redundancy on crucial control circuits. Vibration requirements usually can be met using small board sets with physical dimensions under 5 in. × 8 in. or rigid card cages using pin connectors instead of edge connectors.

Temperature requirements are easily met by using CMOS semiconductor technology. Moisture can be a major problem if packaging allows condensation. Condensate usually causes corrosion and short circuits. Coated boards usually eliminate this problem.

The last point to consider is electronic noise. Noise problems are usually taken care of by using shielding systems and good cabling practices. Simple inspection of these items will aid in evaluating vendor controls for industrial reliability.

Integration

Most factory-of-the-future plans incorporate "islands of automation." The islands are integrated manufacturing cells consisting of robots, machining centers, programmable controls and local area network communications. These planned factories also utilize modern warehousing and AGVS. Still, there is a requirement for manned industrial trucks to perform more flexible material movement functions.

With rising operational costs, hazardous environments, safety regulations and requirements for efficiency, unmanned vehicles present a reasonable alternative to manned units. Figure 4 illustrates an initial phase of integrating these vehicles into operations. The steps illustrated include all kit oriented controls and functions for practical implementation. The application in the example is for narrow-aisle, drive-through warehouse order-picking operations.

Table 3 lists the implementation sequence for this initial phase. Equipment consisting of six guidance kits, four tiller-steered retrofit kits, six on-board functional computers (shown in Figure 1), 12 IR communication controllers and six programmed station controllers is estimated to cost between $225,000 and $250,000.

Table 4 lists future considerations

Table 3: First Phase of Warehouse Vehicle Integration

Warehouse Installation:
  Step 1—Narrow drive-through aisles.
  Step 2—Install wire guidepath.
  Step 3—Install aisle IR communicator.
  Step 4—Install AGVS floor wiring and traffic controller.

On-Board Vehicle:
  Step 1—Decide vehicles to be retrofitted with guidance kits.
  Step 2—Install guidance control on vehicles. (NOTE: IR guidance requires no aisle guidepath, since it guides off the rack.) Aisle access must be determined.
  Step 3—Install on-board functional process (order pickers).
  Step 4—Install AGVS control kits (or retrofit tiller steered vehicles).

Table 4: Second Phase Warehouse

Second phase includes all items of the first phase, plus system advancements. These advancements—expected to be available in 1985—include:

Truck:
☐ Programmable 10-axial controls.
☐ Free ranging robotic vehicle.
☐ Two-way communication.
☐ Image processing.
☐ On-board robotics (for specific applications by other suppliers).
☐ Extended AGVS (non-wire communications) kits.
☐ Location systems (track vehicle through the warehouse).

Warehouse:
☐ Supervisory and traffic controller.
☐ Source location or vehicle tracking modules.
☐ Two-way communication.

New Technical Enhancements:
☐ Digital sensor modules.
☐ Ultrasonic source location verification.
☐ Local microwave ID systems.
☐ On-board vision and parts sort routine.

that will be applicable by 1985 for a second phase of the above example.

In summary, the key to any successful implementation is a well thought out plan. Numerous vendors and experts are available to support users in their pursuit of unmanned vehicles. Technology is advancing at such an accelerated pace that it takes teamwork to integrate facilities at the system level. Phasing in kit vehicle control systems presents the most immediate answer, with eventual evolution into unmanned mobile robotics.

For further reading:

Cauffman, Scott, "An Algorithmic Approach to Intelligent Robot Mobility," *Robotics Age*, May 1983, p. 38.

Kinnucan, Paul, "Machines That See," *High Technology*, April 1983, p. 30. **IE**

Robert E. Smith is President of Integrated Factory Controls, a start-up company serving the market in automated guided control products. He has over 15 years' experience in engineering and marketing of digital control systems. As director of industrial marketing for three major companies, Smith has introduced over one dozen products to the marketplace. These products involve process control, machine tool control, factory data collection, CAD/CAM and industrial communication networks. Smith received a BS degree in industrial technology from California State University, Fresno, and an MBA from Pepperdine University.

*Reprinted from **Industrial Engineering**, December 1982.*

Automated Visual Inspection Systems Can Boost Quality Control Affordably

By John W. Artley
Object Recognition Systems Inc.

On-line factory inspection is a task that human beings probably shouldn't have to perform. Although it is not a dangerous job, it can be tedious and unrewarding, and it is difficult, if not impossible, to maintain concentration on the small details of fast-moving objects for hours at a time, day after day.

Even the most adept human workers simply cannot keep up with the speed of the modern production line. Therefore, several inspectors may be required, or a limited sampling for quality control may be all that's possible.

Ever since World War II, engineers and researchers in the field of pattern recognition have been trying to develop practical, cost-effective methods of improving quality control and industrial inspection through automation. Yet routine on-line inspection has been one of the last factory tasks to resist automation, because it requires qualities of human intelligence: perception and judgment.

Vision for robots

Automated vision is also a desirable quality for many applications involving industrial robots. Without vision, robots usually require that work pieces be delivered to them in a certain position. Any deviation generally confuses the robot, because it cannot determine what is wrong or understand how to correct the problem, since it lacks the human qualities of perception (vision) and judgment (decision-making).

Material handling systems can be automated, but perception and judgment may be required at critical points along the way.

For example, in an automated warehousing situation, certain cases must go in one direction, while others mush reach a different destination. Somehow the various cases must be identified, sorted one from the other and routed to the right locations. Though there are several ways to do that, visual inspection would seem to be a most logical solution.

The principal inventions necessary to reproduce the human qualities of perception and judgment have been around for some time: the television camera and the computer. Unfortunately, combining them into an automated vision system is not as easy as it sounds. Even the smaller and cheaper video cameras produce an extraordinary amount of electronic data every fraction of a second.

To overcome that problem, earlier automated inspection systems required a large and expensive mainframe computer to process all of this electronic information by "brute force" processing techniques. The cost of these techniques was generally prohibitive for all but a few very specific applications.

Cost effective solutions

Recently, technology has finally resolved the problems that held back the development and implementation of cost-effective vision systems. New image-processing techniques can be used to refine and reduce the information received from the video sensor so that the decision-making function can be performed by a low-cost microprocessor.

The result is an automated vision system that is faster, more accurate, more reliable and more rugged and that meets most companies' return-on-investment objectives. Several types of systems are now available in the commercial marketplace.

Some systems utilize a binary processing approach, which is probably the most common strategy in current use. The object to be inspected passes over a translucent plane with lighting underneath and the video sensor above, so that the image of the object appears as a silhouette.

Each point of the image is translated as either black (object) or white (background); therefore the term "binary" processing. However, since the object becomes a black silhouette, illuminated from behind, the surface characteristics of the object, such as printed code numbers or graphic information, are not captured in the image.

Nevertheless, binary processing can be used in industrial applications where the surface characteristics of the inspected object are not critical factors.

Other systems have been designed for specialized applications (such as inspecting computer terminal keyboards). Sometimes these specialized systems can also be adapted to other inspection applications.

Object Recognition Systems Inc. has developed a proprietary technology, utilizing a variety of high-speed gray-scale processing algorithms, which electronically describes a variety of possible features (including surface characteristics) that can be inspected and evaluated. (See diagram in Figure 1.)

The same basic techniques can be used to solve a number of inspection problems in many industries that are otherwise quite different, without the need for custom-designed hardware of software.

Common attributes

Regardless of the differences in processing strategies, automated vision systems share many common attributes and functions. Because on-line inspection has generally resisted automation for so long, productivity gains with these vision systems will be dramatic as the systems are implemented on a broad basis throughout the next decade.

Immediate, practical applications for automated inspection devices in U.S. industry alone are estimated in the hundreds of thousands. Return on investment for such systems is usually estimated at about 18 months, but experience has shown the ROI to be much less for many applications.

The following functions are typical of the kinds of tasks that automated vision systems can accomplish, though not every system is capable of performing all the functions listed:

□ *Identification*—On the production line, vision systems can correctly identify an entire scene or an individual product, components, color, texture, graphics or the object's orientation in order to provide input for such decisions as routing, diverting and auditing.

ORS vision system connected to this Unimation robot enables it to pick up a cylinder from among parts randomly oriented and jumbled together in a bin.

□ *Classification*—Vision systems can recognize and classify different types of objects on the same production line at the same time.

□ *Sorting*—Because the vision system can identify each item and classify it according to type, sorting decisions can be made by the system at high production-line throughput rates.

□ *Inspection*—A vision system can inspect objects on a production line to confirm or deny their conformance to precise production standards. In the broadest sense, inspection encompasses a range of tasks which may include checking for both the total content of a scene and the orientation of patterns or objects within the scene.

□ *Verification*—On a higher level of abstraction, vision systems can validate a particular piece of information in a scene. The information might be a serial number or a lot code. Is it correct? Is it legible? Is it there?

The information could also be the acknowledgment that an assembly is complete and ready to pass on to the next step in the manufacturing process.

Another aspect of verification may be required in solving the specialized problems of gauging. Vision systems will increasingly have a variety of high-speed gauging capabilities for applications where measurements are a critical function of the inspection process.

□ *Control*—Vision systems can direct activities on the production line on the basis of visual information and evaluation.

Visual inspection is not restricted

to the job performed at the final stage of the production line by people with the specific title of inspector. Wherever an object must be seen or checked in order to proceed to the next stage, that function can probably be accomplished by a vision system.

Expanded capability

The basic principles of automated vision technology can also be adapted to industrial robotics. This will extend the capabilities of robots and also stimulate and expand research on vision technology itself. As *New York Times* science correspondent Barnaby Feder wrote recently, the development of robot vision systems is "a key factor in the transformation of robots from valuable but essentially stupid tools into sensitive devices capable of a variety of manufacturing tasks, such as intricate assembly work and arc welding."

For example, Object Recognition Systems Inc. has developed a prototype robot vision system that features a "bin-picking" capability. A bin-picking robot with a vision system can retrieve parts that are randomly oriented and jumbled in a bin or tote box, the condition in which they are probably delivered to the factory assembly area.

This bin-picking capability will be incorporated in an advanced robot vision system, "i-bot 1," which will soon be introduced in the commercial marketplace. It is one example of how automated vision and other sensory capabilities will be used to enhance the range of functions of industrial robots.

Robot vision systems will come into more widespread use in the next several years. It is not necessary to wait for the completely automated factory of tomorrow in order to install automated inspection systems to increase productivity today.

Ruggedness essential

The conventional factory of today is not a sterile environment. There are many hazards that may hinder the smooth operation of new equipment, such as dirt, noise and vibration. Obviously, a practical vision system must be rugged and capable of withstanding such hazards. That's a basic requirement.

In order to survive amid competitive technologies and meet the needs of industry, vision systems must be capable of fulfilling other basic

Figure 1: Diagram of Automated Visual Inspection System

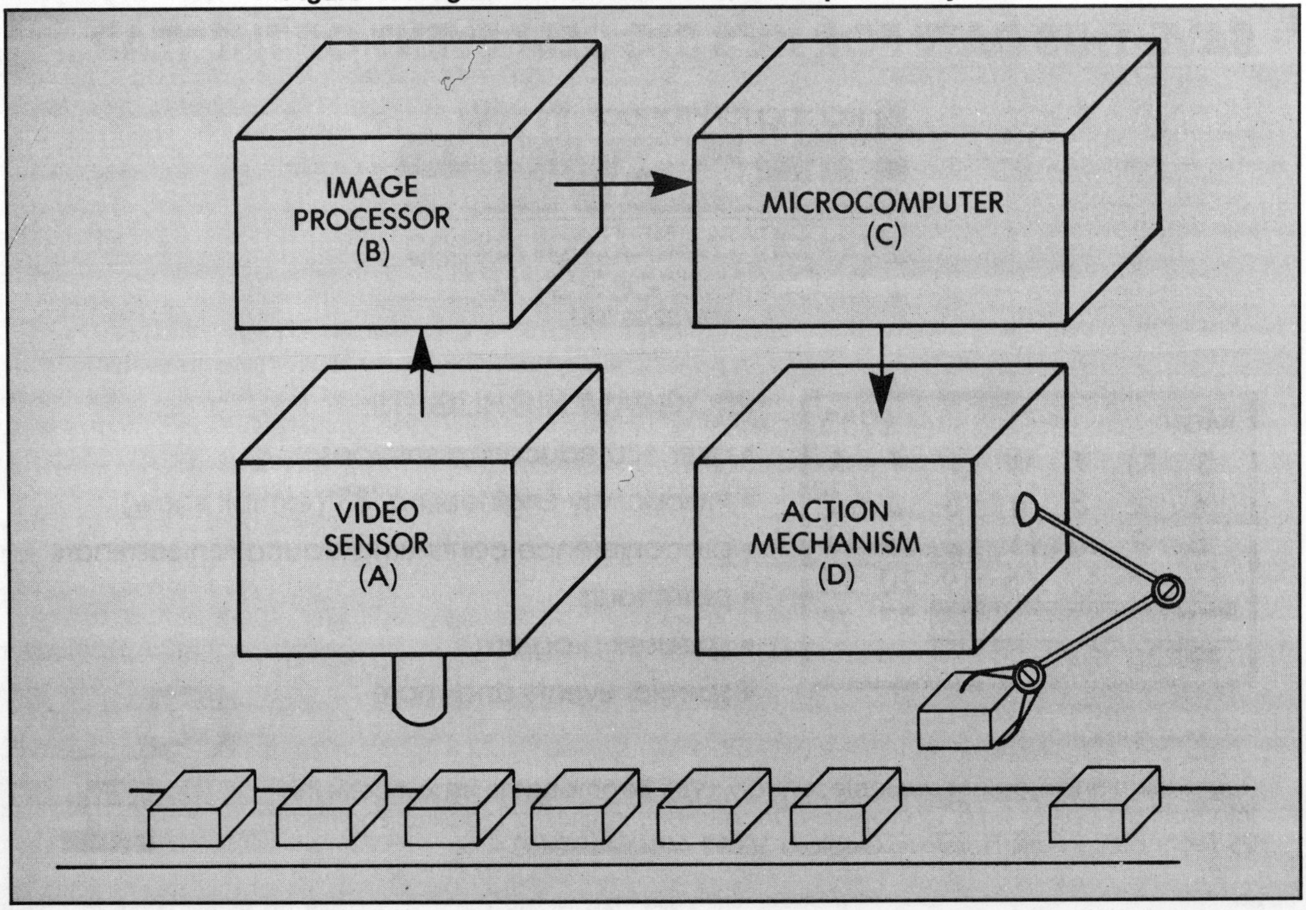

requirements. In evaluating the characteristics of various automated vision systems, some crucial factors to consider are the following:

☐ They must be capable of operating at high speeds. Throughput rates of two to 10 or more decisions per second are required in many production-line applications.

In terms of speed, a complete operating cycle (which takes place in a fraction of a second) would include data acquisition, data processing, initiation of an action response and mechanical handling.

☐ They must be able to recognize a variety of visual features, or differences in scenes, approximately as well as the human eye, including surface features such as color, shape, texture and spatial relations (not possible with binary systems).

☐ In addition, they must be capable of analyzing complex objects or scenes when precise geometric descriptions may not be possible.

☐ They must have sufficient resolution to extract the detailed information necessary to examine complex scenes for small defects or to differentiate between similar objects.

☐ They must be capable of functioning under varied viewing and lighting conditions. A few examples of problems that can arise under these conditions might be labels on curved bottles, serial numbers stamped in metal parts or coding on components.

☐ They should be capable of being "self-trained" or trainable by unskilled or semiskilled workers.

☐ They should be application-independent, capable of solving a wide range of on-site problems with a comprehensive technology.

☐ They must be affordable.

Fortunately, the declining costs and rising functional power of many electronic components will contribute greatly to stable and modest prices in the automated vision industry.

The need now is for IEs and others to understand what automated vision can do and to find cost-effective applications in which these systems can be implemented. That isn't just a marketing or an engineering problem; it is also a matter of education. **IE**

John W. Artley is president and chief executive officer of Object Recognition Systems Inc., which he formed in 1971 with two inventors of a proprietary technology for automated visual inspection systems. Before forming the company, he was assistant to the president of REFAC Technology Development Corp. Artley has bachelor's degrees in commercial art and English and advertising and marketing from the University of Missouri at Columbia.

Machine Vision — The Link Between Fixed and Flexible Automation

David Banks
Object Recognition Systems, Inc.
Princeton, New Jersey 08540

Abstract Modern vision systems can perform useful tasks in current production systems largely based on fixed automation. As the use of flexible automation increases in North America, vision will play a key role.

Manufacturing is currently undergoing wide spread and drastic changes. One Senior GE executive has coined the phrase "automate, emmigrate or evaporate". Others call, the current period the second industrial revolution as computers penetrate manufacturing. John Nesbitt author of Megatrends indicates we're approaching a global economy where information processing, handling and control via computers will revolutionize the entire process of making, distributing and even using goods and services. How does the manufacturing company get from where we are now to the factory of the future; meaning a competitive, efficient enterprise optimizing use of plant, equipment and labor through computer control and communication?

In order to start from a common point we should define terms. Websters New Collegiate Dictionary defines "automation" as "automatically controlled operation of an apparatus process or system by mechanical or electronic devices that take the place of human organs of observation effort and decision". I describe fixed or hard automation as automation which must be changed by physical (human) or mechanical intervention. Flexible automation is automation that can be changed through program control via computer. The most typical description of a flexible automation system is an industrial robot. We will define the term quality as conformance to specification. The term CAD is Computer Aided Design. The term CAE is Computer Aided Engineering. The abbreviation CAM is Computer Aided Manufacturing. CIM is Computer Integrated Manufacturing.

COMPUTERS AND MANUFACTURING

Computers' entry into manufacturing has been from accounting, material control, and data reduction (in other words, number crunching and presentation) to control functions. The number one parameter that has enabled this evolution to take place has been the cost curve on digital computers. Looking at the cost of computers, starting in 1960, a main-frame computer started at $1M and up. By 1970 main-frame computer prices had dropped to $500K and the early mini-computers started at $50K. By 1980, microcomputers costing $1K per unit were available and by 1990, it is speculated that powerful computers will be priced from $5.00 as wafer scale technology comes on stream. In addition to cost reduction, the computational speed and power have become greatly increased over the past thirty years. Computer memories, external storage devices, and communication media are similarly much less expensive and much more powerful.

From the number crunching side of business, computers have entered the control arena for machine tools, process controllers (upgrading from relay electronics), packaging equipment, printing equipment and other tools and instruments.

DESIGN

In the early 1970's computer aided design evolved from the digital plotter and early CRT technologies. Real time computers enabling interactive graphic capability gave the engineer the capability of interacting with this design during the process of design itself. Parts, assemblies, electronic circuits could be specified and designed on a terminal and then detailed drawings, specifications, parts lists could be generated avoiding errors and enabling a data base to be generated which could provide the basis upon which the entire manufacturing process was driven. In the early 1970's only about 200 computer aided design work stations were in use. Now that number is in the thousands.[1]

Later versions of CAD systems allow design/analysis, test, kinematic studies to be done on assemblies without actually building models to under-go these processes. The availability of inexpensive, stand alone workstations that can be networked into larger computers and larger CAD systems via local area networks mean that engineers and production workers can do much more at the production site itself still accessing a central data base. True computer aided engineering is emerging. Thus, the computer's evolution from accounting and material control-the traditional management information systems and MRP systems-has now moved into the production environment via CAD and CAE. We now are seeing more use of CAM.

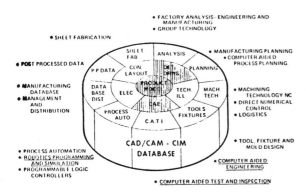

Figure 1 illustrates the data base as the CIM Integrator.

Fixed automation or hard automation typically favors the large runs due to set up and change over expense. The need for flexibility is great. Flexibility even in fixed automation, reduces space requirements, inventory and skilled labor in process change over.

Now that CAD and Computer Aided Engineering can do product manufacturing process design at the same time, the need exists to implement the link from these computerized technologies into production. Both fixed and flexible automation in the form of robotic work cells must tie back to the design and process planning function to enable process simulation, quality monitoring, statistical quality control, yield data, trend analysis for efficient maintenance of the manufacturing process and equipment. Manufacturing companies share a number of common elements; engineering or design, process planning, material procurement and control, manufacturing, quality assurance, and the administrative funcetions. To some extent these functions are automated and interact but not nearly to the extent that utilizes todays technology efficiency. Computer vision is the key technology that links design, production and quality assurance.

VISION - THE LINK

The availability of micro-processors coupled with advanced pattern recognition software makes computer vision an industrial reality. Figure 2 is a block diagram of a typical factory oriented vision system.

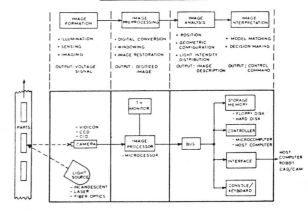

Figure 2

Vision systems, most simply put, are the first available element of sensory artificial intelligence enabling machines to monitor, modify or be guided by visible manufacturing elements on the factory floor.[2] Vision is not a leading automation technology. It fits into processes already in place where some level of automation is utilized.

FIXED AUTOMATION

In fixed automation, the most likely areas for implementation of vision are with machine tools, packaging equipment, printing equipment, painting or finishing equipment and assembly systems. Vision adds quality and process completeness feedback to these systems.

Benefits achieved through implementation of vision are greater than the end of process inspection which most people associate with replacing a human inspector with a machine. The larger value is the ability to monitor in-process (at each critical step) things such as: is the assembly good or bad, properly oriented, are the tolerances correct, has lubrication been applied, is the print accurate, and hundreds of other visually available pieces of information relative to product quality, or completeness? Parameters that need to be considered in implementing a vision system are shown in Figure 3.

```
VISION APPLICATION SPECIFICATIONS
```
° Throughput Rate
    continuous motion
    indexed motion

° Resolution/detail required to detect defect

° Field of view

° Normal - acceptable - variations
    position
    illumination
    appearance

° Machine errors/false rejects

° Gentleness

° Reliability

Figure 3

Previously added to existing systems as a retrofit, vision is now being designed into machine function from the very beginning by forward thinking companies. This leads to greater machine efficiency at a lower overall system cost than retrofitting a vision module. For example, vision can be used for screen alignment on an automated thick film screen printing machine used in fabrication of hybrid electronics. To put a vision system on this piece of equipment costs $28K as a retrofit. Having the vision system designed into the screen printer during the design phase leads to integrated vision costing less than $18K. Another fixed automation example of a vision application is high speed verification of alpha numeric code data on imprinted roll fed pharmaceutical labels. This printing process can be verified accurately using a vision system for a price of $13.5K even as a retrofit. Designed into the label imprinter, the price is under $10K. Thus fixed automation can gain better control, accuracy and product tracking for yield purposes with vision as an integral part of the overall system. Simple train-by-showing vision systems are also capable of changing from one application to another without requiring operator programming skills.

FLEXIBLE AUTOMATION

Programmable flexible automation (robotics) is not truly flexible if part orientation or presentation must be precisely defined for a given operation. Part feeders, molded dunnage and conveyors add to work cell cost and usually are specific to a given manufacturing process thus, robot systems cannot realize their flexibility potential in small batch applications without some level of intelligence which is either visual or tactile in nature.

Vision provides guidance as well as quality measurement or inspection to robotic manufacturing operations such as material handling, welding, spray painting and assembly. Vision with integrated tactile (eye-hand) coordination is a very powerful part of a robot system. A simple application example is a vision controlled work cell to acquire, inspect and place large bolts into an assembly. This process is illustrated in figures 4, 5 and 6.

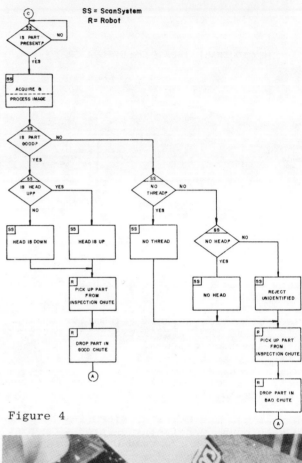

Figure 4

Figure 5 Robot acquiring bolt out of tote bin

Figure 6 Monitor showing process video image for robot guidance

This complete process would be costed as in Figure 7:

BOLT ACQUISITION, INSPECTION ASSEMBLY WORK CELL COSTING		
	Vision Guided	Hard Tooled
Robot	$55K	$55K
Vision Module	$25K	0
Gripper	$2.5K	$1.0K
Bowl feeder and Chutes	0	$28K
	$82.5K	$84K
New part/process change costs		
Replace Bowl feeder and Chutes	0	$28K

Figure 7

If the cell process changes if another bolt size is needed or, the cell is set to do a different manufacturing operation the hardware used to orient this bolt and present it is lost, the tooling must be totally replaced at significant tooling cost. Vision also enables the use of a less precise robot because the vision system can guide can guide the robot to slightly misplaced target parts or target locations.

The evolution of the modern factory - true computer integrated manufacturing - will be just that, an evolution. Few companies can start with a blank slab and build a flexible production facility from the ground up. Machine vision provides the best mechanism to integrate current fixed automation into process planning, process monitoring and process control systems via the CAD data base. Vision enables flexible automation (or robotic systems) to be flexible, to change through software, to adapt to small batch needs. Vision provides benefits far beyond the replacement of the traditional "inspector" at the end of the line by a tireless unblinking machine. It reduces scrap, operates at high speed with very high accuracy, provides feedback to people when a process begins to fall out of spec, gives yield information and generally supplies computers with the inputs necessary to integrate discreet manufacturing operations with the other elements in a business. The pay-back is improved productivity through better machine and people utilization. As the link between fixed and flexible automation vision is helpful in the factory of the present, but is essential in the evolving CIM facility of tomorrow.

REFERENCES:

[1] Computer Aided Design and Computer Aided Manufacturing The CAD/CAM Revolution - John K. Krause; Marcel Dekker, Inc.

[2] IEEE Spectrum, May 1983; Data Driven Automation: Toward a Smarter Enterprise.

[3] Machine Vision Systems - A Summary and Forecast - Tech Tran Corporation.

David Banks, Executive-Vice President of Object Recognition Systems, Inc., received his B.S. in chemistry from Indiana University, the M.S. in chemistry from the University of California, San Diego and the MBA from Purdue.

He is general manager of the Princeton facility of Object Recognition Systems, Inc. responsible for automatic visual inspection and robot vision. Prior to joining Object Recognition Systems, Inc. in January of 1982, he was with Hewlett-Packard for 13 years, most recently as production manager for liquid chromatography and lab automation. His responsibilities before this included various sales and marketing management positions. He is a member of the American Chemical Society and the Robot Institute of America.

Reprinted from the 1983 Fall Industrial Engineering Conference Proceedings.

Automatic Identification and Bar Code Technology in the Manufacturing and Distribution Environment

Edmund P. Andersson
Director, Corporate Relations
Computer Identics Corporation
Canton, Massachusetts

INTRODUCTION

There has been more information published in the last year on bar code technology than in all preceding years combined. Why? Extremely significant developments have occurred that influence virtually every company involved in manufacturing and distribution. If you're not yet directly affected, chances are that it's just a question of time before your company implements systems based on bar code data collection.

The justification for investing in this technology is well documented. The US Department of Defense (DOD) estimates that the introduction of bar code in the various DOD agencies will result in savings of $113 million annually. These are only hard savings; no figures were given for soft, or intangible, savings. The food industry, in its adoption of UPC several years ago, claimed savings averaging 2 per cent of gross sales.

Industrial moves of a voluntary nature are now emerging with dispatch. Perhaps the most ambitious is the Automotive Industry Action Group, an ad hoc committee with representation from all major manufacturers, which has identified the need to select standard bar code symbols and coding systems. In addition, they have targeted many applications for bar code scanning and systems improvement.

But what about you and your company? Maybe you're not involved in industry-wide projects. What do bar codes have to offer you? The buzz words abound: key-entry by-pass, keystroke elimination, factory data collection, increased productivity, inventory reduction, service level improvements...and so on! Let's see what all this means in the context of manufacturing and distribution.

OPPORTUNITIES FOR BAR CODING

The industrial truck will be with us for a long time, but we can make the use of this form of horizontal materials handling more efficient by eliminating many of the time-consuming and error-prone functions imposed on the operators. Hand-held scanners interfaced with controllers can give operators instant instructions on load handling/placement. Moving beam scanners situated in main traffic lanes can provide similar information automatically, without operator involvement, by using video displays for output instructions after loads have been scanned.

Another application is to identify loads at transfer points, either to confirm that a transaction has taken place or to input load identification to another control system that will act automatically upon its receipt.

Product sortation is the most common application for laser scanners in materials handling. A single scanner can increase throughput rates by a factor of 5 (e.g., from 60 fpm to 300 fpm) without the errors associated with human data entry.

Receipts processing is another interesting use of scanners and bar code. Here's a missing link in most MRP systems. How do you inform your computer of what has arrived at the receiving dock? How long does it take, and how many intermediate data handling steps are involved?

Production accounting, one of the applications of interest to the auto industry, can usually be automated simply by adding bar code labels to the products being produced. Most products have some form of identification anyway. Adding bar code information is quite routine today. Again, capturing data automatically from the factory floor eliminates unnecessary data handling functions and errors. Timely data capture is a big bonus.

In-process control can be automated with well-placed laser scanners linked to process control computers, identifying passing loads and their location for the computer.

Shop-floor data collection is made possible through several types of bar code terminals interfaced to a variety of computers and controllers.

Taking inventory is made easier and quicker and is accomplished without unnecessary intermediate data handling steps by capturing information at the source and transmitting that information directly to a computer. Eliminating or bypassing key-entry minimizes transaction time and the potential for error.

SYSTEMS CONSIDERATIONS: SCANNERS

When we think of material handling systems, we most often find ourselves looking at laser scanners. Figure 1 illustrates how these scanners are interfaced with the control system. They are most often used for automatic data collection at very high throughput speeds. From a cost/benefit perspective, the laser scanners offer the most capability for the price. Some new products have been introduced this year that reflect tremendous improvements in performance capabilities, smaller packing and cost. The scanner in Figure 2 scans 2 to 4 times faster than its predecessors and sells for about half the price!

Sometimes the laser scanner won't do the job. If so, it's usually because of some problem with bar code orientation or irregular scanning angles. For these applications, there are dynamic scanners, sometimes called omni-directional scanners. The pros and cons of laser vs omni-direction demand close examination before an intelligent decision can be reached. Sometimes an omni-directional bar code (see Fig. 3) is a solution permitting the deployment of the less expensive laser scanner. In other cases, a cluster of laser scanners is a cost effective alternative to the omni-directional units.

Lightpen terminals are abundant today. There are numerous products on the market featuring bar code data collection supported by keyboard entry and display. These devices (see Figure 4) are operated on-line with a computer and are most frequently used for collecting information from points at which the transactions occur. Figure 5 shows several generic types of devices, each designed to collect data by hand-held scanners. The least expensive is the board set and lightpen for integration within an existing terminal. The next unit is a stand-alone device with its own power supply but without a display or keyboard. It is connected externally to another terminal. The rest of the units are legitimate terminals. The important considerations here are survivability in your environment and ease of use by your operators. You will note that some units are quite rugged, while others are suitable only for light duty applications. The better units will feature an acknowledgement/no

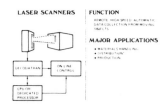

FIGURE 1

acknowledegment protocol that informs the operator that the information has been accepted by the computer to which the unit is interfaced. Some new developments include a device called a "slot reader," which is an alternative to a lightpen. These scanners are designed for reading employee badges, cards, shop travellers, etc. Some are used for access control.

Another new unit is a highly compact, full alphanumeric terminal with current loop and RS232 interface options (see Fig. 6). A seemingly futuristic entry on the market is the laser gun shown in Fig. 7. This is a hand-held moving-beam laser scanner that scans at the rate of about 50 scans per second. In contrast to the lightpen, which must come in contrast with the code, the gun reads the code at a distance of 4 to 6 inches (depending on bar code size). It is a good alternative for reading codes on irregular surfaces (e.g., etched in metal) or bar codes of poor quality that require repeated scanning. These guns must be supported by a decoding terminal similar to the ones shown in Figure 5.

FIGURE 2

Portable terminals (see Figure 8) add an interesting dimension to scanning by reading coded products in remote locations or items that are stationary. They are commonly used for taking shelf inventory and inventory of capital goods. The main improvement in these units is a trend toward larger memory capacity. However, I think the introduction this year of powerful, low cost portable computers will make many users opt for this unit equipped with bar code capability over the more costly dedicated units.

Fixed-beam scanners still have a place in material handling applications, especially in conjunction with permanently encoded tote containers and racks. These units are inexpensive and very reliable. Some units can read bar code, but only once, as the code passes by. They are also being used in conjunction with guided vehicles for automatic vehicle control.

FIGURE 3

SYSTEMS CONSIDERATIONS: BAR CODE SYMBOLS

In this section we will look at bar code generation and the bar code symbologies of importance today. Before we begin however, there are some definitions which must be understood.

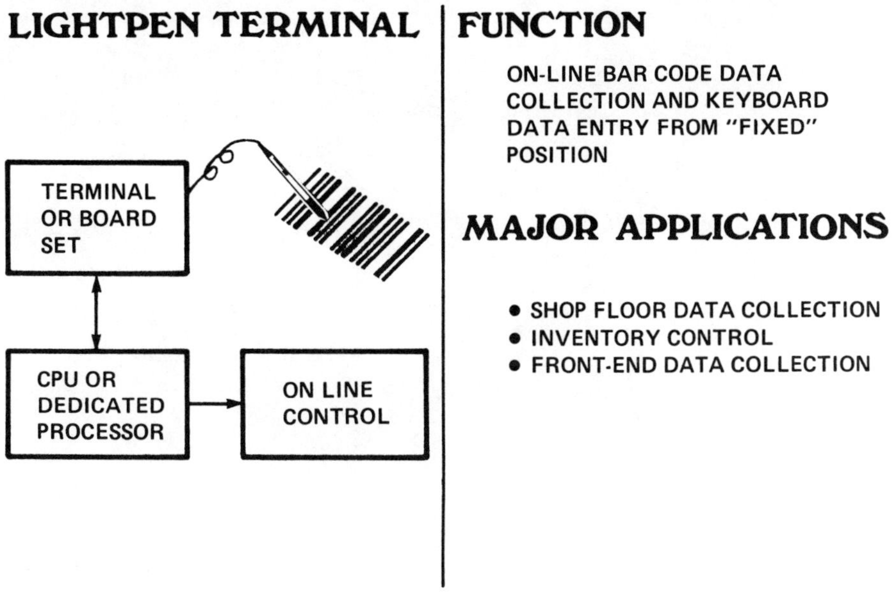
LIGHTPEN TERMINAL
TERMINAL OR BOARD SET
CPU OR DEDICATED PROCESSOR
ON LINE CONTROL
FUNCTION
ON-LINE BAR CODE DATA COLLECTION AND KEYBOARD DATA ENTRY FROM "FIXED" POSITION
MAJOR APPLICATIONS
• SHOP FLOOR DATA COLLECTION
• INVENTORY CONTROL
• FRONT-END DATA COLLECTION

FIGURE 4 BAR CODE TERMINAL INTERFACE

1. "X" dimension: the measurement of the smallest bar or space of a bar code (measured as the width or height of the bar or space, not its length).

2. Code density: high density = 0.010 in or less; medium density = 0.020 in; low density = 0.03 in or greater.

These terms are important because they help to establish which code-genertation and code-reading equipment should be used. For example, the laser scanners we'll discuss later have certain requirements for "X", based on how far away the code will pass the scanner. Generally, the farther the code is from the scanner, the larger the "X" dimension must be. Fig. 9 gives some approximate guidelines for "X", based on distance from the scanner. Hand-held scanners (lightpens and wands) scan extremely small "X" dimensions, the ones we call high-density codes. The benefit of high density codes is that they use less space for the encoded data. They may cause difficulty in printing and scanning if tolerances are not kept at acceptable levels.

Most laser scanner applications involve medium-density or low-density codes. A new term has been introduced for density: STANDARD DENSITY, an important term for those involved in the DOD LOGMARS Program, which we will touch upon later.

Code Generation

CONVENTIONAL PRINTING PROCESSES — There are countless ways in which bar code can be generated. There are the conventional printing processes-offset, letter press, flexographic. Many smaller companies that don't wish to invest in code generating equipment of their own find this a viable option. Large quantities of codes that can be batch-printed are also commonly generated via conventional printing. Other examples include pre-printed corrugated cartons, packaging and product labels.

SPECIALIZED CODE-GENERATING EQUIPMENT — There are several manufacturers offering equipment on which bar code can be generated. These devices are normally available to handle any of the popular bar code symbols; most permit the user to select from at least

FIGURE 5

two modes of operation, batch- or demand-printing. They are capable of sequential numbering and out-put human-readable information with the bar code. Some are equipped with laminating equipment to put a protective coating on the bar code. Some manufacturers even offer on-line applicators that place printed codes on products passing by.

The most popular code-generating techniques in use today are impact printing of pre-formed characters, dot matrix printing, electrostatic printing (non-laser), laser printing (electrostatic), ink jet and laser etch. Each of these methods has its pros and cons. Each has some limitation on its ability reliably to print high-density codes.

For those of you involved in applications that demand high volume (e.g. 40,000 bar code labels per day), I suggest looking into laser printers, because of their tremendous speed. Their higher cost may be justified when certain volume levels are reached.

Permanent laser etching on metals, etc. has caught the fancy of the automotive industry. These codes can survive where paper codes cannot. They are difficult to remove or damage. In the future we will probably see equipment that will laser-etch bar code on products passing by on conveyors or other handling equipment.

Bar Code Symbology

Just when many of us were becoming comfortable with certain bar code symbols (Interleaved 2-of-5, Code 3-of-9), some new ones appear. Our interest here is in bar code symbology for factory and warehouse, so we'll forget about the symbols used in stores, libraries, etc.

I strongly recommend that you purchase a set of Uniform Symbol Descriptions (USDs) from the AUTOMATIC IDENTIFICATION MANUFACTURERS (AIM) Product Section of the Materials Handling Institute (MHI). AIM has published the USDs as a service to users. Inexpensive (about $16 for the entire set), these documents describe each of the popular symbols in a straightforward manner, including all the detail necessary for printers, programmers, etc. Current USD's are:

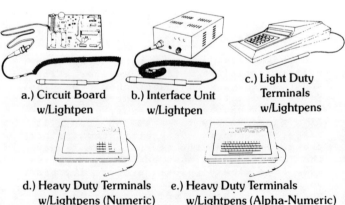

a.) Circuit Board w/Lightpen
b.) Interface Unit w/Lightpen
c.) Light Duty Terminals w/Lightpens
d.) Heavy Duty Terminals w/Lightpens (Numeric)
e.) Heavy Duty Terminals w/Lightpens (Alpha-Numeric)

FIGURE 6

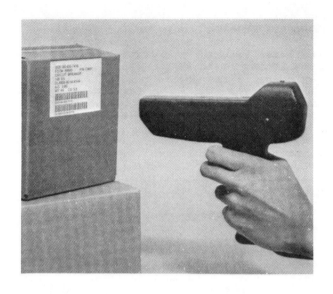

FIGURE 7

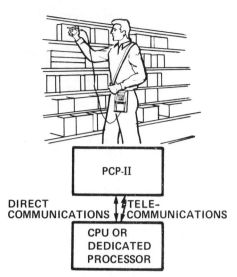

FIGURE 8

USD Number	Symbol
1	Interleaved 2-of-5
2	Code 3-of-9 Subset
3	Code 3-of-9
4	Codabar
5	Presence/Absence Code
6	Code 128
7	Code 93
8	Code 11

The new developments in symbology with which you should be familiar include the previously mentioned DOD LOGMARS Program, the UPC Case Symbol Program, and the introduction of two new codes, Code 128 and Code 93.

DOD LOGMARS PROGRAM As of July 1, 1982 all companies supplying products to US Department of Defense agencies are required to add machine-readable codes to the items and to the containers in which the items are packed. The machine-readable code selected was Code 3-of-9, with the added requirement that the human-readable information would be printed in OCR-A. Two specifications have been published describing the requirements for marking:

MIL-STD 1189: Standard Symbology for Marking Unit Packs, Outer Containers and Selected Documents.

MIL-STD 129H: Military Standard for Shipment and Storage.

LOGMARS means "Logistics Application of Automatic Marking and Reading Symbols". This program represents an enormous effort by the federal agencies involved. They have developed volumes of data relating to bar code technology, including test results for bar codes and scanning devices, and economic benefits for specific applications.

An interesting result of the test program is the revelation that the substitution error rate for bar code is more than 200 times better than OCR and 10,000 times better than human data-entry.

UPC CASE SYMBOL PROGRAM The UPC Case Symbol, formerly called the Uniform Case Symbol (UCS), is the result of years of work conducted by the Distribution Symbology Study Group (DSSG).

The upshot of the program was the requirement that shipping containers for products carrying the Universal Product Code (the supermarket code) must carry the UPC number encoded in Interleaved 2-of-5 (AIM USD-1) on the outer containers. AIM has published a document detailing the code and the recommended practices for creating and placing the code on the containers.

Anyone handling these containers in the future will have the opportunity to use scanners, hand-held or laser, for applications such as productivity sortation, receipts processing, order verification and taking inventory.

AUTOMOTIVE INDUSTRY ACTION GROUP (AIAG) Formed as a non-profit corporation comprised of professionals from the American Production and Inventory Control Society (APICS), AIAG enjoys the support of all major auto manufacturers in the USA. Its objective is to increase the industry's productivity through numerous cooperative projects involving bar code.

The AIAG has identified applications for bar coding as varied as manufacturing process monitoring, broadcast component requirements, assembly verification, production status reporting, shipper generation, receiving verification, warehouse random storage, automatic sortation and routing control, cycle checking and history file development, among others. Bar code also presents great potential in the exchange of information between manufacturers and dealers (e.g., warranty claims, recalls and product quality/availability).

CODES 128 & 93 This year two powerful codes were introduced, capable of encoding the entire ASCII set of characters. They are Codes 128 (AIM USD-6) and 93 (AIM USD-7). Code 128 was recently included in a major trade article and erroneously listed as not being a self-checking code. It is. Both codes have been placed in the public domain by their authors. They share the following features:

Compactness (much data in little space)
Alphanumeric capability
Variable length
Full ASCII set
Ignore ink spread and dot matrix overlap

Why these new codes now? Partly because of competitive pressure from abroad and partly because the marketplace demanded more powerful codes, ones that could encode more information in less space and overcome some of the difficulties in reading poorly printed bar code. Both codes meet these criteria.

OCR AND MAGNETIC STRIPE Invariably, people ask about OCR and magnetic stripe technologies for material handling applications. I believe the LOGMARS Program placed indisputable evidence in the public domain concerning the pros and cons of one technique versus another. Bar code is clearly superior. OCR, at today's state-of-the-art level, can be categorized as being of limited character set (not having a complete alphanumeric capability), requiring careful scanning (substitution rate may be unacceptable), and without a practical remote scanning capability.

Magnetic stripe does not offer a remote capability either; all reading is done by an operator-controlled device or slot-reader, both of which are contact readers. It is also questionable that magnetic cards used in industrial applications can survive operational environments and handling. If you carry a magnetic stripe bank card in your wallet, you have an idea of what I'm talking about. In summary, magnetic stripe holds high density data, BUT cannot be examined or copied, has no remote scanning capability and cannot be printed on paper, etc.

RADIO FREQUENCY TECHNOLOGY The use of passive transponders that are activated by RF-emitting scanners is being proposed for certain applications in industry. These devices may fill certain voids that exist in the capabilities of other identification technologies. They offer a method of identification when line-of-sight is absent. They work well outdoors when radio interference is not a factor. They have appeal for vehicle control applications.

COSTS/BENEFITS

The benefits associated with automatic identification, especially with bar code, are better documented now than ever before. Key-entry bypass, keystroke elimination and automatic data collection are important benefits with high cost advantages for the skilled planner and designer.

The DOD LOGMARS Program produced the most authoritative study to date on benefits. In hard, tangible savings alone, DOD predicts annual savings of $113 million by using machine-readable data-entry rather than conventional human/key entry data collection. DOD has budgeted $66 million for the next three fiscal years for scanners and related equipment. Add imprinters and verifiers, etc., and the total reaches $100 million.

Perhaps your company can achieve a dramatic savings, too. How is DOD going to use bar code? Targeted applications

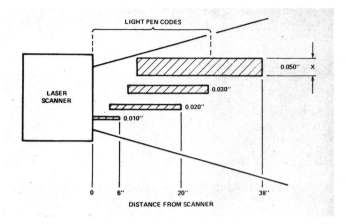

FIGURE 9

The tools are in our hands, and the preliminary studies have been made. Now it is necessary for the leaders in individual industries to implement this new technology in order to achieve the increased productivity and accuracy available for the taking.

1. "Scanning Products on the Move", Automatic Identification Manufacturers (AIM) Handbook.

2. "Automatic Identification--Now It's Really Taking Off", Modern Materials Handling Magazine, September 21, 1982, pp 32-43.

3. "10 of the Most Frequently Used Bar Codes", Modern Materials Handling Magazine, September 21, 1982, pp 35-38.

4. Bar Code Symbology-- Some Observations on Theory and Practice", Monograph by David C. Allais, INTERMEC.

5. "Symbology Fundamentals: Code 128 and Code 93", Paper by Charles E. Mara, Computer Identics at SYMBOLOGY '82, September, 1982, sponsored by Production and Inventory Management Review Magazine.

6. "Uniform Symbol Descriptions", Published by the Automatic Identification Manufacturers (AIM) Product Section, The Material Handling Institute (MHI), Pittsburgh, PA.

7. "Choosing a Format You Can Live With", Bar Code News, March, 1982, pp 6-10.

8. "Improved Bar Code Printing Techniques", Monograph published by the Automatic Identification Manufacturers (AIM), Charles E. Mara, Computer Identics.

9. "Recommended Practices for the Uniform Case Symbol", Specification published by the Automatic Identification Manufacturers (AIM), The Material Handling Insititute (MHI) Pittsubrgh, PA.

10. "Report on Printers", DATAPRO McGraw-Hill.

11. "Improving Distribution Flows With bar Code Technology", by Edmund P. Andersson, Computer Identics, Production and Inventory Management Review, June, 1982, pp 24-44.

12. "Industrial Bar Code Applicaitons Help Control Inventory, Verify Assembly", by Edmund P. Andersson, Computer Identics, Industrial Engineering Magazine, September, 1981, pp 114-120.

13. "How Scanners Add Depth and Versatility to Computer-based Systems", by Edmund P. Andersson, Computer Identics, Material Handling Engineering Magazine, July, 1979, pp 60-67.

14. "Final Report of the LOGMARS Steering Committee with Annexes A-∅.", Stock Number 008-020-00916-L, available from Superintendent of Documents, U.S. Government Printing Office, Washington, D.C. (Price $20).

15. "Logistics Applications of Automated Marking and Reading Symbols," DOD Contractor Seminars Briefing Document, 1982, by M.W. Noll, Chairman, Available from USA Material Development and Readiness Command, Tobyhanna, PA.

Cellular Manufacturing Systems Reduce Setup Time, Make Small Lot Production Economical

By J. T. Black
University of Alabama, Huntsville

A new type of manufacturing system has emerged in recent years. It offers both cost and quality-control advantages over the four traditional manufacturing systems, but it has yet to be implemented on any sizable scale in the United States.

This article presents an overview of this relatively new system—the cellular manufacturing system—and how it differs from the more conventional systems in use. Some guidelines for implementing a conversion to the cellular system are also given, and some of the major impediments encountered in such a conversion are detailed.

Subsequent articles in this section will address specific problems involved in conversions to cellular systems and their integration into a new production system.

Traditional systems

Four kinds of manufacturing systems have traditionally been employed. (See schematics in Figure 1.) The oldest of these is the job shop, a transformation process in which units for different orders follow different paths or sequences through processes or machines.

The major characteristics of the job shop are flexibility, variety, highly skilled people, much indirect labor and a great deal of manual material handling (loading, unloading, setting up and adjusting). General purpose machines are grouped by function and adapted to the special requirements of different orders. The price for this flexibility is paid in the form of long in-process times, large in-process inventories, lost orders and poor quality.

For products built in larger quantities, the flow shop system can be used. The flow shop is a transformation process in which successive units of output undergo the same sequence of operations with more specialized equipment, usually involving a production line of some sort. No back flow is allowed.

Typically, all the units follow the sequence of operations and pass through all machines. Volumes are large, runs are long, and conversions to another product take a long time. The most automated examples of this system for machining are called transfer lines.

The third kind of system, the project shop, is directed toward creating a product or service which is either very large (immobile) or one-of-a-kind with a set of well defined tasks that typically must be accomplished in some specified sequence. The people, materials and machines all come to the project site for assembly and processing. The project is generally backed up by a job shop/flow shop system to supply component parts and subassemblies to the project.

The fourth type of manufacturing system is the continuous process system, in which products—generally gases, liquids or slurries—flow through a series of directly connected processes or operations that link raw material (inputs) with finished products (outputs).

Because this system does not deal with discrete parts, it is generally ignored by the nonchemical community. However, it represents the ideal, or *vision,* of the Japanese just-in-time production system (see Figure 2). Discrete products will ideally flow like water through the system.

The key to making this goal a reality on the plant floor is designing the manufacturing processes and systems such that *small lots* can be produced. The ideal "lot" size is one. Small lots smooth the production flow.

This requires a fifth type of manufacturing system called the cellular manufacturing system—also referred to as the group technology (GT) or work cell. See Figure 3 for an example. The cellular system groups processes, people and machines to treat a specific group of parts. The output typically is completed components.

Tables 1 and 2 summarize the characteristics of all these manufacturing systems and give some examples of each kind. Each of the classical manufacturing systems has a mechanism for handling information and material movement and storage (inventory), purchasing, planning, production control and so forth. (See Figure 4.) Collectively, these activities represent the *production system.* The production system is designed around the manufacturing system, to support it.

The job shop represents the most common type of manufacturing system, however, with 30% to 50% of systems estimated to be of this form. In most job shops, some items will be made in large enough quantity to warrant the use of flow line methods; therefore, it has been common practice in the United States to mix job and flow shop elements.

Cellular manufacturing systems

The emergence of the cellular manufacturing system, and its highly automated form—the flexible manufacturing system (FMS)—has resulted in a new type of production system which is capable of producing high quality products at low cost. The best example of this new production system appears to be the Toyota Motor Co.'s just-in-time (JIT) production system. (See the article by Richard Schonberger elsewhere in this special section.)

Group technology is a systems based rationale for solving the reorganization problems involved in setting up cellular manufacturing systems. It provides a computer oriented data base and tools the manufacturing engineer can use to design the work cell. A GT analysis develops the families of parts which can be manufactured by a flexible, cellular grouping of machines. The machines in the cells can be retooled so that one can rapidly change from one lot of components to another, eliminating setup time or reducing it to a matter of minutes.

Eliminating setup time dramatically alters the economics of lot or batch production to permit the economical production of very small lots.

In equation form, TC (total cost) = FC (fixed cost) + VC (variable cost) $\times$ Q (quantity). Setup is a fixed cost. It does not vary with the quantity made. Labor and material are variable costs. Figure 5 is a graphic picture of this equation.

To obtain the cost per unit, divide the total cost equation by the quantity (Q): TC/Q = FC/Q + V.

The new picture is shown in Figure 6. Note what FC/Q represents. The fixed costs must be spread out over many units to reduce the cost per unit enough to make production of the item economical.

But what if you eliminate or greatly reduce the fixed cost by eliminating the setup time and its related labor cost? Then the picture looks like Figure 7.

Now it becomes as economical to make things in small lots as it is to make them in large lots.

You might ask, "How do you eliminate setup? Isn't it a necessary evil?" It is not, but it takes work to eliminate it. The truth is that American manufacturing never saw the need to reduce or eliminate setup. We have operated on the basis of economic order and production quantities (EOQs and EPQs).

The Japanese have relegated

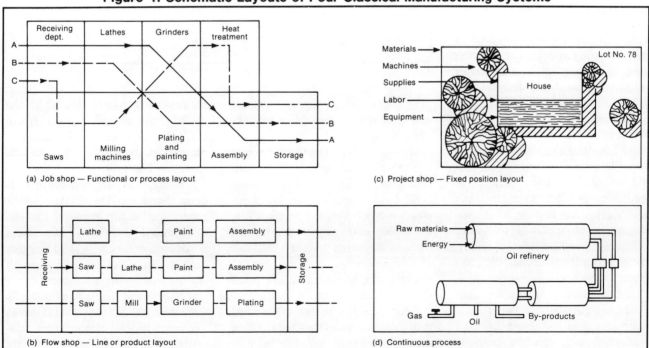

Figure 1: Schematic Layouts of Four Classical Manufacturing Systems

(a) Job shop — Functional or process layout

(b) Flow shop — Line or product layout

(c) Project shop — Fixed position layout

(d) Continuous process

EPQs and EOQs to the archives with their JIT/TQC (total quality control) system, because a primary objective of JIT is to reduce the lot size to the smallest size possible.

In cellular arrangements, one worker can hand a part directly to the next worker for another operation. If the part is defective, the process is halted to find out what went wrong. Quality feedback is *immediate,* and high quality products emerge.

Small lot quantity, coupled with a 100% perfect product (another hard-sought ideal), smoothes the production flow.

After many years and much hard work, the discrete part system begins to look more and more like *a continuous flow process* in which products flow like water through the plant. But the *key* step to transforming a production system based on job shop/flow shop manufacturing systems is the transformation to a cellular manufacturing system.

Designing cellular systems

Cellular manufacturing has existed for many years, but it has not been properly defined or well understood, and it certainly hasn't been recognized as a particular type of manufacturing system. Let us define a manufacturing cell as a cluster or collection of machines designed and arranged to produce a specific group of component parts.

Few rules, and virtually no theory, exists for designing cellular manufacturing systems. However, the first rule is that the design should be as flexible as possible so that it can readily expand to include other components or be modified to handle additional members of the family. The objective is to link the cells into a large manned or unmanned integrated manufacturing system. Cells can be categorized into two general groups: *manned* and *unmanned.*

Manned cells contain machine tools which are conventional or pro-

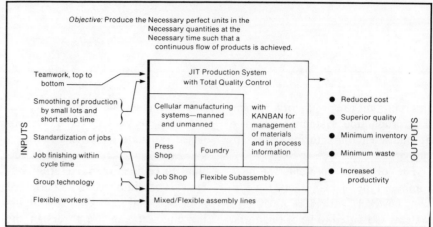

Figure 2: Elements of the Just-in-Time Production System

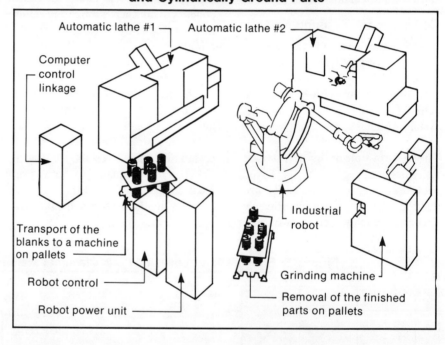

Figure 3: Example of Unmanned Cellular Manufacturing System in Which NC Machine Tools and Robot Work Together to Produce Turned and Cylindrically Ground Parts

grammable (NC or CNC machines) and production workers who have been trained and are skilled in the operation of more than one piece of equipment within the cell. The multifunctional worker is unusual in the typical job shop, but not in microelectronics job shops, where workers have been extensively cross trained with no difficulty. Manned cells are efficient because the number of workers can be adjusted and minimized to meet the desired output. Regarding the design of manned cells, Monden notes that the U shape appears to offer the greatest flexibility (see "For further reading"), because the range of jobs which the multifunctional worker can cover can easily be increased or decreased as needed. A typical manned cell is shown in Figure 8.

Clearly, what these manned cells do is revoke Parkinson's law. As the production requirements of the cell are reduced, the number of workers is reduced accordingly, and workers cannot simply expand the jobs to fill the increased time available.

The cells are tied to each other by a material handling system (JIT used kanban, which Schonberger describes in his article) or are directly linked or combined into larger inte-

Table 1: Characteristics of Basic Manufacturing Systems

Characteristics	Job Shop	Flow Shop	Project Shop	Continuous Process
Types of machines	Flexible; general purpose	Special purpose; single functions	General purpose; mobile	Special purpose
Design of processes	Functional or process	Product flow layout	Project or fixed position layout	Product
Setup time	Long, variable	Long	Variable	Very long
Workers	Single functioned, highly skilled (1 man—1 machine)	One function; Lower skilled	Single function skilled; 1 man—1 machine	Few
Inventories	Large inventory to provide for large variety	Large to provide buffer storage	Variable, usually raw materials	Small in-process
Lot sizes	Small to medium	Large lot	Small lot	Not applicable
Production time per unit	Long, variable	Short; constant	Long; variable	Short; constant
Examples in goods industry	Machine shop; tool and die shop	TV factory; auto assembly line	Shipbuilding; house construction	Oil or chemical processing
Examples in service industry	Hospital; restaurant	College registration; cafeteria	Movie; TV show; play; buffet	Movie; TV show; play; buffet

grated lines. Layouts which result in waiting times for the worker, larger in-process inventories, isolation of the workers and situations in which worker waiting time is absorbed in producing unnecessary inventory are avoided.

Linear manned layouts in which the worker walks from one machine to the next avoid most of these problems, but are not as flexible (in terms of rebalancing the number of workers in the cell in the event of demand changes) as the U-shaped cell.

In the manned cellular system, the worker is decoupled from the machines, so that the utility of the worker is no longer tied to the utility of the machine. (This means that there may be fewer workers in the cell than there are machines.) The objective is to improve the utilization of the people by making them multifunctional, capable of running all the machines in the cell.

Unmanned cells contain machine tools that are programmable (CNC machine tools or other automated equipment), and there are few if any workers within the cell. Unmanned cells have a number of classes or arrangements. These are:

☐ *Fixed automated:* These cells are classically represented by the transfer line, in which the quantities are large (large lots) and the runs long. Such systems are generally arranged in lines, circles or the U shape. They usually have a conveyor which both locates the part and transports it from the machine station, and the line is balanced such that the part spends the same amount of time at each station. The volume of parts is very large and the variety very small. These cells are not very flexible.

☐ *Flexible automated:* These cells are represented by the FMS (flexible manufacturing system) and the robotic cell.

The FMS is generally arranged in a line or a rectangular design with a computer controlled conveyor to transport the parts to any machine in any order. The machines are programmable and therefore can change tools and machining programs to handle different parts (see Figure 9).

Parts can be introduced into the system in any order. Therefore, this system works on a family of component parts with medium to large lot sizes. These systems tend to be rather large and expensive, typically containing five to 12 machine tools; their cost parallels that of the transfer lines. (See article by Zisk.)

The robotic cell generally has few machines and is arranged so that a robot can load and unload the machines and change the tools in them, if necessary. In both the FMS and robotic designs, there should be liberal use of autonomation (automatic inspection) to ensure a high percentage of good parts.

Robotic cells are typically circular in design to take advantage of the range of motion of the robot (assuming the robot has a spherical or circular spatial range), but are not limited to such arrangements. As rectangular and mobile robots become more common, other designs will emerge.

In fact, as robots become more versatile, they will be able to communicate better with each other and will be able to hand parts to each other just as workers in manned cells hand parts to each other.

In most robotic cells in place

today, there is only one robot. If there are multiple robots, they are placed on the floor so that they cannot reach each other or are programmed so that they cannot enter the same space at the same time.

When we have robots which are mobile within cells or can interact with each other, it appears likely that the FMS design will be employed only when parts are either too large (too heavy) for robots to pass from one to another or too small to be properly handled.

The bottom line in these designs, however, is flexibility, with small lot sizes. Figure 10 shows a robotic cell with essentially the same layout as the manned cell in Figure 8.

Designing setups

In setup design, the objective is to eliminate setups between the different component parts in the parts family or, at worst, to be able to use essentially the same fixturing for every part in the family with quick action modification, or "one-touch setup," between parts. In manned cells, this objective can be readily accomplished—as has been demonstrated by numerous Japanese and American manufacturers—with the result that one can economically manufacture in very small lots.

Furthermore, the concept of single setup, which means that setup is accomplished in less than ten minutes, can be extended to presswork and foundry areas which are still essentially in lot-type shop production (see Lesnet's article elsewhere in this section). In the press shop at Toyota, for example, workers routinely change dies in the presses in three to five minutes or less. (The same job at Ford or GM may take four to five hours.) This has the effect of markedly reducing inventory turns. However, the Japanese have found that the *main* benefits are superior quality, worker motivation and enhanced productivity.

Table 2: Characteristics of Cellular and Flexible Manufacturing Systems

Cellular Manufacturing Systems	Flexible Manufacturing Systems
☐ Small to medium-sized lots of families of parts (1 to 200 parts). A special set of parts or products	☐ Medium-sized lots of families of parts (200 to 10,000 parts)
☐ One to 15 machines	☐ Five to 12 CNC machines
☐ Rapid changeover—"Single setup"	☐ Rapid changeover; no setup at machines
☐ Significant reductions in inventory	☐ Significant reductions in inventory
☐ Greatly improved quality control through autonomation	☐ Greatly improved quality control through autonomation
Unmanned:	
☐ Flexible/programmable machines	☐ Flexible/programmable machines
☐ Robotic integration for parts handling (1-5 machines)	☐ Integrated conveyor system for parts and tooling
☐ Network computer control	☐ Networked computer control
Manned:	
☐ A set or group of general purpose machines and equipment laid out in a specific area	
☐ A set of multifunctional workers	
☐ Enhanced worker input leading to job enlargement	
☐ Job enrichment	

Unfortunately, there is no information on how to achieve single setup in any standard American text on manufacturing or tool engineering design. Very little research has been done on the elimination of setup time in the U.S. Why is this? Because we have never seen the need.

Need for new approach

The job shop is, generally speaking, the least productive manufacturing system. It results in products or services which are costly and whose costs tend to keep rising with inflation. In addition, some social and technological trends suggest that the number of and need for small lot production systems will increase in the future. These trends or needs include:

☐ Proliferation of numbers and varieties of products. This results in smaller production lot sizes (as variety is increased) and decreased product life cycles. Setup cost, as a percentage of total cost, becomes greater, as do the problems of managing inventories and materials.

☐ Greater need for closer tolerances and greater precision.

☐ Increased cost of product and service *liability* as consumers demand accountability on the part of manufacturers, which requires greater emphasis on reliability and quality.

☐ Increased variety in materials, with more diverse properties, which requires great flexibility and diversity in the processes used to machine or form these new materials.

☐ Increased cost of energy needed to transform materials and of capital and materials.

☐ The need to markedly improve productivity and reduce the costs of goods and services to halt inflation and meet international competition from those who are using these systems.

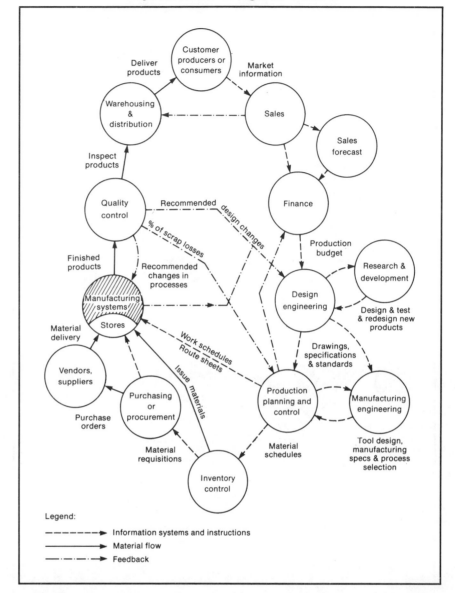

Figure 4: Functions and Systems in a Production System as Related to the Basic Manufacturing System (shaded circle), Which May Have Any of the Layouts Show in Figures 1 and 3

☐ The move away from a labor intensive manufacturing environment toward a service environment.
☐ The need for shorter service times or lead times in production to reduce inventories and allow faster response to changes in demand.
☐ Worker demands for improved quality of working life.

Overlaying these trends is the continued rapid growth of process technology, led by computer technology. No one can forecast the ultimate magnitude of this technological development, but it is clear that no segment of manufacturing and production will escape its impact. Computer-aided design (CAD), manufacturing (CAM), process planning (CAPP), testing and inspection (CATI), and integrated information systems are rapidly being integrated into the factory of tomorrow.

A group technology program

One of the best ways to reorganize a system is by embarking on a group technology or GT program. GT provides a systems approach to the redesign and reorganization of the functional shop. It will have an impact on every segment of the existing system. It is a manufacturing management philosophy which *identifies and exploits the sameness of items and processes used in manufacturing industries.*

GT groups units or components into families of parts which have similar design or manufacturing sequences. (See Figure 11.) Machines are then collected into groups or cells (machine cells) to process the family. Portions of the functional system are converted, in steps, to the cellular system or the flexible manufacturing system.

This implies a number of things. It means redesigning the entire production system and all functions related to it. The change will affect product design, tool design and engineering, production scheduling and control, inventories and their control, purchasing, quality control and inspection, and of course the production worker, the foreman, the supervisor, the middle manager and so on, right up to top management. Such a conversion must be viewed as a *long term transformation* from one type of production system to another.

It is unlikely that the entire shop can be converted into families, even in the long run. Therefore, the total collection of manufacturing systems will be a mix that evolves toward that ideal of a continuous process system over the years. This will create scheduling problems, because in-process times for components will be vastly different for products made by cellular or flexible systems and those made under traditional job shop conditions.

Finding families of parts is one of the first steps in converting the functional system to a cellular/flexible system. There are many ways to do this, but three popular ways are through:

☐ Tacit judgment, or eyeballing.
☐ Analysis of the production flow.
☐ Coding and classification.

The eyeball method is, of course, the easiest and least expensive, but also the least comprehensive. This technique clearly works for restaurants where "chicken" represents a

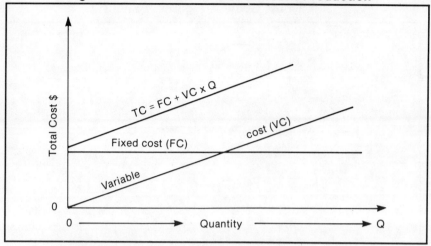

Figure 5: Economics of Lot or Batch Production

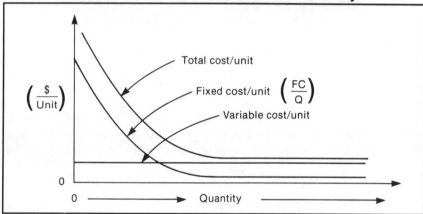

Figure 6: Classical Representation of Production Economics for Cost-Per-Unit Versus Quantity

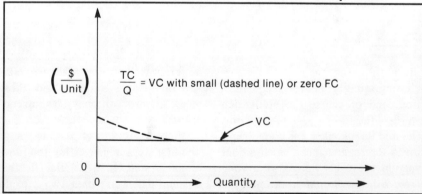

Figure 7: Effect on Unit Cost of Reduced Setup Time

family of parts, and for large lots of similar parts, but not in large job shops where the number of components may approach 5,000 to 10,000 and the number of machines may be 300 to 500.

The second method, product flow analysis (PFA) (see Monden, "For further reading"), uses the information available on route cards.

The idea, which is illustrated in Figures 12 and 13, is to sort through all the components and group them by matrix analysis, using route sheet information to develop the initial matrix. This method is more analytical than tacit judgment, but not quite as comprehensive as the coding/classification method.

PFA can get rather cumbersome when the number of components and/or machines becomes large. Sampling can be used (20 to 30%) to alleviate the size problem, but then it is uncertain whether all the potential members of the family have been identified and how big the family really is. However, PFA is a valuable tool for use in systems reorganization. Part of this technique involves analyzing the flow of materials in the entire factory to lay the groundwork for the new plant layout.

The overwhelming majority of companies that have converted to a cellular system have used a coding/classification (C/C) method. There are design codes, manufacturing codes and codes that cover both design and manufacture.

Classification uses coding to sort items into classes or families based on their similarities. Coding is the assignment of symbols (letters or numbers or both) to specific component elements based on differences in shape, function, material, size, process, etc.

Coding/classification methods are discussed in the article by Dunlap and Hirlinger. Numerous C/C systems have been published (see Ross and Ham, "For further reading"), and many have been developed by consulting firms. Most systems are computer compatible, so that computer sorting of the codes generates the classes or families.

Whatever C/C system is selected, it should be tailored to the production of the particular company and should be as simple as possible, so that everyone understands it. It is not necessary that old part numbers be discarded, but every component will have to be coded prior to the next step in the program, finding families of parts. This coding procedure will be costly and time consuming, but most companies opting for this conversion understand the necessity of performing this analysis.

The composition of parts families, and therefore of the cells that are designed to manufacture them, is dependent upon what characteristics you decide to sort according to. Family generation after coding and classification is not automatic.

Forming cells and FMS

Depending upon their material flow, different families will require different layout designs. In other words, the manner in which the family is determined influences the design of the cell. In some families, every part will go to every machine in exactly the same sequence; no machine will be skipped, and no back flow will be allowed. This is, of course, the purest form of a cellular system, except perhaps for the single machining center in which all parts are made on one machine.

Other families may require that not all components go to all machines, or that the forward sequence of order through the machines be modifiable, or that back flow in the group be provided for. Flexible manufacturing systems are designed to accommodate these situations, as are many cellular systems. This in no way alters the basic concepts, but it does add to the complexity of scheduling parts through the machines.

The formation of families of parts leads to the design of FM systems and cells, but this step is by no means automatic. Design is the critical step in the reorganization and must be carefully planned. Remember, the objective here is to convert the functional system, with its functional layout, into a *flexible* group layout.

We are just beginning to come to grips with the problems of cellular design and how to make these cells truly flexible, but the following guidelines for the formation of machine groups have proved useful:

☐ There should be a sufficient volume of work in the family to justify establishment of a cell. Some families may be too small to form a group of machines, but would be good candidates for processing on a single CNC or machine center, which is the smallest form of a cell.

☐ The composition of a component family should permit a satisfactory utilization situation. In manned cells, this may mean that not every machine will be utilized 100%, or even that the machine utilization rate will be greater than it was in the functional system.

The objective in manned cellular manufacturing is to improve the utilization of the *people*. In designing its manufacturing/production systems, Japan's chief objective is to make them flexible, and this should be our objective as well. In unmanned cells, machine utilization is obviously an important consideration.

☐ The processes in these systems should be technologically compatible.

☐ The needed capacity of the system must be determined from the quantities of parts needed and the production schedule, which determines *when* they are needed. Problems involving balancing labor and machine utilization will be encountered.

☐ The physical reorganization of the manufacturing systems will necessitate the redesign of the production system. The following kinds of problems will have to be addressed:

Product design will be impacted—in terms of both new parts and standardization of old parts—but the designers will be able to see parts being made in total in the cells and to get a better grasp of the manufacturing system and what it can make.

Planning and scheduling for cellular designs will be different.

Scheduling of a mixed shop will involve difficulties with respect to timing components to arrive at assembly points on time. However, scheduling will be easier overall, because the control has moved on to where the work is being done—in the cell itself.

Some people argue that manned cellular processing can be accomplished by simply routing families through machines without ever forming them into cells. This defeats many of the benefits of the cellular system and will obviously never lead to conversion of the functional system to a cellular or flexible system. For example, workers in a cell tend to become more flexible because they become familiar with the entire family and can see the common elements in all the parts in the family.

The manned cellular system provides the worker with a natural environment for job enlargement. Greater job involvement enhances job enrichment possibilities and clearly provides an ideal arrangement for improving quality.

In advanced forms of unmanned cellular layouts, the microcomputers of the CNC machine tools are networked together with a robot for material handling within a cell. It is difficult, if not impossible, to conceive of this kind of arrangement without some method of collecting the work into compatible families. All the machines in the cell are programmable, and therefore this kind of automation is very flexible as well as more economical than traditional job shop processing.

Benefits of conversion

Conversion to these new forms of manufacturing/production systems can result in significant cost savings over a two to three year period and marked improvements in quality.

However, reorganization has a greater, though immeasurable, benefit. It prepares the soil so that the "seeds" of the computer-aided and automated manufacturing of the future will fall on fertile ground. The

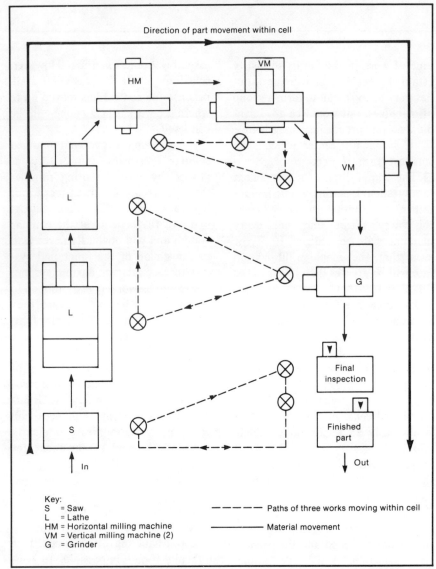

Figure 8: Schematic of a Manned Cell Using Conventional Machine Tools—Cell Laid Out in U-Shape and Staffed by Three Multifunctional Workers

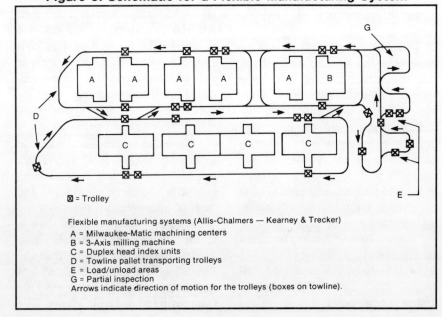

Figure 9: Schematic for a Flexible Manufacturing System

progression from the functional shop to the shop with manned cells to clusters of CNC machines to an entire system of linked cells must be accomplished in logical, economically justified steps, each building from the previous state. Simply adding robots and computers to your existing job shop will not make you as efficient as your competitor who has successfully undertaken such a system conversion.

Since its initiation in Russia in the late 1950s (see Mitrofanov, "For further reading"), the group technology concept has been carried throughout the industrialized world. It is now well rooted in Germany, Russia, England and, especially, Japan, where it is a "way of life" in many manufacturing facilities. Why has this concept not taken stronger hold in the U.S.?

Constraints on implementation

Clearly, such a conversion requires a major effort. Some of the constraints on the implementation of a cellular conversion program are as follows:

☐ Systems changes are inherently difficult and costly to implement. Changing the entire production system is a huge job.

☐ Companies are willing to spend freely for product innovation, but not for process innovation. Few machine tool companies sell integrated and coordinated systems with their machining centers.

☐ Decision making is choosing among the options in the face of uncertainty. The greater the uncertainty, the more likely the "do nothing" option is to be selected. This is because of fear of change and the unknown.

☐ Many companies use faulty criteria for decision-making. Decisions for change should be based on the company's ability to compete (quality, reliability, delivery time, flexibility for product change or volume change), not on the basis of output or cost alone. The high initial cost plus the long-term payback on such conversions equals a high risk situation in the minds of the decision maker. (But is there really an alternative?)

☐ Short-time financially oriented viewpoints conflict with the long-term nature of program.

☐ Unions' fear of loss of jobs and resistance to the multifunctional worker concept act as deterrents.

☐ The general lack of blue collar involvement in the decision making process in companies leads to poor vertical communication.

Summary

Converting to cellular manufacturing is a systems level change. This means that it requires careful planning and the full cooperation of everyone involved. All must understand the problems, costs and limitations of making such a conversion and the long-term effort that will be needed.

However, the magnitude of this change also means that it offers the potential of tremendous savings and meaningful advantages. The just-in-time system, with its simple aim of making goods flow through a plant like the waters of a river, has amply demonstrated the scope of the benefits to be gained.

For further reading:

Burbidge, John L., "Group Technology in the Engineering Industry," *Mechanical Engineering Publications LTD,* London, 1979.

Carrie, A. S., "Group Technology: Part Family Formation and Machine Grouping Techniques, MAPEC Project, Purdue University, 1978.

Flexible Manufacturing Systems, Proceedings of 101st International Conference, North Holland Publishing Co, 1982.

Gallagher, C. D. and Knight, W. A., *Group Technology,* Butterworth and Co. Ltd., London, England, 1973.

Groover, M., *Automation, Production Systems & CAM,* Prentice-Hall Inc.

Group Technology—An Overview and Bibliography, MDC Manufacturing Systems Report, 76-601.

Hitomi, K., *Manufacturing Systems Engineering,* Taylor and Francis Ltd., London,

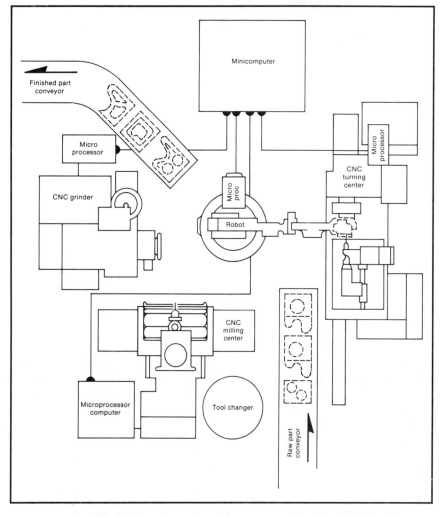

Figure 10: Cellular Layout of CNC Equipment with Robotic Material Handling and Perhaps Tool Changing—Minicomputer Networked to Four Microprocessors on the Four Machines

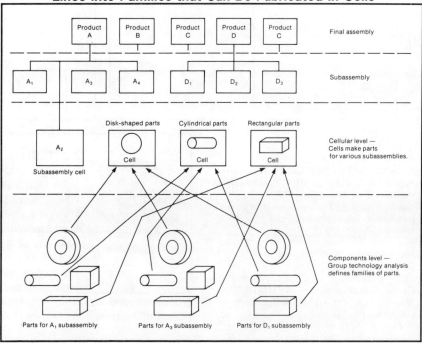

Figure 11: Grouping of Components from Various Product Lines into Families that Can Be Fabricated in Cells

Figure 12: A Matrix of Jobs (by number) and Machine Tools (by Code Letter) as Found in the Typical Job Shop

Job Number	SW	BR	EL	DP	VM	HM	GR	HO	IN	SW
1								X		
2		X	X							
3				X						
4						X	X			
5	X	X	X							
6									X	X
7	X		X							
8						X	X			
9									X	X
10				X	X					
11	X	X	X					X		
12						X	X			
13									X	
14				X	X					
15									X	X
16		X				X	X			
17										X
18	X	X								
19						X	X	X		
20					X					

Figure 13: Matrix Rearranged by PFA to Yield Familes of Parts And Associated Groups of Machines that Form a Cell

Job Number	SW	BR	EL	DP	VM	HM	GR	HO	IN	SW
7	X		X							
11	X	X	X					X (exception)		
2		X	X							
5	X	X	X							
18	X	X								
14				X	X					
3				X						
10				X	X					
20					X					
12						X	X			
4							X	X		
19						X	X	X		
16						X	X	X		
8				X (exception)		X		X		
1								X		
9									X	X
13									X	
6									X	X
15									X	X
17										X

Cell will have 3 machines HM_1GR_1HO for manufacture of 6 jobs.

England, 1979.

Mitrofanov, S. P., "Scientific Organization of Batch Production," AFML/LTV Technical Report TR-77-218, Vol. III, Wright Patterson AFB, OH, December 1977.

Monden, Yasuhiro, *Toyota Production System*, Industrial Engineering and Management Press, IIE, 1983.

Riggs, James L., *Production Systems: Planning Analysis and Control*, 2nd Ed., Wiley/Hamilton, 1976.

Ross, D. T. and Ham, I., "Group Technology Coding/Classification," AFML-TR-77-281, Vol. I, Report, AFML, Wright Patterson AFB, OH, 1977.

Schonberger, R. J., *Japanese Manufacturing Techniques*, the Free Press, 1982.

J. T. Black, P.E., is chairman and professor of the industrial and systems engineering department at the University of Alabama in Huntsville. He holds a BSIE from Lehigh University, an MSIE from West Virginia University and a PhD from the mechanical and industrial engineering department at the University of Illinois in Urbana. He is a senior member of IIE and the former division director of the manufacturing systems division. Black is the author of many technical papers in the area of manufacturing processes and coauthor (along with DeGarmo and Kohser) of the 6th edition of *Materials and Processes in Manufacturing*.

Reprinted from **Industrial Engineering**, *November 1983.*

Cellular Manufacturing Becomes Philosophy Of Management At Components Facility

**By William J. Dumolien
and William P. Santen**
Deere & Co.

Cellular manufacturing at Deere & Co. has grown from its 1975 introduction at the component works facility in Waterloo, IA, to become a major management philosophy. Cellular manufacturing is the transfer of raw material into subassemblies or finished parts within a single organizational entity, or a cell.

Deere has implemented a large number of manufacturing cells containing from 10 to 30 machine tools producing several thousand part numbers. The introduction of these cellular manufacturing systems has coincided with facility reorganization and modernization activities. The practice of cellular manufacturing has evolved into one of John Deere's most important manufacturing strategies.

Reorganization

To reorganize functional manufacturing into a cellular arrangement, the part population must be divided into groups that can be processed as completely as possible within a single cell. Parts are segregated into families. Families are matched with machines to form cells, and the cells are balanced to achieve acceptable machine and operator utilization.

Tacit judgment or manual analysis alone is not feasible for large-scale projects. A computerized "systems approach" is required to review the vast amounts of necessary data. The John Deere group technology system was developed to support the design of cellular manufacturing systems.

This system is a collection of integrated programs for data capture, coding, planning and design. The heart of the system is a classification and coding scheme developed by John Deere engineers for the capture of geometric characteristics of parts.

Data are entered through an interactive classification program which generates a code that references geometric part features. The system also contains production data such as part operation routings, current and future production requirements, cost amounts and machine load information.

All data are available to the engineer through a set of standardized on-line computer modules. These include:
☐ Code—Provides part data entry.
☐ Extract—Isolates a group of parts using desired criteria.
☐ Analysis—Provides a set of analytical routines such as machine loads and part flows.
☐ Modify—Allows changes to data to perform "what if" analyses.
☐ File—Provides file handling programs and interface with statistical analysis programs.

The system is a decision support tool which allows the creation, comparison and evaluation of cellular manufacturing alternatives. The creation of these manufacturing cells is typically approached using the standard phases of analysis, detailed design, implementation and operation.

Data analysis

Establishment of the part data base containing the related geometry and production information becomes the first of three cellular analysis steps. Part numbers, processes, equipment, machine utilization and operating procedures must be determined.

Part families are formed from this part population. Each family is a collection of parts with the same machining features as determined from the coded geometry. Since the code represents the geometric features resulting from the transformation of raw material into finished parts, all parts with the same intrinsic operational requirements are grouped together. This method is used to avoid the proliferation of less than optimal routings rather than grouping according to routing information. An example of random parts before and after the assignment to part families can be seen in Figure 1.

In the second analysis step, part families are grouped into cells for routing evaluation based on one of three themes of cell development. The most general grouping theme is similar geometric features.

The optimum routings are determined for families of geometrically similar parts. The machines needed in these routings become candidates for a cellular design. An example of this may include families of rectangular, flat bar parts with machined slots of similar size and shape, produced as complete within one dedicated machine cell.

A second theme used in cell formation is based on similar process sequencing. Families with similar sequential routings can be assigned to cells which contain the machine tools needed to perform only those operations. Part families with sequential operations of milling, drilling and induction hardening may be candidates for this type of cell formation.

The third theme of cell formation focuses on product lines and subassemblies. These cells include all operations, both machine and assembly, needed to produce completed subassemblies or products. In evaluating a varied part population, all these themes of cell formation may be applied.

The third analysis step is the calculation of the machine utilization for each proposed cell. Cells with relatively high machine utilization will be implemented. Cells with low machine usage must be combined with other cells. Cells with requirements for the same capital intensive equipment are combined to share this equipment. This grouping method provides for effective utilization of expensive equipment.

When combinations of cells will not improve machine utilization, the routing of part families within the cells is revised. Parts are rerouted to similar machines for improved utilization and cell integrity.

Generally, machines capable of producing the larger sized parts are selected as key machines in rerouting the problem part families. This assures adequate machine capacity, but it may cause a shortage of large size machine tools.

It is also necessary to organize the manufacturing cells into departments using the same grouping principles. The objective is to complete all operations, if possible, within one

Figure 1: Random Parts (above); Eight Part Families (below)

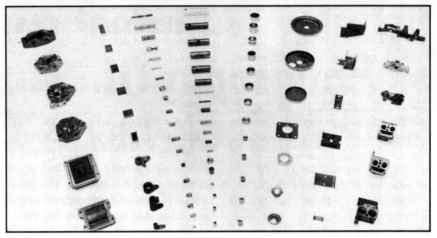

Figure 2: Before and After Analysis of Material Flow

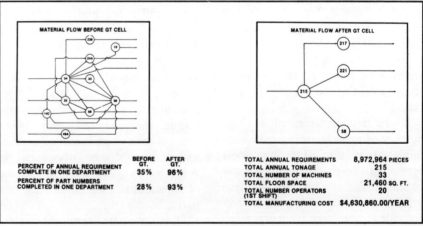

department and maximize machine utilization.

Department size criteria must include the effects of part requirement fluctuation as well as standard manageability factors. Examples of proposed departments are shown in Table 1.

Detailed design

The previous analysis developed the group technology plan of action for cellular development. The detailed design phase must address the specifics of machine layout, common tooling and operator requirements. Also included are evaluations of operational improvements and the introduction of new technology.

The specific machine tools needed in each cell must be laid out. Machine tool performance must match the requirements of each cell. The machine tools are arranged to minimize material handling and eliminate backward flow through the cell. An example of streamlined material flow is described schematically in Figure 2.

One of the basic benefits of cellular manufacturing is the reduction of tooling setups from the use of common part family tooling. Tooling requirements are reviewed and substitutions made to minimize the number of required tool changes.

For example, a family of eight spindle parts which now use common tooling are permanently affixed to a shared machine tool. Previously these parts were produced on five different machines with eight tool changes. Setups, tool crib investments and job change costs are all reduced.

Cellular manning methods are established for effective use of human resources. It becomes advantageous to have operators capable of running multiple machine tools. This provides a greater degree of flexibility within a cell. The operators are now assigned to a group of finished parts instead of to individual operations.

For example, a flexible cell may include four lathes and four hobbing machines run by three operators. While an operator performed a job change on one machine, the other operators would maintain production on the remaining seven machines.

Once the cellular design has been approved, implementation follows basic project management techniques. Physically relocating machine tools is a minor effort compared to the extensive staff manpower requirements associated with a cellular arrangement. Tooling records, mechanical data and job detail standards must be revised. Without these revisions, startup of machine cells will be prolonged and unnecessarily difficult.

The detailed design and implementation phases of cellular manufacturing projects require significantly more effort than the analysis phase. This is exemplified in the manpower percentages shown in Table 2. The project involved the creation of 19 cells producing 600 part numbers.

The cellular core

At the core of cellular manufacturing processes are the machines, tools and floor space required for each operation. Our early attempts at designing cells concentrated effort on the establishment of this cell core.

It is now realized that cellular manufacturing practices offer productivity improvements which are shared between staff and shop floor activities. Cellular manufacturing is part of an overall management strategy. All affected employees should be versed in the cellular concept and the focus and simplification of manufacturing that it provides.

Product engineering can make use of family design practices when introducing new parts. Parts of similar function, when possible, can be designed according to a family standard.

This effort helps to ensure the design integrity and manufacturability of new part introductions. These new parts can then be incorporated into existing manufacturing cells using standard routings.

Process planning and methods engineering must realize the advantages of utilizing part family production. The focus and reduction of routing alternatives simplifies process planning activities. By design, optimal family routings are determined within manufacturing cells. Therefore, many new parts that fall into existing part families will have the optimum routing predetermined.

Process planning and methods engineering efforts are greatly reduced. The prediction of new product manufacturing cost will also be more accurate due to the focused production scheme of cellular manufacturing.

Advantages to production control

Table 1: GT Cell Department Operation

	Department No.							
	1	2	3	4	5	6	7	8
Number of parts	77	107	55	59	57	58	75	103
Number of families	12	20	5	6	7	8	8	9
Number of cells	2	4	2	2	2	3	1	3
% operations completed within department	99.6	91.8	94.7	95.5	99.5	99.9	94.0	100.0
Number of operations	455	442	398	536	694	667	421	497
Number of unique processes	6	6	6	6	5	4	6	5

include part family scheduling. When possible, part families, instead of individual part numbers, should be scheduled for production. Parts should be scheduled within a family before changing to other family groups.

Part families which make use of similar tooling increase productivity by requiring fewer setups.

Quality of finished parts is improved due to the increased visibility of part processing operations. In most cases, one small group of operators is responsible for the complete production of finished parts in sequential operations.

In these cells, quality problems are easily seen by the operators responsible and can be corrected immediately. Fewer inspectors and increased part quality are potential benefits of cellular manufacturing systems.

It must be realized that many staff-supported areas are affected in the development and execution of manufacturing cells. There is no "turnkey" method for the design and execution of cellular manufacturing.

Each cell can have characteristics dependent on the depth of planned and monitored benefits. Only with strong management direction will the potential savings of a cellular manufacturing strategy be realized.

Results

Based on the success of cellular manufacturing systems, the application of group technology concepts has increased dramatically since their introduction at John Deere. Cellular manufacturing is a recognized strategy for improved productivity.

Studies of other manufacturing techniques, particularly the relationship of cells to the just-in-time philosophy, have further demonstrated the importance of cellular manufacturing as a cost-effective management strategy.

The actual benefits of the completed cellular manufacturing systems at three John Deere manufacturing facilities have validated the predicted benefits. Typically, the results are:

☐ A 25% reduction in the number of required machine tools.
☐ A 70% reduction in the number of departments responsible for the manufacture of a part.
☐ A 56% reduction in job change and material handling.
☐ An 8-to-1 reduction in required lead times and a corresponding reduction of inventory.
☐ Shop supervisors who now have more control over processing, with clear delineation of responsibility.

Since their introduction, group technology and cellular manufacturing concepts have been developed, refined, promoted and practiced. Today, all new production and facility improvements at John Deere are analyzed using the group technology system for the application of cellular manufacturing. Current projects at separate facilities are examining cellular manufacturing for cast iron and steel machining, chassis production, subassembly welding and metal forming.

The past success of the implemented manufacturing cells and the continued expansion of these concepts demonstrate the importance of cellular manufacturing at John Deere.

Table 2: Manpower Requirements for GT Cell Formation

	% of Total Time
Analysis:	
Initial cellular groupings	4%
Design:	
Process planning	18%
Process & tool	17%
Industrial engineering—	
Methods & standards	14%
Preset tooling	14%
Manufacturing engineering services	14%
Implementation:	
Facilities redevelopment	13%
Plant engineering	5%
Material engineering	1%

William J. Dumolien, a project engineer with Deere & Co., is currently assigned to the manufacturing projects division. His past experience has been in product engineering, plant engineering, manufacturing research and development and process and tool at John Deere Harvester Works. Most recently, he has been developing cellular manufacturing strategies and assisting in their introduction, design and adoption at various Deere operating units. He holds BSME and MSIE degrees from the University of Illinois. Dumolien is a member of ASME, SME and Robotics International.

William P. Santen is an engineer with Deere & Co. assigned to the manufacturing engineering systems division. Since 1981, he has acted as coordinator of the John Deere Group Technology System at the corporate offices, with responsibility for company-wide promotion, education and application of the group technology concept. Past and current projects include the formation of cellular manufacturing systems for facility and productivity improvements. Santen holds a BS degree in mechanical engineering technology from the University of Cincinnati and is a member of the computer and automated systems association of SME.

Reprinted from *Industrial Engineering*, November 1983.

Well Planned Coding, Classification System Offers Company-Wide Synergistic Benefits

By **Glenn C. Dunlap**
ITT Advanced Technology Center
and **Craig R. Hirlinger**
Garrett Turbine Engine Co.

For several years the prime movers of group technology have been evolving and championing the concept of synergy. As frequently happens, these prophets have been ahead of many of the rest of us, and it has been difficult for them to offer tangible, convincing evidence of the correctness and merit of their ideas.

The basic concept of synergy is that the whole is greater than the sum of the individual elements. This means that working together produces results that are more valuable than the sum of the individual contributions because the unity and connectivity of all the efforts offer advantages that cannot be produced separately.

A few short years ago the primary advantage projected for the classification and coding of information related to machined piece parts was that it would be possible to avoid design replications. Certainly, if the process was properly applied, this advantage was realized, but many times the actual mechanics of the design broke down and the design retrieval process became ineffective.

Surprisingly enough, however, users still claimed that implementing a classification and coding system was advantageous. The advantage they realized was the basic structure upon which a producibility system could be built that would provide input at the critical design stage.

Unexpected advantages were also documented in the classical group technology application of plant layout, this time in regard to the ability to review individual items from a commodity standpoint.

Other paths through group technology history could be traced with the same result of unexpected benefits realized. The common theme throughout all these paths is the ability to analyze geometry, process and material information for purposes other than the original objective. This advantage is provided by the basic classification and coding structure that lies at the heart of group technology.

With the increasing interest in and emphasis on cellular/flexible/integrated manufacturing systems, group technology is proving to be an increasingly valuable, if not essential, tool in a constantly changing environment. The way energy, labor and computerized automation costs have changed over the recent past certainly demonstrates that optimization parameters fluctuate. What constitutes the best possible cell will change over time, frequently in a non-linear or even transcendental manner. Also, the key word "flexible" indicates the need to be able to periodically (or on an event triggered basis) conduct a new analysis and perhaps restructure the cellular system.

This points up the need for a carefully planned flexible coding and classification system that will extend the productive life of an existing cell by providing the capability to readily review the cell's part/quantity mix and dramatically vary this input to maintain maximum cell and plant effectiveness.

Toward a successful code

For both developing and implementing a coding system, the two key words are planning and flexibility. To develop a useful coding system, you must have a plan. This plan must include what you want the system to accomplish and how you are going to use it to accomplish this objective.

For example, consider two companies, both of which produce brake components for the auto industry. Company A's plan is to use its coding system to help its design staff in the production of new brake components. Company B plans to use its coding system to help its production routing department develop production plans for the shop floor.

If each plan is followed strictly, the coding systems developed for the two companies will be considerably different. There will be a central core of information common to each, but Company B will have to know much more about the manufacturing characteristics of its parts than Company A. Company A's code will deal more with the function of the part and its geometric attributes.

If a single coding system is to serve the purposes of both companies, it will have to be very flexible, and compromises will have to be made.

Coding systems are available commercially, and a number of consulting firms specialize in the production of such systems. Regardless of how you acquire your system, it should be an integral part of an overall plan. This will enable you to avoid many of the pitfalls that await those who try to leap into group technology.

Computerization

Group technology implementation would be either very time consuming or impossible without the aid of a computer. Therefore, it is a virtual necessity that the coding system be compatible with whatever computer system is planned for use with the project.

There are two basic areas in which the computer can provide assistance with a coding system: in the application of the code to the data base of parts and in the sorting of parts into groups based on commonality.

If you plan to use existing software in coding application, it must be compatible with the two basic formats of a coding system. One of these, the hierarchical format, applies the code in groups of two digits. As the name implies, it is hierarchical in nature, so that "code pairs" in a long list, or string of code, have meanings that are determined by what code pairs preceded them. Graphically, this code can be shown as a decision tree; it is often referred to as a family tree format (see Figure 1).

The second common format is nonhierarchical. It does not rely on a hierarchy to build the code. Column 3 of a nonhierarchical code will always have the same information stored in its location, regardless of the preceding code's value. Digits of code in this format are usually added one at a time, but also can be paired as with the hierarchical code.

Graphically, a nonhierarchical

The growing interest in integrated manufacturing systems like the flexible manufacturing facility above makes group technology an increasingly valuable tool. (Photo courtesy of Mazak Corp.)

Figure 1: Hierarchical Code Structure

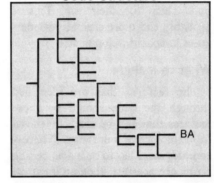

Figure 2: Nonhierarchical Code Structure

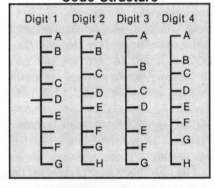

code looks quite different from a hierarchical code (see Figure 2). The key difference is that regardless of what piece of information has been chosen in the previous digit, the next question is the same for all parts.

Most software programs can be easily modified to accept a hierarchical or nonhierarchical format.

The second area of compatibility involves the actual code digits— whether they are numeric, alphabetical or a combination of both. This frequently depends on the type and quantity of information that is being stored in the code. It becomes obvious that there are more possible alpha combinations than there are numerical ones. The program must be compatible with both alphas and numerics, as variables, if the system is a hybrid.

Some misconceptions

There are some misconceptions about coding systems that should be clarified before code development and application are discussed.

First, a coding system with a data base of coded parts does *not* create families or groups of parts. It provides a tool that can be used to develop these families or groups. How these families or groups are to be defined using the code must be spelled out by the plan for the use of

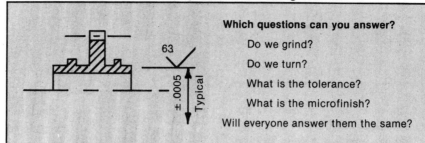

Figure 3: Geometry and Processing Questions

Which questions can you answer?
Do we grind?
Do we turn?
What is the tolerance?
What is the microfinish?
Will everyone answer them the same?

the coding system that was discussed earlier. The list of combinations of parts that can be grouped together is almost endless.

Second, do not be deceived into thinking that it is possible to "buy" a mature coding system that will work for you because it has been used successfully at another comapny. The plan that company followed to develop its code was specified according to its individual needs. Also, the characteristics of its parts may be vastly different, even though at first glance the products may seem very similar.

In one actual case, Company A examined a code developed for one of the large automotive transmission producers. This code addressed the manufacturing characteristics related to the production of all manual and automatic transmissions. However, Company A's application was going to be for gears in the transmissions of small to medium size gas turbines. Whereas automotive gears turn at less than 8,000 rpms, the gears in the gas turbine engine turned at 20,000 to 80,000 rpms. Needless to say, the two were not compatible.

The third point is, do not be afraid to mix hierarchical and nonhierarchical formats in the same code to take advantage of the strengths of both types.

Do it yourself?

A company that has no experience in the development of a coding system might want to use one of the consulting firms that specialize in creating coding systems for various companies. Some of these firms provide a selection of codes that address shape and feature location. They then develop the latter portion of the code to address the features that apply exclusively to a specific company.

Another approach is to develop the entire code in-house. This assures that the entire code is applicable to the company's product and that there is no "wasted" code covering features of no interest to the company.

Either method will require very close interaction with your engineering groups: manufacturing or design, or both. This is necessary to ensure that the information needed by these groups is included in the code in a retrievable form.

The code must be understandable by *all* end users. The quickest and most efficient way to assure this is by involving these people in the actual formation of the code so that it is *their* code, not a tool that is forced upon them for their use. This is probably the most crucial consideration in code development.

What to include

The features that are identified through the code should be those features that can be easily retrieved from the face of a drawing. Wherever a judgment has to be made, coding errors are possible when an identical feature can be coded into two categories by two different people. The primary objective is to capture those characteristics of a part that will not be changing with any frequency.

For example, if our Company A produces a gear with two journals that require a 63AA finish and ±.0005 tolerance, it is possible for these diameters to be either turned or ground. Therefore the code should answer the questions (1) What is the micro finish? and (2) What is the tolerance? not (1) Does the diameter get turned? or (2) Does it get ground? The micro finish and tolerance are available on the blueprint of the part and can be consistently coded the same. Whether the diameters are turned or ground is a judgment decision that introduces unnecessary variables into the coding system.

Figure 3 gives some examples of geometry and processing questions. In addition to the design and manufacturing information, many benefits can be found in capturing information that pertains to the raw materials from which the parts are made.

Classification filters

The formation of a coding system is no small task in itself, but it is followed by the equally important task of determining how to retrieve all the information that has been stored so neatly in the code.

Now the work done in code planning will pay off. Parts will be sorted by selected characteristics using "filters" and grouped together according to their commonality. These filters define a classification and represent groups of features upon which part families are based.

The sorting of a group of parts across a set of filters should define a group of parts that are different from all of the remaining parts. This procedure is continued to create additional families until all the parts have been classified into groups. It is a very tedious, time consuming process.

One problem with using this method to develop families is that two sets of family filters may overlap to some degree. This opens up the possibility that one part may fit filter descriptions for two or more families. If these parts are to be manufactured using the family of parts concept, this will be an undesirable situation.

Also, with several thousand parts coded, several hundred unique sets of filters may be required to subdivide this many parts into usable families.

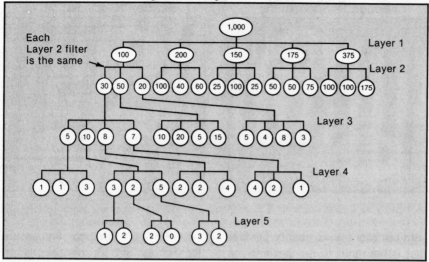

Figure 4: Layered Filters

Table 1: Typical Materiel System Components
- *Scheduling:*
  MRP
  Shop floor scheduling
  Assembly scheduling
  Tooling/gauging
- *Procurement:*
  Automatic order placement
  Supplier load forecasting
  Shipping/receiving posting
- *Quality:*
  Scrap/rejection reporting
  Operator/tool performance
  Dynamic sample sizing/analysis

Again, this represents a time consuming process.

Layered filters

As an alternative to creating individual filters, one family at a time, there is an approach called layered filters. Layered filters are created by, first, taking the parameter types that are covered in the code and listing them according to their level of importance, or influence, in the area for which the filters are to be created.

For example, the code covers material type, OD size, length, ID tooth forms, OD tooth forms and heat treatments. For part groupings for the creation of process plans in manufacturing, the parameters would be ordered so that the parameter with the greatest impact on the process plan was at the top and the parameter with the smallest effect on the process plan on the bottom. Second, groupings are created using only the top parameter. These parts can then be routed through an appropriate manufacturing process.

Each parameter may be subdivided to include any and all codes that could be applied at that level. This will assure that no matter what part is coded in the future, the code for the top listed parameter will be included in a family.

The next grouping is based on the second parameter of importance.

It is essential that all possible codes be covered at each parameter level. When all parameter levels have been addressed, part sorting is initiated.

How layers work

All the parts will be run across the grouping of parameters that were selected for the level one filters. This will begin to split up the parts into groups with at least one common feature.

All of the parts that have already been split once into groups will then be run against the second layer of filters. This will further subdivide the parts into groups that have at least two common features.

This step is repeated until all filter layers have been exercised. The parts will then be very finely grouped into families with many common features. If the families are too small at the end of all of the filters, "back up a level" and look at the groupings at this level.

Once the filters have been created, the priority sequence in which the parts are sorted can be changed. This will produce different groups of parts at the different levels, but the final groupings of parts will always be the same once the parts have gone through all the layers of filters.

Figure 4 shows graphically how a group of 1,000 parts might split over a five-layer filter system. As the illustration shows, parts can be quickly subdivided into very small, finite groups without going through the formation of each individual filter for each individual family. Less time is required to create the layered filters, and the layered filter approach ensures a unique family location for each part. That family location may be vacant at the time, but knowing a part is unique is almost as valuable as knowing what parts are similar to it.

Flexibility essential

As the name suggests, a flexible manufacturing system must be capable of changing as the parts that are produced are changed and as the demanded quantities change. To support a flexible manufacturing system, the coding system must also be flexible. It must be able to expand to absorb new part characteristics and to discern new part groupings as conditions change. A layered filter system provides the ability to change the rules of the families and to view the impact the changes will have.

Synergy in action

If we review a typical manufacturing enterprise, the first level of decomposition would divide it into business, engineering and manufacturing activities. It is useful to examine the interrelationships of these areas with respect to the overall enterprise objectives of:
- Quality.
- Cost.
- Value.
- Delivery.
- Maintainability.
- Performance.

Primary interaction with the customer is the responsibility of the business section. This is usually a two-way interaction involving activi-

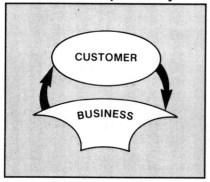

Figure 5: Business Section Responsibility

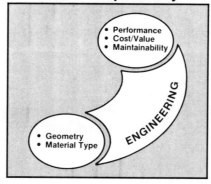

Figure 6: Engineering Section Responsibility

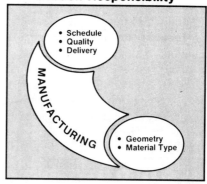

Figure 7: Manufacturing Section Responsibility

ties that initiate as well as respond to customer needs and demands. (See Figure 5.)

Having obtained either a customer or a company-funded commitment for product development, business initiates action in the engineering section.

The primary responsibility of engineering is to respond to fit-form-function requirements to satisfy customer needs. This activity is represented by Figure 6.

Manufacturing is responsible for translating the product engineering requirements into a quality product that is delivered to the customer on time. Its area of activity is shown in Figure 7.

After a customer order is received, there is a common unifying bond among the three areas. This is the part number. It represents the connection between the company and the customer.

Over the past several years, significant advancements have been achieved with respect to the material aspects of this part number. Industry now has available effective, sophisticated tools that address the operational obligations. A few examples of these tools are listed in Table 1.

Since this part number can now be considered a single point, the interrelationship between the customer and each area of the enterprise reaches the ideal. Each area has tangible evidence of its relationship to the customer, and with properly integrated materiel systems there is a universal and clear understanding of the status and timing of that obligation.

Figure 8 illustrates how the customer and the three major sections of the company interact, with the part number at the heart of their relationship.

But this is only half the story. Each part number contains not only materiel type information, but also information on geometric and/or material (ferrous, nonferrous, plastic, etc.) configuration properties. When we receive an order for a specific product, it has (or should have) a uniquely assigned part number, and a good materiel system is well equipped to respond to this requirement.

The materiel information system is unidirectional. If we are given a part number, it is possible to get some idea of what it represents from its subsequent name. However, if we are given only a part name, there are usually many part numbers that would fit the description. At this point a sound group technology system is needed. A properly structured classification and coding system, as previously discussed, can be called on to identify the exact product that is desired.

The implications of the integration of materiel and geometric systems are significant. Each of the overall enterprise objectives is affected by geometry/material/process decisions. Group technology provides a basis for linking this information with the part number data base.

The result is the capability to provide a homogeneous information base for materiel/geometry/material/process based decisions. Genuine synergy is produced by the complete integration of these systems.

Suppose the enterprise is being asked to bid a new part or product. To determine a price quote that will ensure a profit, the following information is needed:

☐ Specific similar parts.
☐ Current/future backlog of similar parts.
☐ Cost of similar parts.
☐ Producibility/quality/performance/maintainability data.
☐ Design/process plan time estimate.
☐ Current suppliers and their backlogs.
☐ Etc.

After it has been decided to produce a new part or product, it may be appropriate to determine the feasibility of creating a cellular manufacturing system. Information and subsequent analysis are required in several areas, including:

☐ Shipment quantity/schedule.
☐ Process/machining requirements

and capabilities.
- ☐ Setup/run time.
- ☐ Tooling/gauging requirements.
- ☐ Floor space/manpower availability.

Proper creation and integration of a materiel/group technology system will provide the synergistic capability to answer these questions in an expedient manner. A flexible GT system will also be able to respond to a variety of geometry/process based questions concerning:
- ☐ Restructuring of cellular systems.
- ☐ Part mix migration in an existing cellular system.
- ☐ Part family augmentation to improve utilization of cellular or noncellular systems.
- ☐ Surge capacity to aid in shop scheduling.
- ☐ Grouping and isolation of quality problem part/operations.

The efforts currently under way in American industry to establish cellular manufacturing systems must not be myopic. With a marginal increase in effort and planning it is possible to create an evolutionary path to much broader productivity improvement opportunities throughout the enterprise. This approach will lead to our next logical step of creating an efficient, responsive, quality-based American industrial base. **IE**

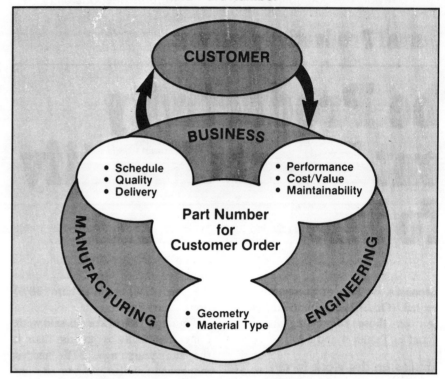

Figure 8: Interaction of Company Sections, Customer and Part Number

Glenn C. Dunlap is director of the CADEM (computer-aided design, engineering, manufacturing) division of ITT. The division's purpose is to improve the company's productivity throughout its computer-assisted areas. Dunlap received his PhD from Arizona State University and has work experience in academia, electronics, aerospace and engineered products. He has also been proprietor of his own manufacturing and sales company.

Craig R. Hirlinger is senior production engineer at Garrett Turbine Engine Co., Phoenix, AZ. He is in charge of production routings of gear line hardware, which includes routings from the blueprint stage to shop floor operations. Hirlinger is a graduate of Arizona State University, where he earned a bachelor of science degree in mechanical engineering technology. He is a member of SME and is highly involved with computer-aided manufacturing at Garrett Turbine.

III. Issues Affecting Development and Achievement of CIMS

Importance of Integration

Design and Software Considerations

Human Factors

Impact on Manufacturing Support Functions

Management Concerns

EDITORIAL OVERVIEW of Section III

Even though the aggressive installation of computer integrated manufacturing systems (CIMS) is necessary for the success of manufacturing companies and, for many, a matter of survival, the overall progress has been less than spectacular. This assessment is made despite a few notable exceptions of substantial progress being made in implementing CIMS in a variety of manufacturing environments. Progress has been inhibited by several technical problems and the lack of comprehensive management strategies for achieving CIMS. Hardware and software vendors, as well as academic researchers, are paying serious attention to issues, such as standard protocols and network structures, to facilitate communication between computers and effective distribution of databases. The establishment of a management strategy for CIMS cannot be assumed and will require more effort than initially anticipated.

The articles and conference papers selected for this section address many of the key issues and factors which affect the development and implementation of CIMS. In the first subsection, Thomas Cornell provides a rationale for integration and explains its importance. He proposes the expansion of the role of systems designer to that of systems integrator in order to enhance the success rate of CIMS implementation.

The second subsection contains six papers which discuss the systems design principles, concepts, and methodologies necessary for the successful introduction of CIMS. Windsor and Nestman propose criteria for the selection of a CIMS database system. Britton and Hammer focus on the importance of systems integrity for the success of CIMS. Devaney's paper examines the CIMS integration task by considering computer language and communication, database boundaries, and organizational relationships. The development of modular, distributed computer architecture by Westinghouse for implementation at its Electronics Assembly Plant is described by David Hewitt. Michael Daugherty discusses the use of IBM's Application Transfer Team approach in planning the high-level design of a CIMS project and identifies pivotal factors in the successful installation of integrated manufacturing systems. In the last article in this subsection, Robert French describes ten mistakes that should be avoided in selecting software for manufacturing control systems.

In the third subsection, Martin Helander addresses some of the human factors and ergonomic problems in automated manufacturing and design. He covers several areas of interest including the allocation of decision-making between humans and computers, ergonomic design of CAD/CAM work stations, and how the human factor affects the design of software. Kelvin Cross presents an approach to job design in an automated environment that utilizes the best attributes of both people and machines.

The two papers in the fourth subsection deal with the impact of CIMS on manufacturing support functions. Richard Shell discusses the impact on work measurement. Kjell Zandin presents the need for computerized work measurement and process planning.

In the final subsection, management concerns are considered. Thomas Klahorst's article discusses the anticipated benefits of flexible manufacturing systems and the importance of identifying responsibility for integration of these systems. In the last article, Robert French presents fifteen issues and realities that face management when considering CIMS. He also provides a prescription for success. Thorough review of the points covered in this article is recommended for all levels of managers responsible for manufacturing and for designers of CIMS.

While additional coverage of the issues influencing the development and installation of CIMS should be undertaken, the concepts and advice presented by the authors of the articles and papers in this chapter should be useful to readers just beginning to consider CIMS, as well as those already involved in the installation process.

Systems Integration Is a Mandatory Component in Achieving an Optimum Systems Environment

Thomas R. Cornell
Vice President
Comp-u-Staff, Inc.

In most organizations today, there is a significant demand or backlog of information systems requirements that are either being inadequately addressed or temporarily shelved. This demand coupled with technology advances in such areas as Office Automation, Computer Aided Design and Manufacturing (CAD/CAM), Automatic Storage and Retrieval Systems (ASRS), distributed data processing, low cost mini-micro computers and robotics present management with a number of complexities and unique opportunities. This article introduces a conceptual framework which is better suited to take advantage of these new technologies and identifies the role of the "Systems Integrator" as the guardian of the critical path.

"This is a man age. The machines are simply tools which man has devised to help him do a better job."

Thomas J. Watson, IBM

Before we begin, it is important that the definition of key elements be established.

- <u>System:</u> The interaction of all components within the systems environment which produce an effective product for the end user.

- <u>Systems Environment:</u> The universe in which all components which effect the processing of a given transaction from origination through conclusion interact.

- <u>Systems Architecture:</u> The framework or constraints in which a system(s) must operate within the organizational environment.

- <u>Systems Integration:</u> The person(s) responsible for producing an effective and productive system which meets both the functional needs of the organization and the humanistic needs of the end user. This is accomplished through integration of all components within the systems environment.

To justify the existence of any system there must be an effective coordination between the functional requirements and the systems objectives. A well defined systems environment that is understood by the organization promotes a better working relationship among the business faction, the system developers and the end users. This must begin at the strategic level and be driven down to the various operational segments. The degree of harmony will become critical to the successful enterprise in the future.

There are three (3) major levels within an organization in which systems address business requirements. These are depicted in Figure 1:

FIGURE 1

Executive management tends to focus on the strategic level which encompass such systems as simulation models, projections etc. They primarily support the planning process requirements of the organization.

The tactical level, traditionally has been the focus of most systems activities. This level provides the information necessary to run the organization and can be equated to such applications as payroll, inventory control, accounting etc.

Technological advances, which have provided new cost effective alternatives point out a blatant need to address optimization at the operational levels of an organization. These recent events have exposed an increased demand or focus on execution level systems which interface and blend with other applications.

Traditionally, systems integration has been addressed in an unstructured, unorganized manner resulting primarily from executive mandates which forced user management to work with other organizations through a Management Information Systems (MIS) function. The MIS organization is usually limited in both resource and charter. Over the past dozen years the corporate structure has tended to support centralized computing environments. The Result has been a classic case of organizational learning curve. Charge-back schemes, data center productivity, capacity planning, and the like have been established to stabilize the MIS function as an effective production unit within the organization.

When third generation computers became generally available during the mid to late sixties, attention was focused on hardware and conversions of applications to operate on the new technology equipment. During the mid to late 1970's, the focus of attention changed from hardware to data resource management. Data resource management concepts clearly place the emphasis on information as the primary asset and emphasized ownership of data to be a key ingredient of control. As a result, new technology emerged in the form of Data Base Management Systems (DBMS) which further supported this notion. With a focus on data and the access to data being of primary importance, new complexities and changes began to be introduced into the system design process. Centralized Data Processing functions were forced to develop new controls and functions such as Data Security, Data Base Administration and Data Dictionaries to support these demands. An increased emphasis for on-line functions coupled with cost effective alternatives further complicated the mission of a centralized data processing organization. Data Communications monitors, networks and the integration of both voice and data communications surfaced more complexity factors.

These occurences have forced most organizations to undergo significant change. The 1980s have brought additional factors and attitudes into focus which directly or indirectly affect the traditional approach to systems solutions. These are somewhat attributable to the following factors:

- <u>Micro Computers</u> have emerged as low cost alternatives which now provide the users ability to automate an increasing variety of applications and functions.

- <u>User Insurrection</u> has emerged due to over control of Data Processing resources, the inability of Data Processing organizations to satisfy systems requirements in a timely fashion. This can also be attributed to an increased level of computer literacy within user organizations.

- <u>The Marketing of Packaged Software</u> has shifted emphasis to the user community where the budget dollars reside.

- <u>Emergence of Decision Support Systems</u> and more user friendly software has provided the ability to provide quicker access of information.

- <u>Multi-Function Work Stations</u> now provide the ability to perform multiple activities such as Application development and Word Procesisng from a single device.

- <u>Flexible Manufacturing Systems (FMS)</u> have become critical in order to change the dimension of a facility to meet market demands quickly and efficiently.

There is no doubt that the information age is upon us. There is an increasing emphasis on Information Systems Technology alternatives in order to optimize the operating environment. This, coupled with the increasing confidence on the part of users that they are qualified to take on more responsibility for the management of systems projects has, created the need for the role of the "Systems Integrator".

In order to build efficient/effective application systems, it is necessary to follow a specific project plan and address each of the activities within a limited phased approach. The execution of this approach with the proper integration of all components is the job of the Systems Integrator/Systems Architect.

In order to provide clarity and consistency to our discussion, the major phases of a systems project are as follows:

- Conceptualization/Justification
- General Systems Design (GSD) - Functional External Design
- Detail Systems Design (DSD) - Technical Design
- Development - Code/Test/Debug
- Implementation
- Post Implementation Follow-up

These are compared to the various phases in which an architect would address the process of building a house:

- Owner Ideas/Concepts about house and Gross Mortgage Consideration
- Preliminary Schedules and Gross Cost Estimate
- Owner Modification and Selection of Estimates
- Blueprints/Detail/Bill of Materials/Refined Cost Estimates
- Construction/Financing
- Furnish/Occupy

Architecture is the art or science of building, which coincides with the systems development cycle from conception through implementaiton. An effort must be made to achieve harmony between the structure and the business function to be performed. For example, the structure might be a home that is being renovated for the purpose of family occupancy and will include such factors as room size, traffic patterns, and the addition of a two-car garage. The function is more related to the human elements (their needs, values, and fears) of the family who will live there. An architect tries to design the home (structure) to address the family's requirements (function). Success is measured by how well both can be integrated and to what degree the major objective is accomplished. If the family needs two bathrooms and the architect provides only one, there is a degree of misfit introduced into the environment.

Information Systems are no different. Here the structure consists of the computers, data communications facilities and data bases which constitute an investment in information technology to the corporation. Function is the business itself. Systems should be designed as structures that best meet the needs of the enterprise, and their success is measured by the degree in which a fit is achieved. For example, if the business requires immediate access to information which is stored outside the system and the architecture does not provide access to the data, then the problem is with the architecture. On the other hand, if the business needs inter-plant communications to manage its open orders and inventory, and the architecture provides a flexible path for a variety of messages to access distributed data, the system is achieving a good fit. This places us into a "Catch-22" situation between the system and the architecture.

Traditional centralized data processing organizations are providing critical function within most corporate structures. They are not going to disappear, but they must be adjusted to effectively utilize new technology and respond to rapid change. Their mission within the business environment is changing to that of a caretaker of information and provider of a utility. Today's challenge is to coordinate and effectively integrate many diversified efforts and activities into a more responsive structure.

The majority of corporations have, at best, an informal architecture. Most do not even recognize the need for consciously designing, building and maintaining an infrastructure that will efficiently support a variety of systems which are difficult to justify and comprehend. In most of the cases where systems architecture is being addressed, the effectiveness level is not good. There are a number of underlying reasons for this, but the primary one is that executive management is not focused on the value of establishing the systems architecture for the total corporate entity. The notion of a systems architecture tends to violate the old maxim that technology must serve the business and not the reverse. The reality of this is that in order for business to accomplish future objectives in a cost effective, competitive manner, a sound foundation and infrastructure must be established to leverage rapidly changing technology properly.

The ability of companies to compete is beginning to depend upon the quality and flexibility of the systems architecture and how effectively it functions. We will begin to see the distinction between infrastructure and business dissolve as the architecture emerges as the enterprise. The architecture is definitely a key factor in accomplishing true Systems

Integration within an organization, and must be recognized by management as a primary vehicle to accomplish optimum results.

With the vast variety of alternatives currently available to the user community it is imperative that a framework be established to not only control project activities but provide them much needed direction. Traditional approaches to designing and developing systems were restricted to a definitive set of variable alternatives. It has been acceptable to rely upon tradition and only introduce a limited number of new alternatives into the design process. Due to the insurge of new technology and the increased literacy of users, design problems are becoming much more complex and traditional approaches are becoming obsolete. This has a significant impact on the design process and the people who are charged to these efforts.

Today's systems designer must not only possess a sound technical background, but also obtain a solid business/functional understanding. The designer's role is changing from that of an agent to that of a creater. In the past, information systems have tended to be the product of replication and have not necessarily required a significant degree of creativity. The scope of most systems has been narrow and was specifically designed to address a finite set of functions. In most cases a designer could utilize proven technology to produce the end product. In a number of other cases, systems were developed by modifying existing systems to retrofit a specific requirement. The role of the Systems Integrator did not exist in the past. Systems concepts and overall design were bestowed upon us from anonymous sources such as the mainframe vendor, the designer's memory of previous like systems, or just simple tradition.

As computing becomes more of an integral part in performing daily operations, it takes on a much different perspective within the corporate entity. Management must make a deliberate effort to establish an overall structure which meets the unique business functions of the corporation.

In a traditional environment, a structure always tends to fit its function because of a constant, automatic and incremental process of self-adjustment. In today's environment, an added burden is placed on the designer to make the process of adjustment happen. Absence of architecture becomes obvious as systems begin to be designed, but the issues are usually not addressed in a comprehensive fashion. The key is to deliberately design systems which force focus attention on the architecture. The systems environment is becoming increasingly unique. No single design suits all companies. Instead, the design solution is intimately interleaved with the context of the problem. Each company must now discover and establish its own information systems architecture.

Design of clearly conceived systems must adapt to a given environment which requires the discovery of the underlying uniqueness between the processes and the structure. For this to be feasible in information systems, both the systems environment and the business activity must be segregated into interacting subsystems. An analysis of major business activities within most corporations can be categorized as one or a combination of the following:

- <u>Function</u> - Across organizational lines such as accounting, purchasing, or manufacturing.

- <u>Business Unit/Product</u> - Markets, customers, or product line.

- <u>Geographic Location</u> - Branch, Plant, Division, Region, Country or other physical characteristics.

Coupled with the business environment is the systems structure which equates very closely to the manufacturing capability of an organization. The various parts within the systems should be categorized as to how they fit within the Systems Architecture. A company's investment in computing technology should be viewed as a system of systems, consisting of three basic components:

- <u>Processing</u> - Sets of algorithms or programs to manipulate or add value to information.

- <u>Storage</u> - Repositories of information and the associated storage management vehicles.

- <u>Flow</u> - Movement of information between processes and/or inventory.

During the design effort, each level of analysis produces a set of components. Each have discrete requirements for integration. Integration is the key measure of success of an architecture. It is both the objective function and the dependent variable. If integration is not necessary, then no architecture is needed. Moreover, the potential contribution to be gained from an architecture is directly related to the overall importance of integration. This greatly varies from one company to another. The importance of integration will vary over time and also from one part of the organization to another. Architecture is determined by the nature and kind of integration required.

Cornelius H. Sullivan, Jr. says in his article entitled "Rethinking Computer Systems Architecture" (<u>Computerworld,</u> November 1982):

"At least six basic varieties of integration are discernable today:

- <u>Horizontal</u> - Across business functions, such as accounting, marketing and manufacturing.

- <u>Vertical</u> - Across levels of control, such as from operational levels to management control and planning.

- <u>Temporal</u> - Through time series, such as from one period or year to another.

- <u>Longitudinal</u> - From one business unit or product line to another.

- <u>Physical</u> - Among physical locations, such as branches, factories or distribution centers.

- <u>Gateway</u> - Between the enterprise and the outside world, such as other companies, customers, suppliers, the government."

Each component of the system structure makes some discrete contribution to integrate, or fit, on two specific levels:

- The <u>logical</u> <u>elements</u> of an architecture are standards. These include the guidelines, suggestions, "standard operating environment" descriptions or other lists of constants to which all parts of the architecture are supposed to adhere. This is the rule book. Examples include vendor standards for different kinds of computers, communications protocols, screen formats and data definitions.

- **The physical elements** of an architecture are its utilities. These are common, shared resources such as hardware, DBMS, and operating systems.

In summary, the framework consists on one side of the structural boundary, of a set of business activities profiled in terms of their integration needs. On the other side is a set of systems elements profiled in terms of the standards and utilities they employ. Once this framework has been established, a picture begins to take shape, consisting of four specific kinds of information:

- Varieties of Integration
- Business Activities
- System Components
- Elements of Architecture

Associations among these four parts are very complex. A given business activity, for example, such as manufacturing, may include several system components in its domain. Likewise, a system component, such as the "purchasing data base", may encompass and support numerous business activities. The relationships between various system components and elements of the architecure are also complex and numerous. The association between business components and integration elements have a high degree of complexities and volume.

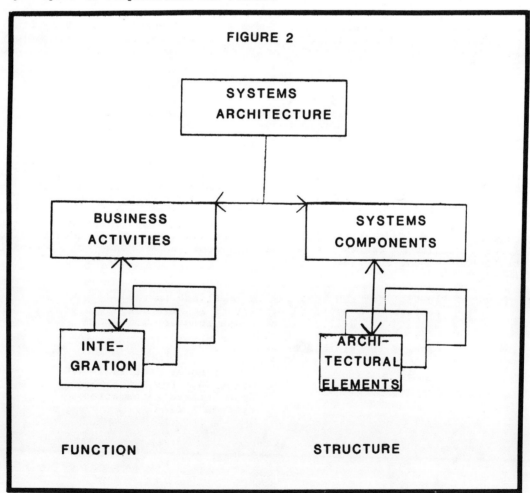

FIGURE 2

Figure 2 shows the various types of information and their interactions, as if they were a logical data base structure. At the next lower level of detail, the data base structure becomes considerably more complex: Business activities, for example, may actually consist of three or more interrelated subsets. Systems components always consist of interrelated processes and flows.

One could conclude that the systems architect and the "Systems Integrator" are one in the same. The role of the Systems Integrator has a broader perspective than that of the architect, and is evolving to that of guardian of the critical path. To some extent that is a true statement in that all the skills, experience and education that are required of the architect must also exist within the role of the "Systems Integrator". Traditionally, once the design was completed, so was the job of the architect. The documentation, blueprints, and responsibility to develop and implement was turned over to the technical staff for continuance of the effort. This is much like an architect would sub-contract the construction phase of a building. In my estimation, this is where the real job of the "Systems Integrator" begins to differ and take shape. Who is better prepared to manage/control the development and implementation phases than the original designer? They have the knowledge to make rapid, decisive decisions and the understanding necessary to assess the impacts of detail-level changes upon the total systems environment. If they have done their homework, they should have established a sound level of rapport and confidence with the users. Systems Integrators must accept the burdens of responsibility for the deliverables they produce. There is no better way to accomplish this than to tie them into the final implementation of the system. Ben Shneiderman, in "Software Psychology Human Factors in Computer and Information Systems" (1980) reaffirms this point:

> "Crudely implemented hospital intensive care systems can cause loss of life; poorly designed Air Traffic Control Systems can lead to disaster; and incorrectly entered credit or police data can ruin careers."

Based on the increasing importance of systems to the corporate entity and the degree in which these systems "gang plank" traditional organizational lines, I believe that the role of the System Integrator is emerging as a unique function within most corporate structures. The person or persons who take on this challenge do not necessarily realize the position that they are fulfilling and in most cases do not even recognize the responsibilities that they are assuming. Their thrust is to deliver quality results in order to achieve a sense of personal satisfaction of the end user.

> "In the long run what may be important is the texture of a system. By texture we mean the quality the system has to evoke in users and participants a feeling that the system increases the kinship between men."
>
> - Theodor Sterling
> "Guidelines for Humanizing Computerized Information Systems" 1974

BIO-SKETCH

Thomas R. Cornell is an Executive Vice-President of Comp-u-Staff, Inc., functioning as chief operating officer and principal consultant for a multi-divisional, multi-million dollar professional services organization. His primary areas of responsibility include branch operations support, account management, and staff management. He has been actively involved with a number of projects supporting such clients as: Westinghouse Electronics and Defense Systems Center, Maryland National Bank, Boeing Computer Services, the Morrison-Knudsen Company, and the Northwest Alaskan Pipeline Company. He is experienced in managing and staffing large scale projects on both a national and international basis. His prior experience encompasses to major corporations such as Bethlehem Steel, Xerox, GTE Sylvania, Carborundum, Marine Midland Banks, Winn Dixie Stores, and several government agencies.

Since 1966, he has actively pursued his career objectives in the field of systems engineering. He possesses a wide range of technical and management skills and is knowledgeable in the areas of manufacturing, material handling, information control systems and the organizational structures that are required to support "state of the art" facilities.

Mr. Cornell's skills and experience include project planning and control, systems development methodologies, efficient design of applications systems, effective management of technical personnel and a solid technical background. He has a broad understanding of business applications systems as they are utilized in both manufacturing and financial environments. He thoroughly understands the functionality of the various disciplines encompassed in major development and production environments.

His applications experience is further substantiated by strong technical skills and a good understanding of current software/hardware technology. This encompasses both large and small mainframes utilizing "state of the art" products in both a centralized and distributed processing mode. He has practical working knowledge of data base management systems and data communications monitors, productivity and development tools, and efficient operations techniques. He has been involved in the analysis, recommendation, procurement and installation of both software and hardware facilities. He has a proven track record of accomplishing cost effective results, and is very skilled in efficient management techniques that motivate technical personnel.

ACADEMIC BACKGROUND

St. Bonaventure University, St. Bonaventure, New York.
Bachelors in Business Administration with a minor in Accounting.

Alfred State College, Alfred, New York.
A.A.S. in Data Processing.

Reprinted from the 1984 Fall Industrial Engineering Conference Proceedings.

Criteria for the Selection of a CIMS Data Base System

Dr. John C. Windsor
North Texas
State University

and

Dr. Chadwick H. Nestman
Senior Member
North Texas State University

Introduction

Computer integrated manufacturing systems (CIMS) have found their way into the modern factory under the leadership of the Industrial Engineer. However, the development of a fully integrated system has long been hampered by the need for improvements in both the hardware and software used by the IE. Although great strides have been made in defining and developing CIMS, the problems of gathering data, and passing and updating information in an efficient manner have not yet been solved. Of the three basic building blocks of a CIMS -- computers, data bases and programmable controllers -- the data base management function addresses the problems of information gathering, storage and retrieval.

The Need For Data Bases

CIMS is built on the philosophy that the Industrial Engineer should optimize the entire business and manufacturing process rather than the individual elements within the organization. This means that the functions belonging to or supported by CIMS include: business planning and support, engineering design, manufacturing planning, resource and personnel scheduling manufacturing control, shop floor monitoring, and process automation (1). The types of computers used by such a system will vary from a host mainframe to a microcomputer, including the use of a programmable controller. Application systems may be found in the areas of CAD/CAM, process automation, and flexible manufacturing. Because of the scope of the function found within a CIMS and the types and sizes of the computers involved, no one data base system will be capable of efficiently or effectively handling all of the requirements of a truly integrated manufacturing system. Willis and Sullivan (2) state that a CIMS will usually include two major data bases: 1) Product Data Base, and 2) Manufacturing Data Base. Although this is a first step in defining the data base requirements, these two data bases do not serve all the functions identified as belonging to CIMS. Discussion relative to the type of data base design required by a fully integrated system was lacking.

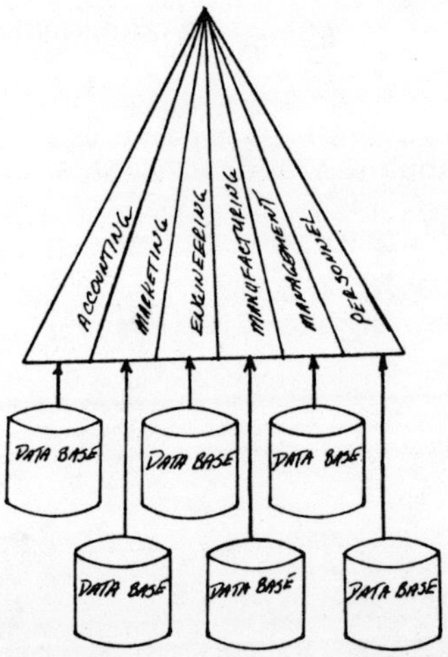

FIGURE 1. FUNCTIONAL DATA BASES

Historically data bases where developed to meet the needs of functional areas within a business. The data bases developed, as shown in Figure 1, were designed to meet the needs of the originating functional area, and the logical and physical design of the data base were determined by those needs.

For example, the functional area of Accounting usually prefers a hierarchical data base because of its 1:m relationship. This classical tree structure closely resembled the structure of accounts already used in that area. Network data bases, with their n:m relationships were generally preferred by other functional areas. This structure, for example, was originally preferred by inventory control because multiple parent child relationships made calculation of materials requirements through the use of Bills of Material much easier than with the hierarchical structure.

The relational data base is the current structural method offering hope for a "true" CIMS. Its use of the mathematics of relationships, ease of understanding, the use of implied relationships, and great flexibility make it the first data base structure capable of handling the requirements of a CIMS.

All of these data base developments were applications oriented, that is, they were implemented to meet the needs of a single functional area. A major improvement began with the introduction of a Data Base Management System (DBMS), and the development of multi-function oriented data bases. These multi-function data bases are usually referred to as Subject Data Bases. Subject data bases, as shown in figure 2, deal with organizational subjects rather than conventional applications areas. An example of a popular conventional approach is described by Willis and Sullivan (2). The Product Data Base should include inventory, order entry, and quality control as well as the engineering information relating to a product. A Manufacturing Data Base, on the other hand would include information related directly to the overall physical manufacturing process. Many manufacturing applications may then be served by the same data base. Designing a stable, nonredundant, and well-documented structure of data will, in the long, run provide for simpler and cleaner control of both data processing and product control. This is more desirable than embedding separately designed data into hundreds of processes. Attaching a form of logic control, such as integrity checks and decision tables, directly to the data structures will allow them to be shared by multiple processes. These "new" data structures will further improve the quality control of the overall processing.

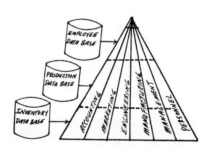

FIGURE 2 SUBJECT DATA BASES

It takes little imagination to employ top-down planning to identify subject data bases which are required by an organization. A few good examples of subject data bases found within a manufacturing organization are: a Personnel Data Base, an Accounts Data Base, and a Vendor Data Base. The purpose of discussing subject data bases is not to list all of the data bases a company might use, or to try to define all of the components of a few selected data bases. The purpose is to establish that the current practice of designing multiple subject data bases under a central Data Base Management System (DBMS) as shown in figure 3, is indeed the most efficient strategy to follow.

To serve all the functions of an integrated manufacturing system, attention must be paid to the concept of the integrated data base system (IDBS). By integration it is proposed that the designer of a CIMS would be interested in not only analyzing and defining the various elements of the system but would want a data base capable of tracking the product from inception to planning, planning to production, and production to delivery. Such a data base is itself a function of many elements:

$$D = \max(f(Ct, Tp, Dc, Ir, L)). \quad (1)$$

Where:
    D is the data base (or data bases) under consideration
    Ct = computer type,
    Tp = task to be performed,
    Dc = data characteristics,
    Ir = information required by decision makers and work stations within the system,
    L = level within the manufacturing environment.

By maximizing the components of this function an optimized data base design will emerge.

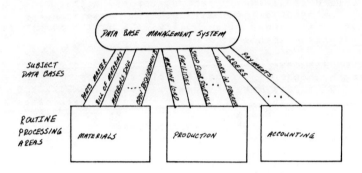

FIGURE 3   DBMS OF MULTIPLE SUBJECT DATA BASES

Specifying the Data Base Interfaces

    Figure 4 shows a multi-layered data base / data path structure. At the top of the structure is the Intelligent Work Center (IWC). This is the location of the "data user" which may be a human processor or, in the case of a fully automated IWC, a microprocessor controlled robot. Data storage capabilities must be present at this location; the amount of storage will depend on the task performed and the type of intelligent processor present.

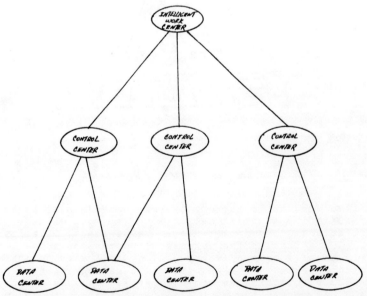

FIGURE 4   MULTI-LAYERED DATA BASE / DATA PATH STRUCTURE

Figures 5 and 6 show the data flows at this center. For purposes of mapping the data flow, the center has been divided into three components: receptor, processor, and exit gate. A receptor is "enegerized" by the arrival of material to be processed. The data flow is first to the localized data base then if and only if nothing is found for the product is the data path extended to the Control Center (CC) requesting processing instructions. A processor performs some task on the material received. The data flow at this point is the receipt of processing instructions from the local data base and the CC. Finally the third component is an exit gate that returns to the CC information about the units processed. Figure 6 shows the same IWC with an inspection center added. The data flow that is added here is a feedback to the receptor without additional interaction with the CC but the local data base is updated.

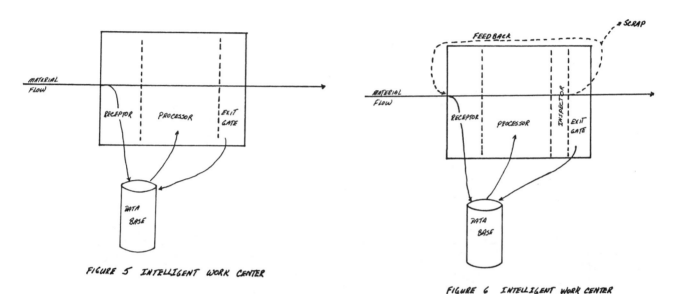

FIGURE 5 INTELLIGENT WORK CENTER

FIGURE 6 INTELLIGENT WORK CENTER WITH INSPECTOR

The Control Center (CC) is the pivotal component in the data flow structure. This component handles the flow of information from the IWC to the Data Center and from the Data Center to the IWC. Its primary functions are the processing of requests for data and the transformation to data from the Data Centers representation to a form usable by the IWC. In order to carry out these functions in an efficient manner, the CC will have its own data base and indirect access to the company level data bases as well as the data bases at the IWC. These functions have traditionally been a part of the DBMS, however, the traditional DBMS is not currently capable of meeting all of the requirements of this multi-layered structure.

The Data Center (DC) is the location of the subject/product data bases used by the manufacturing system. The traditional DBMS, which is tied to the physical data base at all levels, performs the required input/output and updating operations.

In order to interface the IWC with the CC and the CC with the DC three different views or schemas of the data will have to be used. These interfaces are based on the conceptual schema, logical schema and physical schema of data. A conceptual schema is problem-oriented, at least with respect to the data processing and engineering sense of "solution." While it is not geared to a specific application processing requirement, it does depict the natural data and relationships that are shared by many applications. Finally, it assumes nothing about the physical characteristics of the data base. The conceptual schema incorporates the requirements and constraints that must be satisfied by the data base designer in order for the "data user" to perform its task.

The logical interface is based on the logical schema of the data structures and is relation oriented. While the logical schema is not dependent on the physical design of the data base it must transform the physical structure into a conceptual structure understandable by the "data user."

The physical interface is based on the physical schema, it is device dependent and specifies the data structure on the storage media. Although this interface is far removed from the user, it is the basic building block in the multi-layered structure and the physical structure used will effect overall performance of the IWC.

Figure 7 shows the location of these interfaces in the multi-layered structure. When viewed this way it is clear that the conceptual schema is the "data user's" interface with the data base. In fact, using this approach, the "data user's" view of the data base is always relational no matter what the physical design happens to be.

FIGURE 7 INTERFACE LOCATIONS

Because the conceptual design is a relational view of the data the problems of data orientation and complex access-paths are overcome. The "data user" receives the data required in simple relations that are not confounded with complex pointers or branching structures, and can return data in the same format.

The logical interface occurs between the CC and the IWC. It must be capable of transforming the data structure received from the physical interface into a simple relational view that can be manipulated by the conceptual interface. This logical schema will need to deal with the problems of data orientation, cohesion, and coupling but can ignore the complex data access-paths that may exist in the physical data structure.

Transformation of the complex data access-paths into a single usable record is accomplished by the physical interface. This physical interface is the location of the traditional DBMS. Its major functions are the transformation of the physical data structure into a logical representation of the data needed, and the management of the primitive data items.

Passing Data Thou The Interfaces

In order to describe the transformations that take place as data moves thou the multi-layered data path from the DC to the IWC we will use a simple example beginning with a Daily Production Schedule and ending with a single IWC. The Daily Production Schedule is usually generated at the company level after processing data found in the Master Production Schedule, Materials Inventory, and other subject data bases found at the DC. The physical interface through the DBMS will build a set of fixed length records in first normal form that will contain the information needed to implement this schedule. These records will contain such information as what part to produce, how many to start, when to start, and machine routings.

These records, after being built, are passed to the CC where they are separated by part and IWC and the production information is added to the information. The production information added is such data as; materials needed, assembly steps, etc. The data coming out of the control interface will be in fixed length tuples in second normal form based on the part to be processed and the user at the IWC.

The data received by the IWC and given to the "data user" will have passed thou the conceptual interface and be in user dependant form. The data will have added to it the sequence of steps needed to assemble the part that has arrived at the IWC in the form of a variable length tuple in third normal form. For a human processor this could be a simple set of instructions for assembly. These instructions could even be adjusted to match the characteristics of the operator (i.e. a different set of instructions for a left handed operator and a right handed operator). For a fully automated IWC the processor would receive a detailed set of instructions detailing the steps required to assemble the part. This tuple would then be much longer and more complex than the one required by the human processor.

Figure 8 summarizes the data structures and transformations required at each of the interfaces. Notice that the traditional DBMS only handles the physical transformation and is unable to present the data to the IWC in the required format.

Because of the technical flexibility of controllers at each level of the manufacturing process, and because of the ability of Micro-processors to properly handle data base structures, the issues involved with conceptual versus logical versus physical are clouded even more than before. What is needed is an Integrated Data Base Management Control System (IDBMCS). This system would retain the concept of data flows as described but would allow all of the data bases throughout the plant and at each level to become shareable. However certain characteristics must be adhered to: data bases at the IWC level will always be relational and will contain data immediately useful for the particular IWC (and its environment). In the sense of distributed data bases, the CC would contain data moderately useful for several IWC's and the DC would contain data useful for many CC's. The point is that data flow integrity must be maintained and data (in its physical form) must be present at each level.

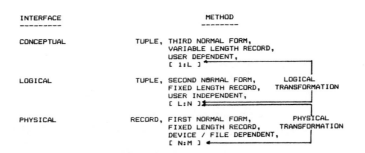

FIGURE 8 INTERFACE DATA STRUCTURES AND TRANSFORMATIONS

Data Base Selection Criteria

Figure 9 shows the structure of data bases resulting form the multi-layered data paths needed for a CIMS. The overall selection criteria for the Integrated Data Base System in this figure remains the same as the selection criteria for an individual data base. However, the characteristics required of a data base at each of the levels within the overall structure change significantly.

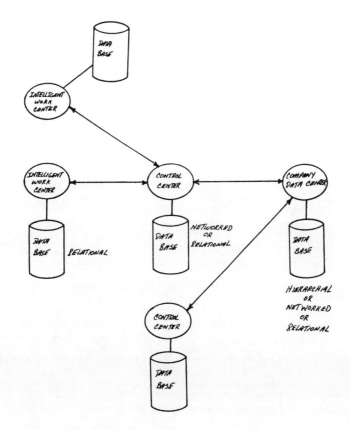

FIGURE 9 DATA BASE STRUCTURE FOR A CIMS

The selection criteria for the data base system should include the following:

1. Data Independence
2. Data Sharibiltiy
3. Non Redundancy
4. Relatability
5. Semantic Integrity
6. Security
7. Access Flexibility and Speed
8. Update Efficiency
9. Ease of Administration and Control

Of these selection criteria data shareability, relatability, semantic integrity, access flexibility and speed, and update efficiency are critical for the overall system.

The characteristics of the data bases within each level of the multi-layered structure become more restrictive as we move up the layers. At the company (DC) level the data bases need to be created independently of specific applications. The data must be designed independently of the function for which they are used. The data must be represented in shared data bases. Examples of the software available for these types of data bases are: IMS, IDMS, IDS, and ADABAS. The operating effects of this type of data base system are many. We could experience a long development period, but lower annualized maintenance costs. Eventually, such a data base system will lead to faster application development.

The data bases at the control level (CC) must be organized for searching and fast information retrieval rather that for high volume production runs. They should employ software designed around inverted files, inverted lists, or secondary key search methods. These data bases must be capable of adding new fields dynamically at any time, have good end user query facilities, and must be capable of interfacing with the DC data bases. Examples of the software available for this type of data base are: IBM STAIRS, ICL CAFS, the relational data bases of NOMAD, MAPPER, MODEL 204, SQL, QBE, and various fourth-generation languages. The operating characteristics of these data bases are that they are easy to implement, and they are more flexible and dynamically changeable than traditional data-base systems.

The data bases used at the IWC level must meet the same requirements as those found at the CC level with the added restriction that they must operate on a micro-processor. The software available at this level is very limited but is growing steadily. What does exist now are the following: MICRO MODEL 204, R4000, KNOWLEDGEMAN (using SQL), some fourth-generation languages, and most common, custom programmer/designed micro-processors. The operating characteristics of these data bases are a loss of some flexibility, limited size.

Goals for the Future

Selection of data base systems will to some degree require a certain amount of isomorphic concern; not that we are concerned with "true" 1:1 mapping but we are concerned with "near" 1:1 mapping, at least as close to 1:1 as we can get. The subject of DBMS and IDBMCS mapping is very important if the CIMS is to function the way it is intended. There are at least four aspects we may want to consider in the mapping process: structures, constraints, operations, and data bases. Of course, structures and constraints when taken together refer to the data model ↑3←. While the mapping process may be within the same data model or with different data models, we will be more interested in inter-data model mapping. However, schema restructuring or translation are complicated and intuitive in nature. The CIMS data base system must strive to define these intuitive processes and reduce the complexity in order for data model mapping to be more effective and useful.

The crux of the data base mapping occurs at the operational aspect. It is the operation of the data base we want to preserve and mappings at this point are easily understood if we are using the same data model, but complexity increases as we introduce additional models. For the CIMS, we are obliged to use multiple models and therefore will face a difficult task in mapping operations.

The last aspect is the data base itself. Data base mappings are important not only in CIMS but in all aspects of the company. Anytime the data base is converted to some other data base we must go through a mapping algorithm. If the new data base is completely isomorphic to the old data base then the process of mapping is trivial. However, if the new data base is a different type of data base with a different structure or if the new data base has different operation the process is very complicated. Because of the nature of the CIMS, mappings between data bases will be complicated sense we are transforming one structure into another.

The future of CIMS rests in the selection of appropriate data bases. Attention must be paid to their relative ease in being transformed into differing structures with differing operations. Selection criteria, while useful under normal conditions, needs to be tempered with the need for isomorphic relationships at the different levels of the manufacturing process: IWC, CC, and DC.

SELECTED REFERENCES

(1) Teicholz, Eric, "Computer Integrated Manufacturing." **Datamation**, March 1984, Vol. 30, Number 3, page 169.

(2) Willis, Roger G., and Kevin H. Sullivan, "CIMS In Perspective: Costs, Benefits, Timing, Payback Periods are Outlined," **Industrial Engineering**, Feb. 1984, Vol. 16, Number 2, page 28.

(3) Tsichritzis, Dionysibs C. and Frederick H. Lochovsky, **Data Models**, Englewood-Cliffs, NJ: Prentice Hall, 1982.

(4) Britton, Harley O. and William E. Hammer, Jr., "Designers, Users And Managers Share Responsibility For Ensuring CIM System Integrity," **Industrial Engineering**, May 1984, Vol.16, Number 5, page 36.

(5) Martin, James, **Strategic Data-Planning Methodologies**, Englewood-Cliffs, NJ: Prentice Hall, 1982.

(6) Curtice, Robert M. and Paul E. Jones, **Logical Data Base Design**, New York, NY: Van Nostrand Reinhold, 1982.

(7) Teorey, Toby J, and James P. Fry, **Design Of Database Structures**, Englewood-Cliffs, NJ: Prentice Hall, 1982.

Biographical Sketch

Dr. John C. Windsor is an Assistant Professor of Information Systems at North Texas State University, College of Business. He received his Ph.D. in Decision Sciences from Georgia State University. Dr. Windsor is co-owner of Information, Decision and Support Systems, a consulting firm concentrating on helping small business. His teaching specialities are in the following areas: computer modeling, design of integrated information systems, microcomputer systems, and general systems analysis and design. His reasearch activities have been in the area of intelligent-based information systems: decision support systems, expert systems. He has several articles in the area and is currently writing a book on the subject. Dr. Windsor has published in **IIE Transactions**, as well as several journals in the field of computers and infornmation systems. He is a member of the Association for Systems Management, the American Institute for Decision Sciences, the Institute of Industrial Engineering, and The Institute of Management Science.

Dr. Chadwick H. Nestman is an Associate Professor of Information Systems, College of Business, Norht Texas State University, Denton, Texas. Dr. Nestman is co-owner of Information Decision and Support Systems a consulting group specializing in the use of small business computers to enhance managerial decision making. His teaching specialities are in the following areas: analysis of data communication systems, analysis to data base systems, design of integrated information systems, and general systems analysis and design. He has been employed by companies such as NCR Data Communications, Harris Data Communications, and LTV Electro-systems. He has served as the chairman of the Richmond Joint Engineering Council and has been active in not only the Computer and Information Systems Division of the IIE but in the Association for Systems Management, ACM, and the American Institute for Decision Sciences. Dr. Nestman has published in **IIE Transactions** as well as several journals in the field of computers and information systems. He is a senior member of the IIE.

Designers, Users And Managers Share Responsibility For Ensuring CIM System Integrity

**By Harley O. Britton
and William E. Hammer, Jr.**
The Duriron Co. Inc.

Systems integrity has been defined as the extent to which a system functions as it was designed to. A term often used to describe systems integrity is credibility. Integrity, or credibility, either is or is not an attribute of a computer-integrated manufacturing system (CIMS); there is no middle ground where user confidence is concerned.

If the users believe in the system they will make an honest effort to maintain the data base and will use the outputs of the system to make decisions.

If, on the other hand, they do not trust the system, they will not be motivated to properly maintain the data base and will not use its outputs in decision making.

A system that lacks credibility is characterized by the following:
□ Users become frustrated and may openly express their distrust of the generated outputs.
□ Users ignore some or all outputs and base decisions upon information from sources external to the system.
□ Parts of the system, if not all of it, are discarded or replaced with "natural systems" created by front-line users to accomplish their jobs.
□ Strained relations between information systems and user personnel make it difficult to resolve current integrity issues and reduce support for future information systems.

To avoid problems with systems integrity, designers must consider a number of key areas. The degree to which these areas are addressed in the development of a CIMS has a direct bearing on the success or failure of the information system.

User involvement

User involvement in the design, maintenance and operation of any information system directly influences the level of systems integrity which will be realized. Because manufacturing information systems are so interrelated with other functions of the organization, the impact of user involvement is great.

A team approach that involves individuals from each functional area, as well as systems and programming specialists, provides for a high degree of systems integrity. The user representatives provide "know-how" in terms of theory and operating practices. System designers act as catalysts in helping users to define objectives, logic, algorithms and procedures. Later in the development process the designer applies the available technologies and/or methodologies to the system.

Conceptual design

The conceptual foundation of a manufacturing information system must be sound for the system to have integrity. For example, a capacity planning system that does not address all of the major capacity constraints will not be credible.

From a purely functional standpoint, the responsibility for integrity falls upon the users involved in specifying the system. However, the entire design team participates during the conceptualization stage by contributing, studying and testing concepts.

Although the conceptual design stage involves mainly a macro-level definition of the proposed system, detail considerations cannot be entirely ignored. The information systems specialists on the team must provide guidance on the feasibility of the concepts developed. Little is accomplished if a system is designed that cannot be implemented.

The appearance of "canned" manufacturing information systems on the market has encouraged many organizations to ignore the conceptual design of a system and accept a software vendor's solution on the assumption that the vendor has completed a thorough conceptual design that is compatible with the actual manufacturing organization.

In general, systems should be designed to fit the organization, but software packages may require a force fit. Compatibility and conformity problems with software packages may be reduced by first developing the conceptual design and then selecting a package that meets the resultant criteria.

The approach used during conceptual design is not as important as the recognition that this step is crucial and must precede any detailed design or package selection. Ultimately, systems integrity depends upon the validity of the conceptual design and its compatibility with the manufacturing organization.

Design considerations

Numerous design considerations influence the integrity of manufacturing information systems. Some of the more important of these are listed below:

☐ *Data sharing:* The data that comprise the information system must be integrated with the data supporting all other associated functions in the company. Traditional department ownership of files must be overcome and the entire organization's data viewed as a shared resource.

☐ *Flexibility:* The system must be designed to be flexible in both the short and long term. A popular technique used in the development of information systems is "prototyping." Prototyping is a repetitive process that allows the design team to program, test and refine an application. By helping identify design flaws early, it enables necessary changes to be made prior to implementation, before costs become prohibitive.

☐ *Maintainability:* Manufacturing systems are dynamic creations that must change as new concepts and techniques become available. For this reason, the system must be designed and programmed with maintainability in mind.

Numerous tools are available to enhance maintainability. These include data base management, data dictionaries, data base directories, fourth generation languages, application generators, modular/structured programming and documentation generators.

☐ *Documentation:* Systems documentation is essential to the ongoing support, maintenance and enhancement of any information system. Without adequate documentation, the integrity of the information system will be marginal and subject to serious degradation during periods of personnel turnover.

Required documentation includes basic program narratives, input/output layouts, restart/recovery considerations, security considerations and current listings of the actual program modules. A narrative describing the overall system and interrelationships of individual jobs and programs is desirable.

A macro-level flowchart depicting the entire system is helpful, but may not be possible in a large and intricate system. Elements of the systems documentation should be combined with user procedures to produce user manuals.

☐ *Input-Output:* The media, format, means and content of reporting provided by a manufacturing information system can have a significant impact on systems integrity. Data collection procedures and man/machine communications must be of ergonomically sound design.

The system should support interactive and conversational communication requiring minimal data entry. For example, bar code and magnetic media might be used to advantage to reduce data entry requirements while improving accuracy.

The system should provide for logical conversation that accomplishes the desired transaction in the language and terminology of the user. User friendliness and reasonable response times are essential.

Where batch applications are concerned, care must be taken to schedule the required computer runs in line with the users' work schedules. These schedules must be considered during the design stage to ensure the availability of required computing resources. Batch reporting should be limited to exceptional data conditions that require action to avoid smothering the user with nonessential information.

☐ *Timeliness:* To the extent practical, on-line entry of transactions should be utilized. It is important that on-line inquiries and batch retrievals provide users with the most up-to-date manufacturing information available.

Data base approach

Perhaps the most important factor affecting the development and evolution of computer-based manufacturing information systems during the last decade has been the advent of data base management systems (DBMS). These software packages, coupled with the availability of low-cost computer hardware, have placed automated manufacturing information systems within reach of most organizations.

This is not to imply that use of a DBMS is the answer to all problems associated with the creation of manufacturing information systems. An effective DBMS is, however, an invaluable tool that provides numerous design and productivity benefits. The general acceptance of this tool has created a new design methodology usually referred to as the data base approach.

The advantages of the data base approach to systems design are:

☐ A DBMS simplifies and standardizes the access to an organization's data. Because all file access logic resides in the data base description module rather than in individual application programs, programmers and designers need not be familiar with a variety of access methods. In addition, the DBMS provides standard data structures which ensure that the data elements have identical attributes and names in all application programs.

☐ The concentration of file access logic in a DBMS offers significant advantages with regard to migrating to new computer hardware or operating systems. Most changes in the computer configuration or environment can be accomplished without impacting existing application programs.

☐ The data independence provided by a DBMS means that programs using it can access only the data items they require, as opposed to reading the entire record. Although the DBMS retrieves the entire physical record from disk storage, only the requested items are returned to the calling program. Physical characteristics of the record, such as length and blocking factors, can change without impacting existing application programs.

☐ One objective of a DBMS is to integrate the corporate data base; that is, to provide a method that will enable access to all associated records. A direct benefit of this integration is the accuracy that is forced by the DBMS in setting up the record relationships. Data items that are key elements in the data base are checked to ensure the integrity of record relationships.

☐ Ideally, the use of a DBMS should reduce data redundancy. In a data base environment, the data are placed in a pool and structured for optimum use. Each unique data element appears only once in the pool, and the problem of non-correlation of reports is eliminated.

☐ Finally, use of a DBMS allows systems to be built in an evolutionary manner. In the data base environment, it is not necessary to predetermine data and structuring requirements for future related systems. The designer can concentrate on the system at hand and be confident that current design decisions will not create serious problems for the systems of tomorrow.

Figure 1 illustrates the data base environment, in which all departments share a non-redundant pool of corporate data. Also depicted is the isolation of the individual systems from the actual computer hardware, with all access to the data resource being processed through the DBMS and on-line software.

The level of integrity that the data base environment ensures is structural only. The DBMS does not guarantee the accuracy or quality of the individual data elements within the various records and files. The old term GIGO, or "garbage in/garbage out," is as applicable in modern data base systems as it is in the traditional systems of 20 years ago.

The responsibility for data input should be moved as close as possible to the source. Data base maintenance should be accomplished on interactive display stations by the personnel who know and understand the data being put in.

The people assigned this function should be aware of the importance of accurate data and of their responsibility for the integrity of the files, records or data elements they maintain. Without a commitment on the users' part, no amount of editing and control will produce the level of integrity required.

Stringent editing procedures for all data that enter the system should be implemented. Occasionally, design teams argue for reduced editing to avoid undue data discipline. There are few cases in which such an argument should prevail. If the data are unimportant enough that any value can be entered, it is questionable whether the data element should exist for reporting purposes.

Data editing practices should include the following tests, where applicable:

☐ *Range checks:* Range is generally tested by setting minimum and maximum values for a particular data element. This type of logic can be used on alpha as well as numeric data. In addition, range tests can be

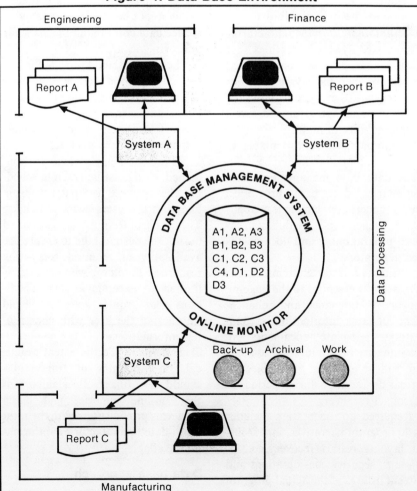

Figure 1: Data Base Environment

made more dynamic by relating the data directly to other data already in the data base.

☐ *Table look-ups:* Many data elements must contain one of a series of known values; e.g., stock room identity, unit of measure, etc. Either internal or external tables can be used to validate the data elements entered. Where possible, external tables should be used to allow edit changes without requiring program modifications.

☐ *Data base look-ups:* Closely related to the table look-up is the data base look-up. Interrogation of the data base for editing purposes can take different forms. A simple edit might consist of reading a part record to ensure that an entered part identity did or did not exist. A more complicated edit could require the retrieval of multiple data base files.

☐ *Required data:* In many data base maintenance applications, especially when record additions are involved, there are certain required data elements. If these data are omitted, the user is notified of the omission, and the data element (or possibly the entire record) is not processed.

☐ *Alpha/numeric tests:* In some cases, the only test that can be applied to a field is an alpha, numeric or combination alphanumeric comparison. This type of edit can be accomplished by either software or terminal hardware.

☐ *Conditionals:* If a data element has a limited and relatively static number of valid options, it may not be desirable to create a table look-up routine. In these cases, simple conditional logic can be written directly into the software.

☐ *Group relationships:* This type of edit is used when the validity of a data element depends upon the content of one or more additional data elements. The source of the data elements making up the group may vary from currently entered values to values already in the data base. The objective of the group edit is to ensure that each data element is valid as it relates to other elements in the group.

Controls

The editing procedures just reviewed represent the most common routines found in data processing systems today. Most manufacturing systems contain them, either alone or in combination. Their primary purpose is to establish a stringent screening process for data entering the system.

Although this process promotes reasonability of entered data, it does not guarantee the required level of accuracy. For the final measure of data integrity to be realized, a comprehensive set of controls must be designed to prevent the following:

☐ *User errors:* Regardless of the extent to which data are edited by the DBMS and application programs, users will make mistakes and enter incorrect data. The error potential is likely to be greater in an on-line system than in a batch system due to the number of users. It does not follow, however, that on-line systems are less accurate. Because data entry is close to the user, resolution of errors and timeliness of input may be better.

☐ *Computer crime:* In manufacturing information systems, at least two areas of security must be addressed. First, since they are generally integral with order entry and shipping/invoicing, they are vulnerable to fraudulent handling of products and money.

Second, since companies become dependent on computer-based systems for information on products, quantities and dates for production, they are open to both internal and external sabotage. Appropriate controls and contingency plans must be designed to limit the company's exposure to security breaches.

☐ *Hardware/software data errors:*

> **"The critical components of the system must be backed up to guarantee that an outage will not severely hamper the operation. Modern manufacturing systems are so computer-dependent that even short outages are noticed, and long outages can be disastrous."**

Sufficient controls must be implemented to ensure that intermittent hardware or software "bugs" do not result in loss of or damage to the data base. As manufacturing information systems become more on-line, the complexity of programs increases and the probability of subtle bugs within program logic increases. These bugs can result in records being damaged or even lost.

Damage caused by hardware is not very prevalent in today's environment; however, data can be omitted or entered twice as a result of hardware problems.

☐ *Computer operational errors:* In most data processing systems, there is considerable room for computer operator error. A variety of controls must be implemented to ensure that the proper procedures are followed and any deviation is detected and corrected before the error has a serious impact. These controls should detect the problem at the data processing level to avoid credibility issues associated with user exposure.

Batch controls

Batch controls are one of the most effective methods of ensuring data accuracy. These controls have been present since the earliest data processing efforts, and although they tend to be more prevalent in batch systems, they can be used quite effectively in modern on-line systems.

Batch controls are implemented by passing the source document through a control point at which a total is computed on an appropriate numeric field. Where a logical field is not present for balancing, a hash total might be generated by adding account numbers, vendor numbers or any other numeric values.

The batch totals established for a group of source documents accompany the data through the entire input process. Totals can be run on the batch of data at any point to ensure accuracy.

Batch controls in an on-line environment are slightly different. The computed batch controls are usually entered into the on-line program, and after all the data are entered, the program tallies the transactions and compares the actuals with the controls. If they match, the batch is accepted; otherwise, the user is prompted through an error location and correction procedure.

Although batch controls tend to be more common in financial applications, they can be used effectively in manufacturing areas such as inventory control, defective material management and work-in-process.

Single transaction controls

Single or individual transaction controls are more commonly used than batch controls. This is due to the real-time interactive nature of state-of-the-art systems. Single transaction controls are designed primarily to ensure that the user submits accurate or reasonable input. These controls employ many of the editing techniques discussed earlier.

In most on-line applications, the editing process is supplemented by passive operator verification. The passive edit usually involves the expansion of coded data fields to include additional descriptive data.

For instance, the user might enter a raw material code in the process of changing material cost. The application should retrieve the raw material record and display the data elements that might be changed, including the raw material description. This gives the user an opportunity to verify that the correct raw material code was specified. In general, cryptic or coded fields should be accompanied by descriptive data where possible.

Another type of passive edit, referred to as a "soft edit," identifies a potential transaction error via a warning message or question. The operator can then make a correction, if there is an error, or confirm the transaction as correct.

> **"Occasionally, design teams argue for reduced editing. There are few cases in which such an argument should prevail. If the data are unimportant enough that any value can be entered, it is questionable whether the data element should exist for reporting purposes."**

Hardware failures

Modern computer systems and associated peripheral devices are generally reliable. However, machines do fail, and this must be considered in the design of manufacturing systems if integrity is to be maintained.

The key to hardware availability is *redundancy*. The critical components of the system must be backed up to guarantee that an outage will not severely hamper manufacturing operations. In modern manufacturing systems, the operation is so computer-dependent that even short outages are noticed, and long outages can be disastrous.

The following ideas on hardware redundancy should be considered:
☐ If a location has a significant number of terminal devices, manage them using two controllers rather than one, with half the devices handled by each controller. If one controller fails, only half the terminals will be affected.
☐ Redundancy in communication lines or the use of dial back-up lines should be considered.
☐ A communication line requires a modem at each end to support data transmission between the computer and terminals. A reasonable number of "hot spares" should be maintained to cover possible modem failures.
☐ Redundancy in the computer system should also be considered. This is an option sometimes ignored because of preconceived and often incorrect assumptions concerning cost.

Attention to system availability issues such as redundancy, disaster back-up and contingency planning reduces, but seldom eliminates, outages. To handle the inevitable system failure, alternate processing procedures are needed to enable work to continue, even if in a degraded mode.

Common approaches include recording transactions manually for later on-line entry or use of a terminal that allows off-line entry and storage and subsequent transmission of the transactions.

Regardless of the method chosen, be certain that sufficient controls are implemented to guarantee that errors are not introduced during system outages.

Software failure

Computer programs, whether on-line or batch, can be terminated abnormally by a variety of errors. Each program in the system must be designed so that termination at any point will not result in lost transactional data or incorrect updating of the data base. To accomplish this, the application module must create a checkpoint at the end of each logical transaction that can serve as a recovery point.

The recovery problem must be addressed system-wide. Each terminal in a network must be capable of transaction level recovery; that is, recovery up to and including the last completed transaction. Systems that require reentry of numerous transactions following an outage tend to lose user support, and therefore integrity.

File diagnostics and scanning

Most data processing organizations have a data base administrator (DBA). One function the DBA performs is monitoring of the various files that comprise the organization's data base. This is done using software routines that scan the files, looking for invalid data conditions or record relationships. Monitoring enables the DBA to locate problems within the data base and implement solutions before systems integrity is seriously affected.

In addition to the ongoing checks performed by the DBA, the application software can be designed to accomplish at least cursory integrity checks on the data base. In the data base environment, records can be retrieved via different paths or linkages. These linkages constitute physical relationships between records.

If the linkages are damaged, it may be impossible to retrieve the record via the damaged path; however, the record may be accessible using another path. This means that two programs could get different results if they utilized different linkages for their retrievals.

Although this characteristic would appear to be a problem, it can be used to advantage by the system designer. By utilizing different link paths across applications and correlating program outputs at appropriate control points, the application software actually audits the data base file.

Computer operations

Control in computer operations begins with well trained and motivated personnel. These people must have a solid understanding of the operational aspects of the systems they run and be capable of dealing with hardware and software failures. This is particularly important when conditions require execution of recovery and/or restart procedures.

Because operators are human and can be expected to make errors, automatic rather than manual operating procedures should be employed where possible. For procedures which must be manual, clear and concise operations documentation should be supplied. The system designer must strive to create an operational environment that affords limited opportunity for human error.

Managing for integrity

Ultimately, the integrity of a manufacturing information system depends upon management. During the design stage the emphasis placed on systems integrity is usually a direct reflection of how important this aspect is in management's view.

Management's support of the additional time and effort required to "design in" integrity cannot be assumed. Often, key managers are uncomfortable with the extra time the design team spends reviewing and testing conceptual designs. Managers, as well as the design team, must be persuaded that the additional time spent on design will be offset by time savings in the programming and implementation stages.

Beyond design and implementation, continued management interest in systems integrity is pivotal. Typically, the integrity of a manufacturing information system will degrade over time.

To offset this tendency, management should require feedback and control mechanisms to monitor the overall integrity of the system. This can be accomplished by measuring specific outputs, correlating actions recommended by the system with actual decisions made and determining user confidence. When a significant drop in systems integrity is observed, corrective action can be taken.

Systems integrity is a fundamental feature of a successful manufacturing information system. Integrity or credibility does not occur automatically as a by-product of the design and implementation process. Management and the design team must recognize the importance of systems integrity and plan for its development and control.

Integrity is indeed critical to the success of a CIMS.

For further reading:

Booth, Grayce M., *The Design of Complex Information Systems,* McGraw-Hill Inc., 1983.

Cardenas, Alfonso F., *Data Base Management Systems,* Allyn & Bacon Inc., 1979.

Haleui, Gideon, *The Role of Computers In Manufacturing Processes,* John Wiley & Sons Inc., 1980.

Martin, James, *Security, Accuracy and Privacy In Computer Systems,* Prentice-Hall Inc., 1973.

Orlicky, Joseph, *The Successful Computer System,* McGraw-Hill Inc., 1969.

Tonies, Charles C., *Software Engineering,* Prentice-Hall Inc., 1979.

Wolf, Arthur E., *Computerized Plant Information Systems,* Prentice-Hall, 1974.

This article was developed through the cooperation of CIMS series editor Randall P. Sadowski and CIMS committee member Edward L. Fisher.

Harley O. Britton is the manager of programming at the Duriron Co. in Dayton, OH. He has expertise in the application of data base management technology. Britton holds an associate degree in data processing from Sinclair Community College and has completed additional studies at Miami University in Oxford, OH. He is a member of DPMA.

William E. Hammer, Jr., is the director of information systems at the Duriron Co. He received his BIE from the University of Dayton and an MSIE from the Ohio State University. A senior member of IIE, Hammer is the immediate past director of the computer and information systems division and past president of the Dayton chapter. He is a member of the Engineers Club of Dayton and is included in Who's Who in Engineering.

Building the Bridge Between CAD/CAM/MIS

C.W. Devaney
Price Waterhouse
Houston, Texas

ABSTRACT

This paper begins with a description of the challenges to manufacturing management resulting from shorter product life cycles. The significance of quick-response, flexible manufacturing capabilities, along with a planning and control system that exhibits similar characteristics, is discussed.

Current efforts to build this capability appear to be fragmented in CAD/CAM and MIS efforts. The CIMS acronym is an expression of the ultimate goal and not necessarily an approach to achieving such a goal.

This paper examines the CIMS integration task from three points of view that seem critical and basic; namely, computer language and communications, data base boundaries and organization relationships. The paper does not, however, discuss group technology coding and classification.

INTRODUCTION

Manufacturing management today is challenged by competitive pressures from abroad, cost pressures from within and shorter product life cycles from changing markets.

These challenges are being addressed, in part, by the computer. The computer is changing the way we design, manufacture and distribute products. In many companies these efforts to utilize computers have been underway for twenty years. In most situations these efforts have not been coordinated between engineering, manufacturing and accounting as well as they might have been.

More recently, the concepts of computer integrated manufacturing have emerged. CIM has stimulated interest in linkage between existing CAD/CAM/MIS installations and interest in planning integrated systems. Integration promises benefits far in excess of isolated applications of computers in either design, manufacturing, or management planning and control.

THE PROMISE OF CIM

The manufacturing person may look at Computer Integrated Manufacturing from two perspectives:

- computers driving, supporting and controlling design activities, processes, material handling and measurement functions.

- computers acquiring, storing and reporting information to plan and control these functions; in other words, as information handlers.

Since CIM is a relatively new term, and computers have been supporting manufacturing with planning and control information for many years, what is the uniqueness of CIM? It is probably in the word "integration." CIM promises that computers will integrate process support and information handling, efficiently and effectively, for order-of-magnitude improvements in product quality and productivity.

CIM today defines a goal or objective more than an approach. Since the word integration is the most vague term in the acronym, this paper attempts to clarify the integration issue and offers an approach to the information integration challenge.

Webster's New Collegiate Dictionary defines the verb "integrate" as follows: to form or blend into a whole, to unite with something else, to incorporate into a larger unit. I think we have the right word.

INTEGRATION IN CIM

There are at least the following possibilities:

- integrating organization elements.

- integrating engineering functions with manufacturing functions.

- integrating manufacturing processes with one another.

- integrating automated material handling with manufacturing processes.

- integrating computer control of a process by linkage with design specifications.

- integrating information or information systems (horizontally or vertically).

Consideration of the possibilities described above means a matrix of significant complexity in the manufacturing situation where the product is engineered and component part manufacturing is involved. The kinds of integration will include, at least:

- organizational or functional integration (where responsibilities or duties change or combine)

- process or methods integration (where machine or tool sequences change or combine)

- data or information integration (where data elements or representations change or combine).

This paper, as mentioned previously, examines some of the information or data integration issues and possible approaches to practical solutions. This does not suggest that the task of the data integrator is any more or less important in CIM development than the task of the facilities integrator or the process integrator. I am just trying to develop an approach to the information dimension.

THE CAD/CAM/MIS ISLANDS

The "islands of automation" term has been used to describe isolated or non-integrated computer applications that co-exist in a manufacturing environment. Integration suggests bridges that connect these islands. You cannot build a bridge unless you know what's on the other side. A basic impediment to integration has been the lack of understanding and cooperation between the individual sponsors and architects of the CAD/CAM and MIS installations.

Perhaps the most common reason for this is the lack of incentive to cooperate. The payoff in benefits has been high enough in individual CAD/CAM/MIS projects that sub-optimization is perfectly acceptable. Classic organization relationships aggrevate the problem, CAD in the engineering organization, CAM in manufacturing and MIS in the finance and accounting shop.

Steering committees help, but usually the only result is communications are improved. You already know the one about the system designed by a committee.

INFORMATION DIMENSIONS

Input-Process-Output-Mechanisms-Controls

CIM data travels horizontally and vertically. CAD data may travel across a bridge into CAM and across another bridge into MIS. Part information travels this way; but, in addition, we need to consider the hierarchy that may exist. Typically, information systems in manufacturing can be categorized into three layers or levels:

- a top management information level.

- a functional planning and control layer.

- a process driver layer or level.

Both within and between these levels are threads of information consisting of data describing the product (technical data) and data describing a process or the results of operations or an operations plan or schedule (management information).

Simplistically, we can assume two (2) major threads (technical and management) that must flow horizontally (across organizational boundaries and functional areas) as well as vertically (some data elements and/or their aggregation flow from the shop floor up to the board room).

Bridging islands of automation in both dimensions is a big challenge if the bridge traffic (the data) is uniform; but, of course, it isn't.

The integrator must address the following kinds of problems or issues:

- data representation
  Textual data may be ASCII or EBCDIC

  Floating point numbers may be 8 bit or 7 bit exponent with a 23 bit or 24 bit mantissa and either base 2 or base 16.

- programming language
  CAD software may be programmed in PL/1, FORTRAN IV, FORTRAN 77, Pascal, Assembler or C.

  CAM software may be programmed in any of the above plus COBOL, RPG II, RPG III or others.

MIS software is usually programmed in COBOL but one may find several of the other languages mentioned above.

- operating systems and file access methods vary from vendor to vendor.

- communications protocols vary from vendor to vendor.

In addition to these dimensional issues caused by the use of computers, there is a more fundamental problem in bridging or integrating:

- do the existing engineering documentation standards, process specifications and cost accounting standards or conventions adequately and accurately describe our data? If so, are these existing standards or conventions uniformly adopted and used throughout the organization? Fortunately we have tools to cope with both kinds of data dimensional issues.

REDUCING THE COMPLEXITY OF THE BRIDGE

The basic tools available that make the integration task feasible are:

- the data dictionary
- the data base management system

The dictionary is an information set that describes or models the enterprise. It includes entities, attributes, relationships, sources, destinations and the rules governing the use of this data.

A discussion of the features of either software capability is beyond the scope of this paper. It is difficult to visualize a CAD/CAM/MIS integration effort that was not initiated in a dictionary/data base environment. The dictionary makes it possible to standardize and control at the data element level and the dbms provides control over the organization, size and scope of the islands of automation, plus isolating the data management task (as defined by the dictionary) from the application programs.

Unfortunately, state-of-the art dbms/dictionaries do not completely solve the data representation problem. What we really need is software that has universal data representation capabilities, where the user can define the requirements for using each data element.

BUILDING THE BRIDGE

The bridge building task can be generalized into an approach, if one accepts the notion that there is probably no single situation where the approach will be optimal. We can speculate that the worst case would be a free-standing, mixed-computer vendor environment, with top managment not recognizing the need for a top-down, comprehensive requirements study or design effort. Management climate and behavioral issues are important, but outside the scope of this paper.

The integrator facing a challenge such as this might pursue the following course of action, once the need is recognized and a top-down approach is adopted:

PHASE I

a. Evaluate the engineering standards for documentation product description and configuration control. Are they comprehensive, sensible, understood, in use? Can they drive a major thread of information? Is data described as well as the products?

b. Establish key data element standardization throughout the company. This will probably require redefinition of data elements in the MIS system and in manufacturing.

c. Implement a data dictionary and dictionary support of all data processing applications. Construct the enterprise data model.

PHASE II

a. Establish consolidated data base boundaries and an orderly plan for conversion to a data base environment.

b. Document and optimize all data entry points to add/change or delete information.

c. Initiate a data administration function and an orderly conversion to the data base environment.

PHASE III

Establish minimal system entry points for add/change/delete updates. A single entry point with automatic, concurrent updating in all data bases would be the ideal. Some data base bridges will probably utilize data communications, either initially or later on.

At present, there is not a local area network available to support most multi-vendor environments. Many vendors do, however, support the ISO Open Systems Interconnect Model, which currently is adhered to by the IEEE 802.3 Local Area Network Standard and the X.25 Packet Switching Network.

The worse case communications bridge will require communications protocol conversions, which usually are accomplished by one of three methods:

- mainframe resident software

- a hardware converter box

- add-on boards (in a workstation)

Each method has its place. Software provides the greatest flexibility and usually works out to be the least cost per terminal or workstation (where a number of workstations are involved).

MANAGING THIS APPROACH

As mentioned earlier, islands of automation and the bridge problem exist for rational reasons:

- the availability of short-term improvements and attractive payback periods

- the organization problem and different priorities for computer support in engineering, manufacturing and accounting

- the focus on annual budgets and short-term planning

If the promise of CIM is to be realized, top-down direction and planning is required over individual automation projects in engineering, manufacturing and management information. Some specific individual must be made responsible for making it happen. The approach described here for integrating the information thread must be planned and executed as part of an overall CIM plan including processes, process control, automated material handling, facilities, etc.

The benefits of building the information bridges are significant. We have much work to do in assessing the impact of information integration on the knowledge worker. I believe this is the big payoff area. I suspect that about 50% of the knowledge worker's time, in a manufacturing environment, is devoted to interpreting or defining information received or created in the course of normal efforts.

If CIM today, as a concept, is where MIS was twenty years ago - and it is, in my opinion - there are lessons to be learned from the MIS experience. Those companies that addressed MIS strategically, with top management, have systems that contribute more to the business than those who left lower or middle managers alone to automate those tasks that seemed attractive to them.

Bridging CAD/CAM/MIS and supporting CIM is a complex undertaking, but with the better tools available today and our experience of the last twenty years, the job is certainly doable. Let's do it and lets keep looking for a better data dictionary to make the job easier. Finally, don't overlook the importance of the engineering standards and documentation function. It is absolutely essential that a comprehensive standards program and configuration control system be in place and understood. This is the engine that will drive the product data train and keep it on the track from design all the way to cost accounting.

For further reading:

Hales, H. Lee, "How Small Firms Can Approach, Benefit From Computer-Integrated Manufacturing Systems"
Industrial Engineering, June 1984

LeClair, Steven R. and Hill, Thomas W. Jr., "Functional Planning Approach Maps Out Connections Between People and Systems"
Industrial Engineering, April 1984

Tompkins, James A., "Successful Facilities Planner Must Fulfill Role of Integrator in the Automated Environment"
Industrial Engineering, May 1984

Katz, Tony, "Protocol Conversion"
Computerworld on Communications, May 2, 1984, Vol. 18, No. 18A

Proceedings, Seventh Annual ICAM Industry Days 5-9 June 1983
Aeronautical Systems Division
Air Force Wright Aeronautical Laboratories
Material Laboratory
Manufacturing Technology Division

Messina, Andrew, "Automated Factory, Automated Office"
Computerworld Office Automation, June 13, 1984, Vol. 18, No. 24A

"Manufacturing Technology: A Report to Management"
DATAMATION, February 1984, Vol. 30, No. 2

C. William Devaney is a principal in Price Waterhouse Management Consulting Services. His current assignment is partner-in-charge of a manufacturing services group specializing in computer integrated manufacturing. He is a Certified Data Processor, Certified Management Consultant and Registered Professional Engineer. Mr. Devaney received BS and MBA degrees from Temple University and is a senior member of IIE.

Reprinted from the 1982 Fall Industrial Engineering Conference Proceedings.

Distributed Computing in the Manufacturing Environment

David G. Hewitt
Vice-President
United Research Co.
Morristown, N.J.

ABSTRACT

Expanding utilization of computers has required the development of a modular architecture for manufacturing systems. Westinghouse defense's Electronics Assembly Plant will first utilize this architecture in 1983.

The Electronic Assembly Plant (EAP) Project will establish modernized state-of-the-art manufacturing processes for the manufacture of printed wiring assemblies of analog, digital, and flat pack types that are required for systems produced by the Westinghouse Defense and Electronics Center. These modernized processes, equipment and systems are to be implemented in a new 180,000 square foot facility to be located in College Station, Texas. The initial EAP technology will incorporate computer-integrated-manufacturing advancements that will achieve substantial savings over PWA manufacturing technology currently in use at the Westinghouse Defense and Electronics Center.

From the initial concept, the EAP Project has been designed to evolve into an automated "Factory of the Future" that will incorporate the planned integration of Integrated Computer Systems, Material Control/Distribution Systems, and Process Automation. The initial EAP system elements have been designed explicitly to include interfaces for the future integration of the advanced flexible manufacturing systems for robotic assembly and material handling/control that currently are being developed under the Westinghouse/Air Force Technology Modernization effort.

The key critical design elements in the "Factory of the Future" are:

o   Integrated Computer System(s)
o   Material Control/Distribution Systems(s)
o   Process Automation

The EAP INTEGRATED COMPUTER SYSTEMS are being designed to integrate the quality, test, process control, business and material handling sub-systems. The Computer Systems within EAP are broken down into three logical operating system levels as shown in Figure I.

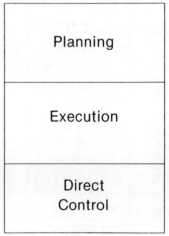

Figure I

Planning Systems are used to prepare for the manufacturing activity and to monitor and analyze the results of the manufacturing process. These systems include: Manufacturing Engineering, Quality and Visibility, Finance, Industrial Engineering and Plant Business Systems.

Execution Systems are tactical in nature and support the factory floor with real time information required for short term decision making. These include: Manufacturing Control System, Test Control System, Material Handling and Control, and Work Management System.

Direct control systems are normally real-time in nature and control physical processes embedded with the manufacturing process. These include: Automated Manufacturing Equipment, Automated Test Equipment, and Material Handling Equipment.

Further analysis of this structure indicates that the nature of the computing performed at the various levels is quite different (See Figure II). Movement upwards in the structure leads to very complex processing such as product design and Material Requirements Planning. Movement down to the lowest levels of the structure leads to conveyor control and automation control. Processing becomes more and more time critical at the lower levels of the system structure.

A further partitioning of the architecture is shown in Figure III. Rather distinct differences exist in the nature of business and technical processing at all levels of the architecture. Business computing deals primarily with data movement, transaction (terminal) processing and rather straightforward mathematical processing. Technical computing typically involves heavy computation and relatively little data movement and terminal processing. For this reason the architecture is partitioned along the lines noted. Computer languages, data handling techniques and staff skills tend to be quite different between the business and technical portions of the architecture.

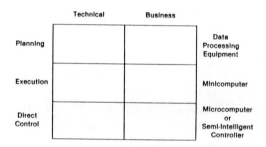

Figure III

An additional analysis of the architectural levels reveals significant differences in the planning horizon (or data lifetime) at the various levels in the architecture. Figures IV, V and VI summarize the processing requirements at these levels.

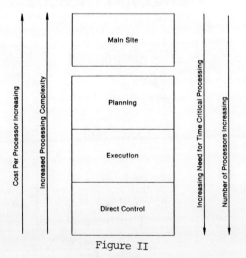

Figure II

Planning

- 30 Day Horizon

- Large/Complex Processing
  - Factory Scheduling
  - Process Planning
  - Postprocessing

- Redundancy
  - Unaffordable
  - Unnecessary

Figure IV

Execution

- 7 Day Horizon
- Simpler Processing
  - Program/Data Download
  - Collection of Machine Status
- Rapid Response Requirement
- Redundancy
  - Affordable
  - Necessary

Figure V

Direct Control

- 1 Day or Less Horizon
- Simple Processing
  - Sensor/Actuator Mgmt
  - Program Download
- Time Critical
- Redundancy
  - Unaffordable
  - Generally Unnecessary*
  - Spares/Maintenance Philosophy

* This May Not Be True in a Continuous Processing Environment

Figure VI

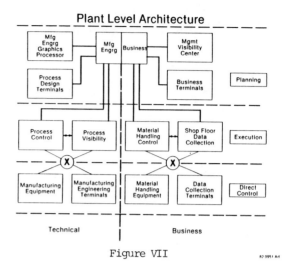

Figure VII

Figure VII shows the computer architecture to be utilized in the Electronics Assembly Plant. Note that the three levels in the architecture and the technical/business partitioning have been preserved.

The major CPU's are identified on the diagram for the Planning and Execution levels. CPU functions are shown generically for the execution level. Approximately 35 mini and microcomputer CPU's and 16 dedicated microcontrollers will be used at the execution level.

The execution level of the architecture is a major focal point for data communication. As a result of this critical pathway, it is necessary to provide a practical method for circumventing hardware failures at this level. Figure VIII outlines the general systems approach to be used for providing failure recovery. Critical processors will be saved with shared disc resources. Each processor will monitor the others. If a critical resource fails, an alternative hardware device or data pathway is provided.

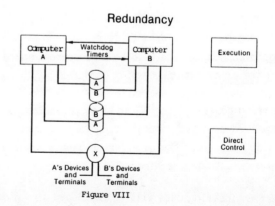

Figure VIII

The basic computer system architecture and data bases have been designed to evolve into the projected CAD/CAM system of the future. The modularized system design being undertaken will provide for maximum flexibility and assure minimum revisions other than enhancements to the initial systems being implemented.

The following is an overview of each major Sub-System identified above as planned for EAP.

Manufacturing Engineering

The Manufacturing Engineering System accepts data from Design Engineering in the form of drawings and electronic data. The data is cataloged and organized in a data base.

The coordinates/parameters file which will describe, in electronic terms, all of the technical data regarding a given assembly will become the source data for all post processing for factory machinery.

Post processors will use the coordinate/parameter information in order to develop the Machine Control Data (MCD) required to operate the various automated manufacturing machines.

As machine control data is developed and proven, approved MCD will be held in an electronic vault in the plant data processing computer.

Working copies of the MCD will be periodically downloaded to the Manufacturing Control System which will store the data for on demand download to manufacturing machines.

Test Engineering data will be developed in Baltimore, by Test Engineering, using the CAD/CAT system. As programs are developed and proven in Baltimore, they will be electronically transmitted to the remote computer facility. Then programs will be categorized and placed in the Electronic Vault and periodically downloaded to the Test Control System Computer.

Manufacturing Control System

The Manufacturing Control System will be the central node in a computer network which will deliver data to manufacturing machinery and continuously monitor the states of the machines.

All manufacturing machines will be electrically connected to the Manufacturing Control System and will possess the ability to exchange data on a bidirectional basis.

Test Control System

The test control system is the central node in a network of computer and microprocessor controlled electronic test stations.

This node will receive test programs from the main plant site via the Manufacturing Engineering System and store programs for on demand download to lower level test stations in the network.

All test stations will be electronically connected to this node and a bidirectional data exchange pathway will be supported.

Automated Manufacturing Equipment

Individual workstations will operate under micro or minicomputer control to perform various manufacturing activities such as component placement and insertion.

The programs to operate these individual stations will be called selectively by manufacturing operators as required.

As the new generation of robotic workcenters comes on line at the EAP facility, they will be incorporated into the network at this level.

Automated Test Equipment

Computer and microprocessor controlled test equipment will be employed at EAP in testing and in performing fault analysis on Printed Wiring Assemblies (PWA's).

Programs required to operate the automated test stations will be available from the Test Control System.

Factory Business Systems

The Factory Business Systems at EAP will, in general, be a subset of the same modules which are to be used at the BWI site. Copies of these modules will be installed and operated in a small business computer located at the EAP facility.

Schedule data will be supplied to EAP from the BWI site MRP process. This data will be entered into a local order processing module at EAP (one of the few additions to the BWI Manufacturing System) and a local factory schedule will be developed with the delivery constraints mandated by BWI's MRP process.

As work orders become ready for release, these orders will be sent to the Work Management System (WMS).

All material ordering, processing, kitting, etc. will be performed at the Material Acquisition Center (MAC) located in Baltimore.

As work is completed, materials will be packaged and shipped back to the MAC facility for distribution to the other DESC facilities.

Work Management System

The Work Management System (WMS) will be the primary tactical information system supporting the factory operations at EAP.

The Factory Business Systems modules at EAP will prepare and plan all orders. As orders come due, the data will be released to the WMS system.

WMS will organize the work queue for all work centers and stations based on the critical ratio schedule number supplied by the Factory System modules.

As work is completed in workcenters, operators will pay off jobs via barcode wands and terminals managed by WMS. WMS will then determine the next job to be performed in a workcenter and instruct the Material Handling and Control System (MHC) to perform the physical moves required.

Material Handling and Control

The Material Handling and Control System is the primary control system for material location control (both on and off the shop floor).

As orders to move material are received from the Work Management System (WMS), MHC will look up material and bin location data stored within its memory and direct the mechanical devices in the Automated Material Handling group to perform physical material movement.

Automated Material Handling Equipment

The automated material handling equipment consists of a computer controlled carousel storeroom, conveyor systems which flow through the workcenters, receiving and material dispatch areas; and material handling devices to interface the carousel and conveyor systems.

All material movement within the EAP facility will occur via the conveyor and carousel systems. All factory operators and material control personnel will have convenient access to this system within the factory work cells.

Quality and Visibility Systems

All relevant data regarding factory performance will be electronically collected and sent to the Quality and Visibility Systems for analysis.

Each of these systems will feature regular, standard analysis and reporting which will be done on a scheduled basis; and the ability to perform and hoc analysis at the direction of plant management.

Data from these systems will be accessible to plant management in a visibility center located in the plant facility which will feature color graphics computer terminals, which have access to data bases, coupled into a large screen (4' x 6') color television projector.

Groups of management personnel will have the ability to quickly review, graphically, performance data regarding their operations.

Finance

Financial Systems at the EAP facility will consist primarily of "front ends" to the basic financial systems at BWI.

A copy of the employe personnel data system containing data regarding EAP employes will reside at EAP. This system will update the larger BWI version on a periodic basis.

Contract charging data will be collected from the Work Management System and sent to BWI for incorporation into the DESC wide cost distribution systems.

Corrections to the salaried payroll will be processed via a local front end to the BWI payroll system.

All standard accounting reports will be produced as a part of the normal financial data processing operations at the BWI site.

The EAP MATERIAL CONTROL/DISTRIBUTION SYSTEM is being designed as an integral part of an overall Integrated Material Handling Systems and Technology (IMHST) Strategic Plan that is addressing the following key objectives:

° Dock-to-Stock/Stock-to-Ship material tracking,
° insure product integrity,
° maintain product identification, and
° provide a totally automated integrated material handling/management system

There is considerable synergism between material control/distribution systems irrespective of the facility, manufacturing process and/or product. A generic modular design approach is being followed on all IMHST Projects in work and being planned to assure maximum flexibility and total integration with both the overall integrated computer system(s) and process automation objectives described above.

The initial EAP Material Control/Distribution System is based on the combination of several sub-system modules. These modules are to be integrated through physical and logical interfaces that allow the components to function reliably as one system.

Centralized Work-In-Process (WIP) storage and retrieval is the basic material handling and material management concept to be used in this facility. Work-in-process will be automatically stored and dispatched to and from the work centers under real-time computer control. In this way, WIP storage on the production floor will be minimized and positive tracking of job progress will be maintained.

The total system will be very responsive to the real-time needs for material in the work centers and be designed for an extremely high level of reliability. The system will incorporate revisions for back-up modes of operation so that work can still be provided to the work stations during periods when the system is not fully operational.

Conclusion

The development of the modular, distributed computer architecture which will be implemented by Westinghouse at it's Electronics Assembly Plant in 1983 will set the stage for the full scale development of the Factory of the Future.

A major new series of robotic workcenters will begin arriving at the facility by 1984. The systems (both hardware and software) which are being designed for the EAP facility include the software and hardware attachments for the next generation manufacturing technologies.

Biographical Sketch

DAVID G. HEWITT is Vice-President of United Research Company, a management consulting firm located in Morristown, NJ. He specializes in improving performance through integrated systems. Previously he was the manager of Satellite Plant Systems Development for Westinghouse Defense and Electronics Center located in Baltimore, Maryland.

Achieving Integrated Manufacturing Systems

Michael J. Daugherty
The Duriron Company, Inc.
Dayton, Ohio

ABSTRACT

Success in manufacturing is, to a large extent, determined by the old axiom of being able to make the right thing at the right time. When the product is complex, expensive, highly engineered, and often made-to-order, this simple statement can be extremely difficult to accomplish. In many companies today, this problem is being addressed by the implementation of integrated Manufacturing Resource Planning Systems. This paper discusses the planning and design of such systems. Many of the issues discussed are relevant to all areas of Computer Integrated Manufacturing (CIM); however, the primary focus is on integrated Manufacturing Resource Planning (MRP) systems, as opposed to Industrial Automation Systems.

COMPANY PROFILE

Since many of the concepts, ideas, and examples presented in this paper were developed in the context of manufacturing systems at The Duriron Company, Inc., it is appropriate to give a brief company profile.

The Duriron Company, Inc. is a world leader in the design and manufacture of fluid movement and control equipment for the Chemical Process Industries. Classes of products include centrifugal pumps, valves, pipe and fittings, filters, and high pressure equipment including valves, compressors, reactors, and gauges. In addition, the Company manufactures various specialty corrosion resistant materials and components.

Duriron began manufacturing a corrosion resistant material in Dayton, Ohio, in 1912. Today, Duriron is a vertically integrated, multi-division, multi-plant corporation employing 2000 employees. The Company consists of six (6) domestic operating divisions, nine (9) wholly owned foreign subsidiaries, and twenty (20) sales offices located throughout the United States. Duriron has many Manufacturing Resource Planning systems installed throughout the corporation.

PLANNING INTEGRATED MRP SYSTEMS

Integrated MRP systems do not just "happen." Integration does not occur naturally - it has to be forced. If a company does not actively and visibly plan to have integration, then, by default, it will not have it. If each particular department in an organization is free to act on its own, the systems developed may be optimized, but only from the originating department's point of view.

Unfortunately, optimization of all the various isolated systems of the organization will not necessarily lead to optimization of the performance of the entire corporation. Only a completely integrated system, designed with a full understanding of inter-functional relationships, will meet this objective.

Integrated systems will be developed only if they are part of a Corporate plan firmly based on CIM concepts. The direction for this plan must be defined, supported, and understood by Senior Management. Only after this direction is set can detailed planning and design proceed in a top-down fashion.

The importance of this support extends to all levels of management; furthermore, to have true support you must have involvement and understanding. There are many methodologies for obtaining this involvement. The best choice for a particular company depends on many factors, the most important of which is the amount of resources the company can afford to commit to the project. The following discussion summarizes a technique that has been used successfully at Duriron.

The example used is a recent project to develop a new system for: Scheduling, Capacity Planning, Shop Floor Control, and Purchasing. From the beginning it was recognized that this system would need to integrate with essentially all existing manufacturing systems (Figure 1). The job functions of a large majority of employees would be affected in some way. Instead of involving only select managers and employees, it was decided that essentially everyone functioning within the scope of the project should be involved.

To plan the high-level conceptual design of the project, the IBM Application Transfer Team approach was utilized. Using this approach, the planning was divided into the following seven steps:

1. A "conceptual design team" was established. This team had at least one representative from each division involved. The following functional areas were also represented: Material Control, Manufacturing, Manufacturing Services (representing Industrial Engineering, Manufacturing Engineering, and Computer Aided Manufacturing), Accounting, and Information Systems.

The following objectives were established:

. To develop planning and measurement methods that will allow better utilization of resources to improve both effectiveness and profitability.
. To identify and remove those obstacles which cause customer orders to miss their required or promised dates.
. To develop the means by which long-range production modeling can be used to determine future resource requirements.
. To integrate all new systems with a major portion of existing systems.

2. The Group Vice President responsible for all the Divisions involved was asked to be the project sponsor. Having the complete commitment of the sponsor is key since he is the link to Senior Management. The plan was to put together a high-level conceptual design fully supported by him and by Divisional Management. Final approval would have to come from the Information Systems Steering Committee.

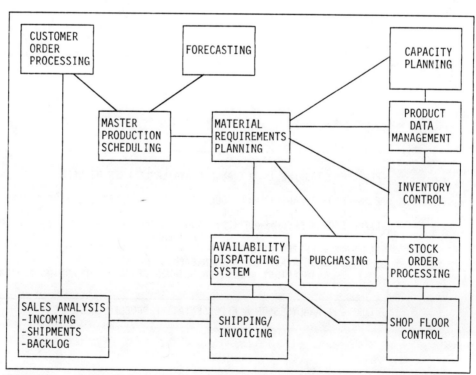

FIGURE 1. INTER-RELATIONSHIPS OF MANUFACTURING SYSTEMS USED AT DURIRON.

3. The next step was to interview the potential users of the system. The interview was approached from a user-oriented perspective. They had to be convinced that the design team was genuinely interested in their problems and ideas. Instead of asking what features they wanted to see in the new Manufacturing Resource Planning system, three basic questions were asked:

   - What are your objectives and how are you evaluated against them?
   - What problems do you face in trying to meet these objectives?
   - What kinds of tools could you use to help improve your performance?

   In order to reduce pressure, the users were interviewed in small departmental groups. Representatives from all the divisions and functional areas involved were interviewed. This included everyone from Shop Foremen to Division Presidents.

4. The interviews generated several hundred problems which were combined first into 39 categories, then later into 12. For each of these problem areas, the following was identified:

   - the cause of the problem
   - the impact of the problem
   - the benefits of solving the problem
   - a set of recommendations

   Figure 2 gives an example.

5. The recommendations that were within the scope of the original project were identified. A proposal to implement the recommendations was written. The general specifications for the new system were defined. By the time this step is reached, the need for integration becomes obvious. The interdepartmental relationships and dependencies can only be addressed through total system integration.

6. The proposal must next be presented and sold to the Group Vice President sponsor and the Steering Committee. A particular problem in this step is that management must be convinced to make major commitments in time, money, and personnel for systems where the payback is often long term and intangible. It can be very difficult to quantify in dollars the long-term advantages of installing sophisticated and costly CIM systems, over less sophisticated and costly discrete systems. For example, normally a stand-alone Purchasing System can be installed faster and at less cost than a system which must be integrated with several manufacturing systems.

7. Once the project is approved, the detailed design must begin. Up to this point the conceptual design team has only written general specifications for the system. The detailed design team will expand on these specifications by defining the actual data base and the inputs, outputs, and processes involved.

PROBLEM: TOOLING IS NOT ALWAYS AVAILABLE WHEN NEEDED.

EXAMPLE OF A CAUSE: ALL TOOLING IS NOT INDICATED ON THE ROUTINGS.
EXAMPLE OF AN IMPACT: POOR UTILIZATION OF MACHINES.
EXAMPLE OF A RECOMMENDATION: ADD TOOLING REQUIREMENTS TO THE ROUTING DATA BASE.
BENEFITS: BETTER UTILIZATION OF RESOURCES BECAUSE TOOLING WILL NORMALLY BE AVAILABLE WHEN NEEDED.
INVENTORY REDUCTION BECAUSE OF BETTER PLANNING AND EXECUTION.

FIGURE 2. ANALYSIS OF A REPRESENTATIVE PROBLEM AREA.

This team must analyze the software make versus buy decision, monitor the software development, and direct the implementation of the system.

CRITERIA FOR SUCCESS

The capability to install successful Integrated Manufacturing Systems will vary in different companies depending on several factors.

- The Type of Business

    Integrated Manufacturing Systems will improve the performance of any manufacturing company. However, the more complicated the organization and product structure, the greater are the relative benefits. For example, a company making only a few products with a very simple bill-of-material, a one-step routing, a very large SOQ (standard order quantity), and no variations, will probably have little to gain by installing sophisticated Integrated Manufacturing Systems.

- Management Policy

    As discussed previously, it is critical that all levels of management understand and support the implementation of CIM. The involvement must continue through the development and implementation of the project.

- Communication

    The development of sound integrated systems requires excellent interdepartmental communication. In fact, improved communications is often an intangible benefit of the development process.

- Education

    The need for across-the-board education cannot be overstated. It is vital that all users understand the particular sub-system with which they interact and also, how that sub-system interfaces with all other sub-systems. It is critical that users understand not only what they should do, but how their actions affect other departments. It is equally important to know the effect of actions taken by users in other departments.

- Existing Systems

    In the move to integrate MRP systems, an important consideration is whether it is necessary to start over from scratch, or can existing systems serve as a base and be enhanced to fit the CIM master plan. This is an extremely important decision that depends to a large extent on the design of the existing systems. If existing systems have been installed as stand-alone systems, the conversion to Integrated MRP II is going to be very difficult. In this position most companies should probably elect to start over from scratch, possibly by using a packaged system. The problem of fragmented systems has been increasing recently due to the introduction of Personal Computers and low cost computers in general. Too many departments are installing stand-alone systems on PC's without giving any consideration to integration to larger systems.

    If the current systems have been designed with ease of integration as a criteria, the evolvement to CIM will be practical.

- The Data Base Policy

    The most important technical factor necessary for the development of effective integrated manufacturing systems could be the use of a data base management system (DBMS). Britton and Hammer, in a recent _Industrial Engineering_ article, made the following six points in support of this argument:

    . "A DBMS simplifies and standardizes the access to an organization's data.

    . One objective of a DBMS is to integrate the corporate data base (by structuralizing formats, access techniques, and data relationships).

    . The use of a DBMS should reduce data redundancy.

    . A DBMS allows systems to be built in an evolutionary manner.

    . The data independence provided by a DBMS means that programs using it can access only the data items they require.

    . The concentration of file access logic in a DBMS offers significant benefits with regard to migrating to new computer hardware or operating systems."[1.]

In order to develop integrated systems, the data must be well structured, non-redundant, easy to modify, and easy to access. It must be possible to easily "navigate" through the data base to collect or update all relevant data in a real-time mode. There is a tremendous growth in "computer literate" people throughout the organization. If we expect them to develop and use integrated systems, we must provide an easy technique or vehicle for integration. A DBMS is that vehicle.

MAKE VERSUS BUY

In recent years, many excellent articles have been written which analyze the packaged software "make versus buy" decision. This paper does not discuss the pros and cons of this decision. However one criteria that must be emphasized is the effect on integration. Assuming that the company only considers well designed software, the primary source of problems in this area comes from trying to "mix and match" software. It is often possible to <u>interface</u> an order processing system from Company A with the manufacturing scheduling system from Company B but it is unlikely that they can be <u>integrated</u> without major modifications.

This problem is just as real when one of the companies is your own. It is very difficult to buy just a Capacity Planning system, then try to integrate it into in-house developed Bill-of-Material, Routing, and Order Scheduling Systems. Obviously, there are some exceptions to this rule, but a general guideline is if you decide to install packaged manufacturing software, then go with it all the way. Other important criteria are flexibility, cost, compatibility with the company's and user's requirements and method of operation, and capability and availability of in-house development staff.

EXAMPLE

Many of the issues involved in integrated manufacturing systems can be best illustrated by an example. Let us consider a Customer Order Processing System. Suppose the system is to have the following features:

. Processing of Quotations
. Entry and Maintenance of all customer orders.
. Generation of all order paperwork including the customer acknowledgement, shipping document, bill-of-lading, and invoice.
. Entry to the manufacturing scheduling and work-in-process system.
. Entry to the corporate sales/profit analysis system.

This system and associated systems are represented schematically in Figure 3.

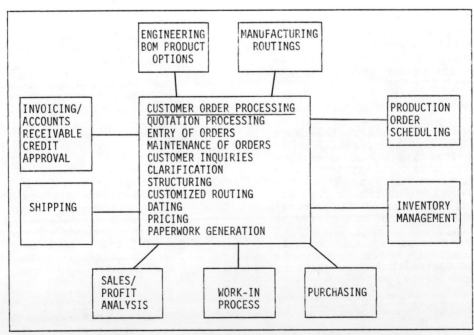

FIGURE 3. RELATIONSHIP OF CUSTOMER ORDER PROCESSING TO OTHER SYSTEMS.

In a true CIM installation these systems must be integrated, not just interfaced. For example, there must be one order master file which is fundamentally interwoven into the order processing, manufacturing scheduling, accounting, and sales/profit analysis systems. All updates to the file should be real-time. An example will illustrate this point. Suppose the Sales Department does an availability check on Part A at 9:00 AM and finds one available. They then enter an order which reserves the part. If a different Sales office tries to order one Part A at 9:05 AM, the system will show, of course, that none are available. However, if the Material Control Department, at 9:03 AM enters a schedule receipt for 5 pieces of Part A to be received October 1, then the second order inquiry must show that 5 are available on or after October 1.

Suppose the order is for a non-stocked item. Just to give a promise date on this order, the system must be able to check all features and options requested, material and capacity availability at possibly several plants, purchase material availability, order priority, and special manufacturing requirements. In this type of environment, nightly downloading of data is simply not acceptable.

Next, consider the Credit Approval System. Many times the Sales Department will want to enter a "ship immediately" order for an off-the-shelf item. Since the order cannot be entered without credit approval, the two systems must be integrated so that all the data relevant to this customer can be accessed. This includes the customer's credit limit and all current open orders, including any that may have entered 5 minutes earlier through a different sales office. By following similar arguments, it can be shown that most of the other systems shown in Figure 3 must be completely integrated with the Customer Order Processing System.

"User-friendliness" is another benefit of integration. In the example above, the order processing clerk is in the Customer Order Processing system when it becomes necessary to do the availability check. He must be able to check availability of Part A without leaving the system, preferably by just hitting one key. Exiting Customer Order Processing to enter a Material Status System is not acceptable.

SUMMARY

The use of computers and information systems pervades all aspects of manufacturing, from instantaneous control of robots and NC machines, to the development of long-term business plans for Senior Management. Optimal performance of these systems in aggregate can only be obtained through integration. This integration must be planned and directed from the top. It must be fully understood and accepted by all levels of management and staff.

REFERENCES

1. Harley O. Britton and William E. Hammer, Jr., "Designers, Users, and Managers Share Responsibility for Ensuring CIM Systems Integrity," Industrial Engineering, May 1984, pp. 36-48.

BIOGRAPHICAL SKETCH

Mr. Daugherty is Manager of Systems for The Duriron Company, Inc. He has been directly involved in planning and implementing a wide range of information systems, especially in the area of manufacturing.

Mike received his BS degree in Physics from the University of Dayton, and MBA degree in Management from Wright State University. He is active in the Association for Systems Management and is a past President of the Dayton, Ohio, chapter.

Ten 'Cardinal Sins' To Avoid In Selecting Software For Manufacturing Management Control Systems

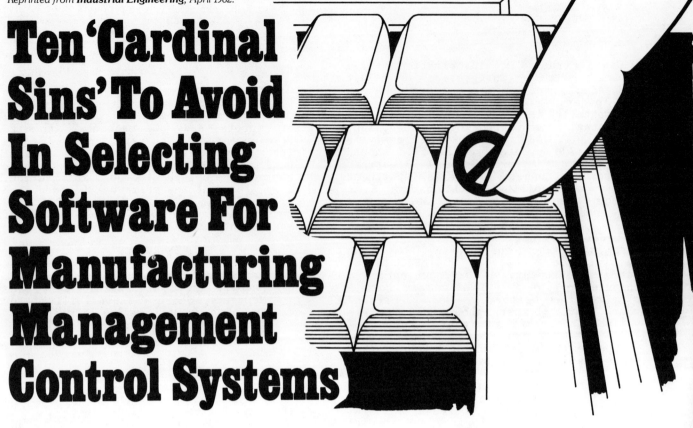

By Robert L. French
R. L. French and Co. Inc.

The most important decision a company introducing a computerized system of manufacturing management control must make is probably that concerning the selection of application software that is reasonably compatible with company needs. The following "ten cardinal sins" must be avoided to prevent a serious and costly error in software selection:

1.) *Mismatch with basic manufacturing process*—The major modes of manufacturing are:
☐ True job shop—make to order, little or no repetition.
☐ Job lot—make to order with repetition.
☐ Continuous process—make to stock.

Amazingly enough, manufacturing software packages are often sold to companies with a complete mismatch in this fundamental area.

2.) *Mismatch with the cost accounting system*—This is similar to the preceding example except that the company may truly need to reevaluate what control procedures would best serve the needs of the business.

For example, a manual job cost system may need to be converted to a computerized combination of standard and job cost. In any event, the matter needs to be carefully considered before a software package is selected.

3.) *Failure to provide for alternate path manufacturing*—The software package may assume fixed path (or relatively so) manufacturing when in reality the process routing may vary a great deal.

Unless the system to be adopted provides a mirror of the real world of the manufacturing environment it is to be used in, it will fail or be marginally successful at best.

4.) *Mismatch of the degree of job lot control required*—The two extremes are complete accountability after every processing operation, on the one hand, and an almost complete lack of work-in-process accountability on the other. In the latter situation, control is exercised only upon issue of materials to work-in-process and with the reporting of completed finished goods.

In connection with consumer protection and product liability, the emphasis is shifting more toward precise lot accountability. To some extent, the availability of computerized shop floor control systems contributes to this trend. For products such as drugs, medical devices, aircraft parts and so on, lot control is required by government regulation.

This very important area needs to be considered on both a short-term and a long-term basis in selecting manufacturing software.

5.) *Mismatch of the degree of specification and configuration control*—Closely related to the question of job lot control is the matter of controlling product configuration and engineering changes. While this may be primarily a matter of product liability protection, the degree of control exercised over engineering changes and decision rules on the effective dates of changes and the "using up" of materials can have an important influence upon company losses to obsolescence.

6.) *Improper provision for required quality discipline*—To be a useful tool for those charged with the responsibility of making commitments to customers regarding expected delivery dates—and for a balanced work load in the plant—the manufacturing software package should give visibility to the current status of all parts in the process of being manufactured, including those that have been given a quality hold status. In addition, parts that are in the process of being reworked need to be properly identified. The software package should provide an "inventory bucket" for every different part status and location.

7.) *Improper provision for "designed" yield loss as opposed to unplanned scrap*—For certain manufacturing processes, the amount and use of byproducts have an important influence on the entire planning and control process. For example, certain plastics and rubber compounds can be reground and "worked back" into new products on a percentage basis.

While the distinction between "yield loss" and scrap may be a gray area in some situations, the software package must be adaptable enough to recognize and record what actually happens, or the system will lose credibility because of inaccuracy.

8.) *Requirement of an incompatible degree of shop floor reporting discipline*—In this area, the state of the art is ahead of practical application. Many software packages are written for "on-line interactive" input. With this type of system, the files are automatically updated immediately after the transaction is entered. The alternative is a "batch" mode of entry with on-line inquiry.

Shop disciplines are far less demanding in the batch mode than in the on-line interactive mode. Therefore, before selecting a software package requiring on-line interactive processing, the customer should be certain the shop environment is ready for the required discipline.

These same considerations must be applied to some other potential "automatic" system parameters. For example, some available software provides finite work center loading, but few users are ready for this degree of sophistication.

9.) *Improper provision for man-machine relationships*—Variables such as a crew working at a work station vs. one person running several machines must be checked. In addition, in some plant environments, the machine-hour rather than the man-hour provides the basis for scheduling, quoting, costing and control. Obviously, compatibility is needed.

10.) *"None of the above are important. A standard package is available that is adequate for use without change"*—Proponents of this approach argue that making the software fit the manufacturing environment is not important. The small business must either change its way of doing things in manufacturing to fit the package or, if doing that immediately is totally impractical, put the package in anyway and wait a year or so to make changes.

This approach overlooks a very important ingredient—acceptance of the system by the users. If the results of the system are totally unreliable because the software is incompatible with the real world of manufacturing, *it is extremely difficult to regain acceptance and the dedication necessary for a second effort.*

The successful manufacturing control system developer will very carefully do the necessary "homework" by documenting exactly what is needed (and desired) in a manufacturing control system and searching the software field for the package that most nearly meets these needs. He or she will also be prepared to enhance that package as circumstances warrant. Good systems result from careful work "up front." **IE**

Robert L. French holds BS & MS degrees in business administration from the University of Missouri and has held positions in industry as an industrial engineer, manufacturing engineering manager, vice president-manufacturing and vice president-general manager. In addition to heading R. L. French & Co., he is president of Computerized Manufacturing Management Corp. and chairman and CEO of On-line Data Inc.

Human Factors Problems in Automated Manufacturing and Design

Martin G. Helander, Ph.D.
Department of Industrial and Management Systems Engineering
University of South Florida
Tampa, Florida 33620

ABSTRACT

The use of computers for manufacturing, product design, and engineering analysis is expanding very rapidly. There are predictions that by 1990 practically all U.S. engineers will occasionally interact with CAD/CAM work stations, and that approximately 35 percent will conduct most of their work at CAD/CAM work stations. There are therefore good reasons to analyze and improve the design of CAD/CAM work places. During the last ten years there has been considerable interest in the human implications of automation. Mostly, the discussion has centered on social issues such as unemployment and the utilization of human skills. In comparison, fairly little attention has been devoted to human factors issues. This article gives a brief overview of some human factors and ergonomics problems. There are several issues of interest: allocation of decision making between humans and computers, ergonomics design of CAD/CAM work stations, choice of optimum input devices, training/retraining of employees, and human factors design of software.

INTRODUCTION

The opportunities for automating manufacturing are often misunderstood. The emphasis has been placed almost exclusively on the production process, and automation has become symbolized by the use of industrial robots. Actually, the work involved in assembling a product is not where automation is likely to have its greatest impact; assembly accounts for only 10-25 percent of the cost of manufacturing. Of greater consequence is the organization, scheduling, and managing of the total manufacturing enterprise including functions such as product design, distribution, and customer service. For this reason, the most important contributions to increase in productivity will be obtained by linking information networks and thereby provide a common pool of data on design, management, and manufacturing.

A computer integrated manufacturing (CIM) system could be able to handle just about all aspects of the work including forecasting, cost planning and control, purchasing, manufacturing, inventory control, quality control, and plant maintenance. CAD/CAM systems at different plant locations can also be tied together through the use of communications satellites. This extended information network makes it possible to coordinate the design and manufacturing of products, and standardize product design features. For such functions, productivity increases of several hundred percent have been mentioned. The 50-100% increase in productivity generally obtained for drafting and engineering design is by comparison fairly modest. The management of information may therefore have a greater impact on work in an automated plant than the automation of the assembly, and it is likely that these changes will alter far more white collar jobs than blue collar ones.

DECISION MAKING IN COMPUTER INTEGRATED MANUFACTURING

Several large U.S. corporations are presently working on the problem of integrating new information resources into their existing CAD system. Programs for drafting and engineering design have already been used for several years. The challenge is to integrate features like:

o Storage and retrieval of information about the parts and the products being manufactured.

o Management and control of available sources, such as labor, machines, and materials.

o Materials handling including inventory management, purchasing, and retrieving.

o Design of machine tools.

o Control of robots and N/C machines for manufacturing.

In the future, other functions such as cost planning and control, forecasting, and plant maintenance may be added. Perhaps the most remarkable aspect is that the only processes which may be difficult to automate are decisions made by senior management with respect to long term planning, including market analysis for new products and the decision to manufacture those. Figure 1 shows a computer integrated manufacturing planning system.

From a human factors point of view there are at least two problems involved:

o Change of job assignments from manual work to monitoring and control of the system.

o Allocation of decision making between computers and humans.

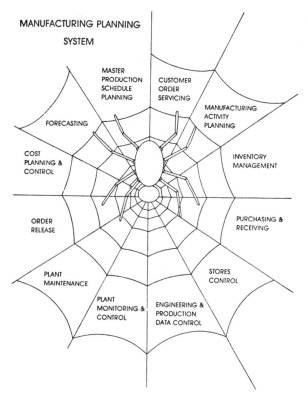

Figure 1. A manufacturing planning system relies on a centralized computer. It may be difficult to decentralize authority and decision making. If so, job satisfaction may suffer.

In the factory of the future, human operators will rarely be used for manual work. However, even highly automated manufacturing needs human beings for supervision, adjustment, maintenance, expansion, and improvements. The recent increase in interest in human factors among engineers reflects the paradox that the more automated control system used, the more crucial are the contributions of the remaining operators, and the more human factors/ergonomics tools are needed for the design of the human/machine interface (Bainbridge, 1982). For example, a process control operator is often a highly skilled individual with an engineering degree. This is required in case the process breaks down. The normal work condition, however, is one of mental underload, with monotonous supervision of instrument panels. But when the process breaks down, the work becomes extremely stressful. It is often difficult to recruit and retain skilled individuals for such tasks. However, by incorporating human factors design it is possible to create a more meaningful task that provides greater job satisfaction.

The other main problem has to do with allocation of decision making. By its design, a CIM system requires a centralized computer in order to integrate information, see Figure 1. The question is then to which extent individuals will be given the opportunity to influence decisions. For example, corporations like IBM and General Electric are presently implementing CIM networks. Although these corporations have a company policy to decentralize decision making and provide many employees with personal computers, the design of the CIM network is not particularly well suited to decentralized decision making.

Table 1 illustrates that decision making is a fairly complex procedure involving preparation of several decision alternatives, and communication and implementation of the decision.

Table 1. LEVELS OF AUTOMATION IN DECISION MAKING
Adapted from Sheridan (1980).

100% Human Control	1.	Human considers decision alternatives, makes and implements a decision.
	2.	Computer suggests set of decision alternatives, human may ignore them in making and implementing decision.
	3.	Computer offers restricted set of decision alternatives, human decides on one of these and implements it.
	4.	Computer offers restricted set of decision alternatives and suggests one, human may accept or reject, but decides on one and implements it.
	5.	Computer offers restricted set of decision alternatives and suggests one which it, the computer, will implement if human approves.
	6.	Computer makes decision and necessarily informs human in time to stop its implementation.
	7.	Computer makes and implements decision, necessarily tells human after the fact what it did.
	8.	Computer makes and implements decision, tells human after the fact what it did only if human asks.
	9.	Computer makes and implements decision, tells human after the fact what it did only if it, the computer, thinks he should be told.
100% Computer Control	10.	Computer makes and implements decision, does not tell human.

At level 1 computers are not used for aiding or making decisions. At level 10, the other extreme, a computer makes and implements the decision without communicating to the human operator. It is reasonable to assume that a compromise between the two would enhance both productivity and job satisfaction. For several tasks levels 4 or 5 might be appropriate. However, the desired level of automation in decision making depends upon the task. Highly automated routine types of tasks which do not require human supervision may utilize level 10, whereas expert systems for aiding management decision making may only go as far as level 2.

There is a need for developing policies. Ideally, there should be company guidelines governing the allocation of decision making between computers and human beings for different types of tasks. This information is particularly important for those who develop software for CIM. Lacking such guidelines, there is a great possibility that the needs of the individual employees are pushed aside; the technological challenge in developing CIM might be overpowering to most organizations. Sooner or later it will, however, be recognized that the delegation of authority, responsibility, and decision making are imperative to job satisfaction, and that software should be deliberately designed to consider the needs of the individual operator.

Some researchers maintain that the delegation of decision making should be made in a dynamic fashion, so that sometimes the human operator has full control and sometimes the computer (Greenstein and Rouse, 1982; Rieger and Greenstein, 1983). According to this principle, the computer continuously monitors the process and simultaneously evaluates the appropriateness of the control performance of the human operator. When there is a need, the computer may take over decision making. Such schemes have been proposed for some specialized tasks including air traffic control and the operation of nuclear power plants. Dynamic decision making leaves much freedom to the human operator who, in most cases, retains the authority to delegate work to the computer. The work is therefore more satisfying and improves the freedom of the operator.

ERGONOMIC DESIGN OF CAD/CAM WORKSTATIONS

As CAD/CAM workstations become increasingly used by managers and engineers, there will be an increased demand for ergonomic design features. Many have already been incorporated in visual display terminal (VDT) workstations, and include features such as: height adjustable table and keyboard, wrist-wrist and foot-wrists, and ergonomic chairs with height adjustability, lumbar support and a high back rest.

As with other CRT terminals there are also problems with veiling reflections (mirror-like reflections of surrounding objects). Both of these problems may be reduced by the deliberate design of the office illumination system.[1] For example, an illumination level of about 300 lux allows visibility of both the screen and the source document. However, when the screen is not used, the operator might prefer to increase the illumination level. Similarly, when the operator is interacting exclusively with the screen, with no need to refer to printed documents or drawings, it might be advantageous to lower the illumination level below 100 lux. It is therefore suggested that a graphics workstation with a CAD/CAM system should have easily adjustable illumination that can be individualized for each workstation.

Another important problem is display flicker. Unless the screen is refreshed at a rate greater than 60-70 Hz, there might be problems with perception of flicker. Flicker is perceived easier

[1]These and other VDT design issues are addressed in Helander, Billingsley, and Schurick (1983).

and is more annoying the larger the light area of the display is. As a result, some CAD operators, in order to keep the display as dark as possible, avoid crosshatching of drawings and similar details which would increase the flicker.

It seems likely that alternative display technologies will be used in the future. Flat panel displays including plasma, LCD, and electroluminescence displays require much less space at the workstation and would avoid most of the problems with flicker and reflections.

Work with CAD/CAM equipment may require considerable concentration. The design of the work area must therefore permit concentration by allowing individuals privacy at their work. It is recommended that the background noise not exceed 50dB. Computers, plotters, printers and other noise producing machines must therefore be located in special room(s) accessible to the design room. A high level of security, including fire detection and extinguishing equipment must also be built into the facilities.

Finally, most designers of present CAD/CAM equipment seem to disregard the fact that there is an unusually high demand for space for drawings and other computer generated graphics and documents. Information from such documents is used as input in the CAD/CAM design process, and operators need to refer to them constantly. As a result, there is a need for tables where operators can spread out and store drawings. An example is given in Figure 2.

Figure 2. In a couple of years there will be considerable design improvements of the CAD/CAM workplace (Source: Helander, 1982).

GRAPHIC INPUT DEVICES

A graphic input device should enable the operator to quickly and easily specify the contents of the display. Appropriate selection of a graphic input device has the following benefits:

a) reduces time to alter the display

b) reduces errors in selecting or specifying symbols

c) reduces operator fatigue.

There are several different methods for graphics control:
- alphanumeric dialogue
- light pen/light gun
- mouse
- trackball
- joystick
- digitizing pads/tablet
- cursor control keys
- touch sensitivity display
- keyboard coordinate entry
- scanner/digitizer
- touch sensitive pad

Research has been performed to evaluate the relative advantages of different input devices (Parrish, Gates, Munger and Sidorsky, 1981). For example, a light pen is usually preferred for sketching and drawing lines on the display when the need for accuracy is not great, or for placing, moving and deleting symbols. On the other hand, the main advantage with a mouse is the speed by which it is possible to move symbols between different locations on the display. A joystick is particularly advantageous for selecting or specifying graphics commands where the input rate is high. It is therefore possible to optimize the choice of input device depending upon the task. It seems likely that as CAD/CAM workstations become more frequently used, there will be greater versatility in the choice of input device. One word of caution, however, is not to mix the use of several input devices at the same time. Alternating the use of, say, a light pen and a mouse may produce much confusion and decreased productivity (Chapanis, Anderson and Licklider, 1983).

THE INEXPERIENCED USER

With the increased use of computers at the workplace there will also be a wider recruitment of employees to handle computerized tasks. It should be recognized that the potential users represent a wide range of the general population: males and females, 20-65 years of age, minority groups, and different educational background (Muckler, 1981). Although most everyone has had some contact with computers in our society, very few have much direct experience. The major human factors challengs is: how can we design a computer system that may adapt to the range of skills and backgrounds of people that use them so that we can accommodate both experienced and inexperienced users. There are two major issues: the problem of negative attitudes and acceptance of innovations, and the design of training programs.

The Problem of Attitudes

Most computer designers have experienced a wide range of attitudes, both positive and negative towards their systems. Many of the negative attitudes seem to stem from that fact that prospective users might perceive the computer as a device that will depersonalize and dehumanize their work. However, these attitudes, regardless of whether positive or negative, typically change with increased experience with the system. O'Dierno (1977) cited an illustrative example of attitude change during the conversion from a batch-oriented data collection system to an on-line, terminal-oriented system. As can be seen in Figure 3, attitudes differed markedly as a function of organizational level and phase of implementation.

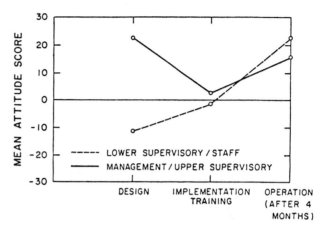

Figure 3. The formation of attitudes towards a computer system depends on organizational level and phase of implementation (Source (Source: O'Dierno, 1977).

A potential user's initial positive or negative attitude depends on several aspects:

- perceived features of the innovation
- prior experiences with similar developments
- estimates of relative advantage
- compatibility and complexity
- perceived personal risk.

The process of acceptance or rejection is, as illustrated by O'Dierno, a dynamic phenomenon that might take place over days, weeks, or even months. The initial inclination toward acceptance or rejection is more refined as further information is received during during the course of the development of the system. In the case of automation of the factory, it is particularly important that the individual be informed about all aspects of the new system. This includes the following:

- information must be comprehensive and factual
- misconceptions must be recognized and corrected
- problems associated with other similar systems must be addressed
- design inputs from users must be considered

Computer Aided Training

The flexibility of computer systems allows for human aids to be a basic part of the system.

It is not uncommon to see training modes in the system itself. This is particularly frequent in advanced control systems in which inexperienced users learn to perform complex tasks. The choice of computer aids for human operators, however, should depend upon what the user needs. In some advanced systems for computer aided training, the computer can make intelligent decisions by monitoring the progress of the trainee. Computer aids may therefore be used less and less as the operator learns the new task. It might be worthwhile to develop a short computer based diagnostic routine for new operators. This kind of exchange information could bring new meaning to the term interactive computer systems (Muckler, 1981).

SOME USER CONSIDERATIONS FOR COMPUTERS

It is likely that many basic human factors considerations found to be fundamental to good systems design in other applications will be equally important in the design of computer-based systems. The human factors principles that seem to be particularly important in specifying a human-computer dialogue include the following (Williges and Williges, 1982):

o Compatibility of computer input and output. This suggests that the input required of the user should be compatible with the output of the computer and vice versa.

o Consistency in the design of the user interface. This permits the user to develop a conceptual model of the operational system. Since computers are complicated machines, it might be necessary to write special software in order to present an interface that is more readily understood. The system should then perform in a generally predictable manner without exception.

o Brevity in computer input and output. The limited capacity of a human shortterm memory (about 7 items) makes it difficult to handle much information at the same time. For example, the number of choices in a menu presented on the screen should be limited to 7 items; thereby the user does not need to read the menu several times before making a decision.

o Flexibility to accommodate different users. Individual differences among users necessitate systems which are flexible enough to accomodate both inexperienced and expert users. In computer-based dialogues the input required of the user and the output provided by the system should depend upon the user's expectations and capabilities. For example, specially designed software might be used in order to evaluate the operator's errors; both the type of errors and the frequency. Such information might then be used by the computer in order to modify messages so that they suit the individual user. One example is the use of a HELP function which provides different explanations, depending upon the experience of the user.

o Immediate feedback from the computer to the user. In computer-base systems, users should at all times be aware of whether their own actions were successfully interpreted by the computer and if so what the computer is doing. One example of immediate feedback is the use of a small blinking symbol on a CRT screen that appears as soon as the operator has logged on. Without that information the operator cannot know whether his/her attempts were successful.

o Operator workload kept within reasonable limits. If this is not the case, there is a greater probability of human errors. This applies both to operators subjected to mental overload and those who are subjected to underload. In the latter case, monotony and boredom makes it difficult to maintain performance at a high level.

Principles like these have been used to formulate guidelines for design of human-computer dialogue. A considerable amount of information is available. Table 2 presents, as an example, guidelines for operating systems recently published by the U. S. Department of Defense (Hendricks et al, 1982).

TABLE 2. SOME HUMAN FACTORS REQUIREMENTS FOR OPERATING SYSTEMS

o LOG ON appears automatically
o A response is made to every input
o Errors are easily correctible
o Error messages begin with error ID
o Error messages are useful
o Error messages are polite and neutral
o Error messages are brief
o Error messages are appropriate
o Error messages are documented
o Levels of HELP are available
o Escape function is provided
o User error will not destroy data

CONCLUSION

Several principles for human factors design of CAD/CAM work stations have been discussed above. For some of the issues there is already a substantial amount of research; for example, ergonomic principles for design of the workstation. In this case it is easy to suggest design improvements. Other issues are more complex and require further research. This includes allocation of decision making between humans and computers, and development of human-computer dialogue. Unfortunately, most designers of computer systems are unaware of the existing research in this area. "User-friendliness" is a popular word now but few people realize exactly what it implies and how existing and future research can be used to support the development of user-friendly design concepts. The editorial column of the BYTE (April, 1983) gives a typical example of such failure to understand: "the air is filled with claims and promises about the merits of each company's products, but nobody knows what makes software easy to use; the final answer will be in what the people buy."

Fortunately, the situation is not all that bad. Research can suggest affirmative answers to some issues, and there is plenty of existing information that may be applied to the design of user-friendly systems. I am hopeful that such information will be incorporated to a greater extent in the design of computer hardware and software in the future.

Acknowledgement

This article is based upon a paper published in the IIE Ergonomics Newsletter, Vol. XVII, No. 4, 1983. I am grateful for the help of Mrs. Martha Pratt and Ms. Debbie Shelor for help in the preparation of the manuscript.

REFERENCES

Chapanis, A., Anderson, N. S., and Licklider, J. C., User-Computer Interaction. In: Research Needs for Human Factors. National Research Council, Washington, D.C.: National Academy Press, 1983.

Helander, M. G. Ergonomics in Automation. Blacksburg, VA: Virginia Polytechnic Institute and State University, Department of Industrial Engineering and Operations Research, 1982.

Helander, M. G., Billingsley, P. A. and Schurick, J. M. An Evaluation of Human Factors Research on Visual Display Terminals in the Workplace. Human Factors Review, in press, 1983.

Hendricks, D., Kilduff, P., Brooks, P., Marshak, R., and Doyle, B. Human Engineering Guidelines for Management Information Systems. Aberdeen, MD: U. S. Army Human Engineering Laboratory, 1982.

Rieger, A. C., and Greenstein, J. S. The allocation of tasks between the human and computer in automated systems. Proceedings of IEEE Conference. Seattle, 1983.

Muckler, F. A. The Inexperienced User. Paper presented at Symposium on Human Factors Considerations in Computer Systems for Nonprofessional Users. AAAS Annual Meeting and Exhibit. Toronto, 1981.

O'Dierno, E. M. Designing Computer Systems for People. In: R. E. Granda and J. M. Finkelman (Eds.). The Role of Human Factors in Computers. New York: Human Factors Society, Metropolitan Chapter, 1977.

Parrish, R. N., Gates, J. L., Munger, S. J. and Sidorsky, R. C. Development of Design Guidelines and Criteria for User/Operator Transactions with Battlefield Automated Systems, Volume IV. VA: U. S. Army Research Institute, 1981.

Greenstein, J. S., and Rouse, M. E. A model of human decisionmaking in multiple process monitoring situations. IEEE Transactions on Systems, Man, and Cybernetics, SMC-12, 182-193, 1982.

Sheridan, T. B. Workshop Documentation. Human Factors Society Annual Meeting. San Francisco, 1980.

Williges, B. H., and Williges, R. C. User Considerations in Computer-Based Information Systems. Blacksburg: Virginia Polytechnic Institute and State University, Department of Industrial Engineering and Operations Research, 1982.

BIOGRAPHICAL SKETCH

Dr. Martin G. Helander is an associate professor at the Department of Industrial and Management Systems Engineering, University of South Florida. He was previously employed by Canyon Research Group, Inc. as a senior scientist. He is presently establishing graduate programs in Human Factors at USF. He is a member of the Human Factors Society, IIE, Ergonomics Society, Society for Information Displays, Robotics International, and IEEE.

IEs Must Address Automation's Effects On Worker Motivation And Capability

By Kelvin Cross
Wang Laboratories Inc.

The information age is upon us, and its most significant impact will be found in the work environment of both the manufacturing and service sectors. Unfortunately, in our rapid quest toward the automated work place, the impact on the work place and on workers is often overlooked or discounted.

As the number of workers diminishes per unit of output, their role in the work place will increase. Using ever-advancing automation and information technologies, each worker will handle a greater volume of work and/or a greater number of tasks. Workers' new role should not be the haphazard secondary result of automation, but rather an integral part of the transition to the automated work place. Job design is as important as the design of products, manufacturing equipment, office automation and the production/service systems themselves.

Workers must have not only the capability of doing the job, but also the incentive to do the job. As automation/technology alters the nature of work, both issues—capability and motivation—must be addressed. Recently there has been much discussion of training to enhance worker capabilities. Yet not enough attention has been paid to motivating the capable worker. The emerging automated work place creates an increasing need for a well thought out relationship between the work and the worker.

Good jobs must be designed; they usually do not occur by accident. The key is to structure jobs around the needs of the worker as well as the needs of the company. The needs of the company usually boil down to delivering a quality product or service on time and profitably. Generally, the division of work is somewhat discretionary as long as the work is done and company goals are met.

Good job design must consider the following: capability, security and pay, objectives and definition, challenge and opportunity, and understanding and involvement.

Capability

The first step in job design is ensuring that the worker is capable of doing the job and doing it well.

The interface between the person and the machine must be designed to utilize the best attributes of each. In other words, jobs for humans should be designed to take advantage of the tasks at which they excel, and jobs for machines should take into account what they do best.

In Table 1, the relative merits of task assignments to humans and machines are illustrated.

Primarily, capability must be designed into the job itself. Training cannot be relied upon to provide workers with the desired capabilities if expectations are unreasonable or unrealistic. The tasks must be within the capabilities of those who will perform them. Using sociotechnical systems analysis, the job designer can ascertain what is causing errors, quality lapses or similar problems.

For example, in a material handling situation, material may often be chipped or otherwise damaged while being removed from a shelf or conveyor. A better designed handling device may eliminate this. Or a worker may have to extract material from a small opening. The worker might be able to do this without damaging the material if he or she is slow and careful, but a larger, or differently designed, opening might speed up the extraction process.

The issues of job security and pay are obviously major concerns of the work force. The introduction of any new process technology can create apprehension unless these issues are dealt with up front. A successful implementation will depend upon making the workers comfortable with their future by assuring them that their basic needs will be met.

Objectives and definition

In order for anyone to do a good job he or she must have some comprehension of what is expected. This does not mean that every job must be defined in minute detail, but rather that the objectives must be made clear. These objectives could change significantly with the introduction of a new automated process. Care should be taken to ensure that the new scope and responsibilities of the job are communicated to those who will be doing it.

Challenge and opportunity

Most, but not all, people perform best in a job which is demanding, but not too demanding. A problem with the automated work place is a tendency to create jobs in which a worker "observes" rather than "does." The result is boredom, which in this case is created by a bad job, not a bad worker.

A job shouldn't be allowed to stagnate to the point of no further learning or growth. Again, the influx of automation can stifle the on-the-job learning process unless jobs are well designed.

Jobs should be designed to include some challenge. For example, a plant may install robots to perform dangerous or routine jobs previously performed by several workers. Instead of doing these tasks, one worker may be assigned to "observe" the robots in case something goes wrong. In order to relieve the boredom of constant observation, the job designer might rotate this task among several workers. Or it might be possible to design an alarm system which would eliminate the need for this observation altogether.

In an assembly-line situation, workers may be made responsible for a whole job rather than just a very small piece of it. Or the team concept could be utilized—with a group of workers responsible for the whole chassis of an automobile instead of just a small part of it.

Worker involvement

Most individuals feel a need to be a part of their work. This means that methods changes must be communicated to the workers at an early stage and their ideas solicited. Every effort must be made to involve workers in the change. They are the ones who will have to make the change work.

Workers should also be informed about the purpose of their work. They must perceive the role their jobs play in the context of the whole operation. They must recognize the impact of their productivity and the quality of their work on the rest of the production process. Feedback of information regarding productivity and work quality is essential to maximize worker performance. Such feedback should be designed into the production/service operation.

Of course, in order to utilize this information, some degree of control is necessary. Along with responsibili-

Table 1: Man-Machine Abilities

	Man	Machine
Analytical abilities		
Reasoning	Inductive Can improvise	Deductive
Computation/Analysis	Single channel of data Subjective Use of judgment Adaptable to changes	Multi-channel Objective Fast, accurate Consistent
Intelligence	Learns from experience	Even "artificial intelligence" is programmed
Retention	Perceptions of principles and strategies	Photographic memory short and long term
Sensitivity	Distinguish patterns Distinguish a wide variety of stimuli occurring simultaneously	Measure minute change Undistracted Measure accurately Designed to perceive inputs well beyond human capability
Physical Abilities		
Speed	Slow, inconsistent	Fast, consistent
Power	Limited	Consistent per design
Dexterity	Very mobile Very versatile/flexible	Fixed by design Can handle fixed multiple high complexity tasks
Endurance	Varies (by day, week, lifetime, etc.) Usually predictable Rapid replacement	Steady pace Instant, unexpected breakdown Long replace or repair lead times possible

Adapted from Singleton, *Man-Machine Systems*

Work space may be closely related to a worker's job satisfaction and performance.

ty for productivity and quality must come the ability of workers to control their portion of the operation.

Reinforcement and incentives

Good work must be rewarded in both monetary and personal terms. A conscious effort must be made by management, supervisors and support people to verbally recognize and reinforce the efforts of others.

Care must also be taken to avoid reinforcement of bad behavior. For example, a high volume of output may be rewarded while quality is overlooked. Obviously, the worker will have every incentive to sacrifice quality in the pursuit of quantity. When a new automated process is implemented, the incentive/reward structure, whether official or unofficial, must promote good performance with regard to all relevant criteria, not just the major benefit of the automation.

Relevant rewards

Rewards must be relevant to the worker. A worker should be able to see a tangible connection between extra effort and extra recognition.

This connection does not always occur. For example, profit sharing in a large company may not always promote extra effort by the individual worker because it is so far removed from him or her. (This may not be the case in a small company with fewer than 1,000 employees.)

There must be a tangible relationship, whether on an individual or a small group basis, between the work and the recognition. Workers need a "piece of the action" over which they have some control.

Group pay incentive plans, such as gainsharing, foster this tangible relationship by rewarding productivity, quality and improvement ideas. To some extent, piecework incentives foster individual output. Unfortunately, traditional piecework systems can actually squash worker participation, improvement ideas and quality by rewarding only output.

Social structure

A work environment is a social structure, with an official and an unofficial set of activities, beliefs and interactions. The implementation of a new method of working can significantly alter the existing social structure. The successful implementation of change must address both the formal and the informal social structure.

For example, an employees' job satisfaction and therefore performance, may be related to the size of his or her work space. Or perhaps a worker's craftsmanship and pride are inextricably related to the "old" method. Each case represents a social problem which must be solved. An effective and mutually beneficial solution will usually require a cooperative effort between the worker and management.

Another social consideration is the external perception of one's job. There is some satisfaction in being able to relate work to social life. Jobs should be designed to encourage pride, especially when explaining a job to others outside the company.

Conclusion

All too often a new automated system is implemented with little or no regard for the kind of job being created. Yet with a little consideration of human factors and motivation, the effectiveness of both the automated process and the work place can be maximized.

For further reading:

Argote, Goodman, Schkade, "The Human Side of Robotics: How Workers React to a Robot," *Sloan Management Review,* Spring 1983.

Birn, Serge A., "Productivity Slump: Broader IE's Needed," *Industrial Engineering,* February 1979.

Cherns, A.B., "Helping Managers: What the Social Scientist Needs to Know," *Organizational Dynamics,* Winter 1972.

Connell, John J., "Managing Human Factors in the Automated Office," *Modern Office Procedures,* March 1982.

Emery, F.E. and Trist, E.I., *Toward a Social Ecology,* Plenum Press, London, 1972.

Herbst, P.G., *Sociotechnical Design,* Tavisock, London, 1974.

Singleton, W.T., *Man-Machine Systems,* Penguin Education, Harmandsworth, England, 1974.

Kelvin Cross is a senior planning analyst for Wang Laboratories Inc. He is primarily involved in medium to long range manufacturing and distribution planning. Cross holds a master's degree in industrial engineering and an undergraduate degree in psychology from the University of Massachusetts, Amherst. He is a member of the Boston chapter of the Institute and of Alpha Pi Mu.

The Impact of Automation on Work Measurement

Richard L. Shell P.E.

Department of Mechanical and Industrial Engineering
University of Cincinnati
Cincinnati, Ohio 45221

ABSTRACT

This paper provides a brief overview of automation and computer aided design and manufacturing developments with estimates of future growth. Common benefits and problems associated with automation are discussed along with future developments predicted for the automated factory. Work measurement and methods engineering activities and techniques that will likely remain unchanged and those that will likely change in the future are discussed. In addition, some of the major industrial engineering considerations likely to be associated with the automated factory of the future are identified.

AUTOMATION OVERVIEW

Computer aided design and manufacturing (CAD/CAM) is perhaps the heart and brain of automation and may be defined as the integration of the complete factory system from order entry through design engineering, manufacturing/industrial engineering, production, assembly, inspection and testing, to final storage/distribution of the completed product. The exact boundaries of CAD/CAM are not well defined for either hardware or software because of the widely diversed number of industries that produce segments of the total system, e.g., data processing/computers, machine tools, and material handling. Also, some CAD systems run on general purpose computers, and with CAM it is sometimes not clear where the computer controlled system begins or ends. Consequently, it is difficult to estimate the present market size of CAD/CAM and even more difficult to precisely forecast future growth and development.

The principal components of CAD include systems and applications software and computer hardware. All sizes of computers (micro, mini, and main frame) have been utilized for CAD activities. Estimates of the 1981 CAD market varied from $275 to $590 million sales. Forecasts for 1985 range from $1.5 to 2.0 billion sales.

The principal components of CAM include systems and applications software, computer hardware, machine tools, NC, DNC, CNC systems, robots, sensors, and automated material handling equipment. Estimates of the 1981 CAM market varies from $6.5 to $7.0 billion sales. Most forecasts for 1985 exceed $10 billion sales not including automated material handling equipment. The most conservative estimates predict that CAM sales will increase in excess of 10 percent annually.

A recent market research report by International Data Corporation surveying only CAD/CAM systems and applications software illustrates the rapid growth of the CAD/CAM industry. Sales in 1981 were reported to be $35.8 million. Sales for 1982 are expected to more than double and sales for 1986 are projected to be $543.5 million. It is interesting to note that low cost software configurations (below $100,000) accounted for about 5 percent of 1981 revenues. It is estimated that by 1986 low cost configurations will account for almost 20 percent of CAD/CAM industry revenues [6].

COMMON BENEFITS AND PROBLEMS OF CAD/CAM AND AUTOMATION

It is generally understood that computer based technologies are expanding rapidly and improving manufacturing productivity through more cost effective utilization of materials, labor, energy, improved inventory control, product quality and ultimately better customer service and marketplace performance [3,4,13,16,17,18]. While not true in all cases, when one considers the manufacturing cost equation, increased levels of automation usually improves overall productivity [9].

Benefits of CAD/CAM and automation include the following:

- Reduced through-put time for the design and manufacturing functions.
- Increased use of "standard" parts and tooling.
- Reduction of errors and changes in design and processing.
- Greater design and production flexibility permitting more product variation at affordable cost and response to market demand.
- Improved management control and understanding of design engineering and manufacturing.
- Greater utilization of equipment and physical plant.

Common problems and concerns associated with CAD/CAM and automation include the following:

- High capital investment.
- Rapidly changing technology.
- Failure to develop a master plan (system architecture) for all CAD/CAM elements.
- Hardware and software selection (purchase or lease) versus internal development versus complete turnkey system.
- Personnel problems, e.g., resistance to change, training older technical employees and production personnel, and properly developing management awareness and understanding.

While most of the problems cited above can be resolved, a few suggestions are offered:

- Work toward developing a combined CAD/CAM system by selecting elements that can later be interfaced with minimum cost and difficulty. Obviously, the combined system reduces duplicate technical activities, further lowers total through-put time and improves response to marketing, and in general increases productivity.
- Emphasize training for all employees, technical, manufacturing, and managerial.

- Interface CAD/CAM with the remaining management information system within the firm.
- Review and develop as needed any remaining support functions, e.g., improved basis for cost/benefit evaluation for future automation, material requirements planning, group technology, and specialized maintenance.
- Continue research and development for CAD/CAM to permit ongoing improvement in the system "intelligence" through human creativity.

FUTURE DIRECTION OF AUTOMATION

Clearly the future holds higher levels of automation for both the factory and the office. In addition, we have become a service oriented economy. The shift of employment from industrial manufacturing to service industries has steadily continued during the past three decades. It is projected that by 1985 less than one-fourth of the work force will be employed in the industrial manufacturing sector [16]. Figure 1 shows the projected distribution of an estimated United States total work force of 114.1 million in 1990.

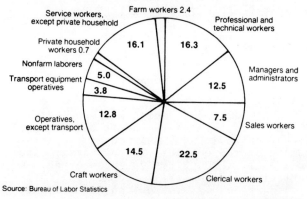

Figure 1. Projected Distribution of Employment by Occupation in 1990 (Millions of Workers)

Before discussing the factory of the future, clarification recently offered by John White is useful. "The automated factory is not the same thing as the automatic factory. The automatic factory is a peopleless factory. In the automated factory, automation and mechanization dominate, but people are still needed to perform a limited number of direct tasks and a greater number of indirect tasks" [19]. Humans will also be required in the automatic factory for indirect support and managerial functions. The automated manufacturing factory is comprised of the following major areas: machining, metal forming, assembly and testing, quality assurance, material handling, robotics, computer control, and support functions.

The trend in automated manufacturing is the ability to process different components, models, or families of parts on the same machining system as contrasted to high production single purpose transfer machines. As Thomas C. Kennicott recently put it, "The dedicated machine of the past is rapidly becoming an anachronism. Increasingly, hard automation is becoming flexible"[7].

For most American companies, assembly accounts for over one-half of all the direct labor costs in manufacturing. Consequently, for the cost motivation alone, assembly automation will probably increase dramatically during the 1980's. With the exception of noncontacting sensors, no major breakthroughs in automated assembly equipment are forecast. However, it is predicted that productivity gains will be made through improved workplace designs. As Frank J. Riley recently stated, "Using ergonomic principles and simple automated processes, these work stations will integrate the operator with the simple machines"[15].

Material handling will become more important in the automated factory [19]. It has been predicted by Richard Polacek that the use of robots and pallet shuttles that operate around the clock will become commonplace. Even the job shop with small lot sizes and great varieties of parts will have unmanned operations [14].

The future development of factories has been very well summarized by Russel M. Loomis. "Business is preparing for a fundamental change in production techniques and management that, by the end of the 20th century, will rival the Industrial Revolution of the 18th century. This transition will herald the manufacturing facility of the future - the fully automated factory. These factories will operate at optimum levels of efficiency and productivity with virtually every operation, from ordering and accepting raw materials to manufacturing, to assembly, to quality control inspections, to shipping, and customer billing performed, monitored, and controlled by intelligent machines" [8].

FUTURE DIRECTIONS OF WORK MEASUREMENT AND METHODS ENGINEERING

Introduction

During the next decade, CAD/CAM and automation will significantly change the nature of the work force. The decrease of jobs for minimal skilled direct factory workers will be accompanied with a rapid increase in demand for highly trained (and perhaps better educated) specialists. There will be a large number of clerical/office support, technical, professional, and managerial personnel performing irregular, complex, long cycle work activities. Most of there time will be devoted to mental processes with much less time expended doing physical work.

Recently reported data on sources of productivity loss in high production manufacturing operations illustrates the importance of work measurement in the highly automated factory. For dedicated automatic equipment where set-up and tear down time losses are minimal, studies show that actual output is only 59 percent of the optimum capacity or losses of 41 percent. The losses are categorized as follows [2]:

Source	Percentage
Equipment Failure	42
Machine in Wait Mode	34
Work-force Control	16
Other	8

Based on the above percentages, effective work measurement to improve maintenance, scheduling, and support personnel utilization could greatly reduce productivity losses. For job shop operations where equipment changeover becomes a major activity and consumer of human resources, work measurement of the set-up procedure would produce additional benefits.

The major objective of work measurement and methods engineering will remain fundamentally unchanged as levels of automation increase and we move toward the automatic factory. Work measurement and methods engineers will still function to increase productivity and lower unit cost, thus allowing more goods to be produced for more people [11]. How industrial engineers realize this objective will indeed change. The following sections summarize what will likely remain unchanged, what will likely change, and what will be some of the major considerations of work measurement and methods engineering in the future.

What will Remain Essentially the Same

While numerous changes will occur in the theory and application of work measurement and methods engineering, it is believed that several traditional practices will remain. In the future CAD/CAM environment, the following practices will be conducted as in the past and present time.

- Perform the production engineering function to review product designs, and select raw materials.

- Perform the manufacturing engineering function to determine processes, manufacturing sequence, tool/fixture and test equipment design, quality assurance planning, and cost estimating.

- Review the product design and total manufacturing planning for possible safety/health problems and/or product liability exposure.

- Interact with other functional areas of the business, e.g., sales/marketing, finance/accounting, and general management.

The equipment utilized to conduct the above work activities as well as the work time expended will change as the automation level increases. For example, there will be a greater availability and use of computers and interactive graphics [5, 8, 20].

Examples of specific work measurement and methods engineering functions that will remain essentially unchanged include the following:

- Design of work place and human-machine interface considerations.

- Evaluation and measurement of indirect and support personnel, e.g., data processing, professional/technical staff, clerical/office, and maintenance.

- Evaluation and measurement of equipment change over.

- Determination of personnel allowances.

- Planning for manufacturing capacity and utilization.

Most of the existing work measurement and methods engineering techniques will be utilized in performing the above functions.

Probable Changes

The CAD/CAM automated factory of the future will cause several changes in the practice of work measurement and methods engineering including the following:

- The requirement to conduct direct observation timing and the associated rating or leveling of performance will decrease because of the increased machine and computer controlled processing cycles. Many cycle times will be automatically determined from internal clocks and calendars of the digital control device.

- While the determination of personal, fatigue, and delay allowances will be done as in the past, there will be changes. In general, physical fatigue will decrease while mental fatigue of workers will increase. Unavoidable delays in the highly automated factory will be different from traditional manufacturing plants.

- Work sampling will be simplified because of the ability to electronically monitor and record the activity state of most elements of the automated factory.

- While predetermined time systems will still be used in the traditional form for certain indirect and support personnel measurement, there is a need for further development in computerized synthetic time systems to facilitate application to the activities of knowledge workers. Robot Time and Motion (RTM) is an example of a special predetermined time system that will be useful in the highly automated CAD/CAM factory [12]. The trend to computerize predetermined time systems is likely to continue [10].

- Standard data will probably be developed for major segments of the automated manufacturing system as contrasted to small elements of an individual processing cycle. With major segment manufacturing times built into the CAD/CAM data base, one could easily simulate cost comparisons of alternate methods of manufacture and rapidly complete product cost estimates before production.

Other Considerations

The advent of the highly automated CAD/CAM factory will present a number of questions or difficult challenges for the work measurement and methods engineer. Among these are:

- The ability to easily compare the total cost effectiveness between various levels of automation to produce a given product. It is possible that certain direct production activities in the automated factory could be best (lowest cost) performed with a more labor intensive process.

- Should the automated system be provided with a "manual" backup during extended periods of breakdown? One must consider both the CAD/CAM system as well as the automated factory equipment.

- Inaccuracies in work measurement systems occur at every level of the total system, starting with the basic technique(s) [1]. Consequently, accuracy requirements must be determined for those techniques employed in the automated factory.

CONCLUSIONS AND RECOMMENDATIONS

The following conclusions and recommendations are offered for improving the practices of work measurement and methods engineering in the highly automated CAD/CAM factory of the future:

- The development of manufacturing automation is proceeding at an increasing rate, and the automatic factory is predicted to arrive by the year 2000.

- The ongoing development of CAD/CAM will greatly influence the practice of work measurement and methods engineering. The work activities of industrial engineers concerned with work measurement and methods in the automated factory will be significantly different, e.g., less time devoted to direct observation time study, more concern with indirect and support functions. Because of this, work sampling may become the most widely utilized of the presently known work measurement techniques.

- The overall work assignment of the industrial engineer in the automated or automatic factory will be fairly comprehensive and broad based. Consequently, the future engineer should be well educated and trained to understand and properly interact with the total complex CAD/CAM factory.

- There is a need to upgrade and expand existing work measurement techniques. Examples cited include specialized predetermined time systems, and standard data for automated manufacturing segments.

- Because of the increase of knowledge workers, there is a need to accurately assess mental output performance and the resulting mental fatigue.

REFERENCES

1. Brisley, Chester L., and Fielder, William F., "Balancing Cost and Accuracy in Settling Up Standards for Work Measurement", Industrial Engineering, Vol. 14, No. 5, May 1982, pp. 82-91.

2. Carter, Charles F., Jr, "Toward Flexible Automation", Manufacturing Engineering, Vol. 89, No. 2, August 1982, pp. 75-78.

3. Classen, Ronald J., and Malstrom, Eric M., "Effective Capacity Planning for Automated Factories Requires Workable Simulation Tools and Responsive Shop Floor Controls", Industrial Engineering, Vol. 14, No. 4, April 1982, pp. 73-78.

4. Engelberger, Joseph E., "Turning America Around with Robots", Manufacturing Engineering, Vol. 89, No. 2, August 1982, pp. 112-114.

5. Hosni, Yasser A., "Time Standards Using Micro Computers", Proceedings, Annual Industrial Engineering Conference, Institute of Industrial Engineers, 1982, pp 694-699.

6. "IDC Forecasts Low Cost CAD/CAM Systems Market", Typeworld, August 20, 1982, p. 15.

7. Kennicott, Thomas C., "Transfer Machines in the '80s", Manufacturing Engineering, Vol. 89, No. 2, August 1982, pp 111-112.

8. Loomis, Russell M., "The Programmable Controller Today and Tomorrow", Manufacturing Engineering, Vol. 89, No. 2, August 1982, pp. 117-119.

9. Malstrom, Eric M., and Shell, Richard L., "Projections for Future Manufacturing Work Force Productivity", Proceedings, 25th Annual Joint Engineering Management Conference, The American Institute of Industrial Engineers, 1977, pp. 41-47.

10. Martin, John C., "Program, Portable Microprocessor Allow Direct On-site Input of MTM-1 Data", Industrial Engineering, Vol. 14, No. 8, August 1982, pp. 50-53.

11. Niebel, Benjamin W., Motion and Time Study, 7th edition, Irwin, 1982, p.7.

12. Nof, Shimon Y., and Lechtman, Hannan, "Robot Time and Motion System Provides Means of Evaluating Alternate Robot Work Methods," *Industrial Engineering*, Vol. 14, No. 4, April 1982, pp. 38-48.

13. Ottinger, Lester V., "Questions Potential Robot Users Commonly Ask", *Industrial Engineering*, August 1982, pp 28-31.

14. Polacek, Richard, "Machining Centers: Today and Tomorrow", *Manufacturing Engineering*, Vol. 89, No. 2, August 1982, pp. 97-99.

15. Riley, Frank J., "The Look of Automatic Assembly", *Manufacturing Engineering*, Vol. 89, No. 2, August 1982, pp. 79-80.

16. Shell, Richard L., and Malstrom, Eric M., "Measurement and Enhancement of Work Force Productivity in Service Organizations", *Proceedings, 25th Annual Joint Engineering Management Conference*, The American Institute of Industrial Engineers, 1977, pp. 29-35.

17. Van Singel, Gary, "Tips on Selecting the Most Suitable Equipment Type", *Industrial Engineering*, Vol. 14, No. 5, May 1982, pp. 32-36.

18. Vasilash, Gary S., "General Electric: Bring the Factory of Tomorrow to Life", Manufacturing Engineering, October 1981, pp. 72-75.

19. White, John A., "Factory of Future Will Need Bridges Between its Islands of Automation", *Industrial Engineering*, Vol. 14, No. 4, April 1982, pp 61-68.

20. Wright, Marc B., "Work Measurement System Monitors The Output of A Word Processing Operation", *Industrial Engineering*, Vol. 14, No. 7, July 1982, pp. 70-72.

Computerized Work Measurement and Process Planning — Vital Ingredients in a Computer-Integrated Manufacturing (CIM) System

Kjell B. Zandin
Senior Vice President
H.B. Maynard and Company, Inc.

One of the most popular topics today in magazines covering the business world is the vision of the totally automated factory. But with a very few exceptions, it is a dream that exists only in the minds of some long-range planners and industrial prophets.

There is no doubt that eventually factories will be more highly automated than today. Steps have already been taken in that direction with such devices as computer numerical control (CNC) and robots. However, these are only small parts of the total system for which there needs to be an overall and highly integrated master plan.

There are several questions that need to be examined with regard to total manufacturing automation. First, why automate? Second, what are obstacles to automation? And third, what practical steps do managers need to take to start their companies on the path to an operation that is essentially controlled by computers.

Competition is the biggest reason for modifying the present factory system common in the United States. European and Japanese companies are already following a course that will result in totally automated factories within ten years. United States companies cannot afford to lag far behind. The second reason for automating is productivity. There is no doubt that machines are more efficient than humans who do not perform consistently and who need rest and lunch breaks. Third, is the declining workforce pool in the United States. Population growth is slowing and the blue collar ranks are thinning as education levels rise and more workers aspire to white collar jobs.

OLD WAYS ARE NOT BEST

Despite these trends, there are only sporadic moves toward more automation. American technology leads the world in computer development and yet United States industry is in the dark ages regarding the application of that advantage to Computer-Assisted Manufacturing (CAM). We have robots that can weld parts and perform other manufacturing tasks. We have numerically controlled machines that cut metal precisely time after time.

More complex automation, such as the economical and efficient preparation of a production plan or automatic cost estimating for a job does not exist in most United States companies. Product design is still performed by row after row of draftsmen toiling over blueprints. It recalls scenes of monks drawing illuminated manuscripts in the monasteries of medieval Europe. Cost estimating still stumbles about in the darkness of the "best guess", based partly on past experience and partly on a look in the crystal ball.

A majority of industrial engineers still march out to the shop floor with stopwatches in hand to time the work, then spend long hours toiling with calculations and paperwork to produce standards that are out of date almost as soon as they are set.

ROADBLOCKS TO PROGRESS

Why are United States companies still bogged down in these old practices? Why, with our superior knowledge of computer technology, don't we program our computers to do much of this routine, repetitive work for which these machines are so well adapted? One of the major reasons is cost. It is certainly not inexpensive to design and build machines that will take the place of men. The recurrent recessions that have plagued the world's economy over the last several decades add to the reluctance of industrialists to make such long term investments. Wickham Skinner, a professor in business administration at the Harvard Business School, wrote in the Wall Street Journal recently that: "Investments in new equipment and processes usually take a good three to five years to perfect, but in too many United States companies, capital budget systems demand an earlier payback -- with the result that essential new technology is turned down. Management rewards structures that discourage risk-taking also hold back the adoption of new technologies and equipment, particularly as needed investments grow even larger in cost, scale and interconnectedness."

Another reason progress on automated factories is so hesitant is that most managers do not know how to proceed in putting together such a system. This leads us to a discussion of what the practical steps are in developing a computer-integrated manufacturing system.

THREE BASIC ELEMENTS

There are three building blocks in the first stage of a Computer Integrated Manufacturing System. These will then lead to other links in the chain of assembling the automated factory. We believe the heart of any system is measurement. You must have good methods for determining how much labor depends on this calculation, as does the cost estimate.

There are essentially two ways to go with standards calculations. You can take manual data you have already developed and put it on the computer, or you can buy a developed software program that does these calculations for you. The tradeoffs here are cost versus maintenance. It will probably cost you more initially to put a usable system of data operation and management on the computer. On the other hand, by transferring manual data, you will have to update it at some point, which means redoing the measurement program.

If you purchase software, you should be sure that it contains a mass update feature to avoid process plan delays or inaccurate data feeding into the process plan from this source. And, of course, the system itself should be interactive and on-line. This means that the computer operator can talk to the machine and receive answers in real time. Batch processing is not the solution for the automated factory. Data management and retrieval on such a system should be equally convenient and simple to facilitate rapid assembly of the elements that make up a process plan.

While work measurement gives the labor input, a separate system is needed for machining and welding operations, known in the industrial engineering language as process time or machine time. This system is related to work measurement because the worker is often involved in the setup or other operations surrounding process time on the machine or with the welding unit. Even in the case of numerical control machines, it will be necessary to know the operation time in order to calculate the total process time. Most of the process time formula is knowing what machines are available and assigning jobs to machines that can do the work most effectively.

PART CLASSIFICATION

A third ingredient that goes into the CIM system is a classification system. This is a means of defining the part or parts to be manufactured and grouping these parts with other similar parts. By classifying parts into families, it is possible to find out quickly if any similar part has already been manufactured. Another feature of the classification system is quick retrieval of component design.

There are a number of classification systems in existence today and to say that one system is better than another would be misleading. A company looking at such a system to augment its manufacturing operations would have to consider which system would be best for its application. Even then, the best system would have to be tailored for a company's particular needs.

CIM SYSTEM INTEGRATION

These three elements, work measurement, process planning, and classification form the basis of an integrated system that contains all the elements needed by the manufacturing engineer to develop process plans automatically and put them out to the shop floor as a routing which tells the worker how to do the job. It can also serve as a source of cost estimates for accounting which generate the lowest feasible bid that can be put into a proposal or the expected cost of a job for budgeting purposes.

What this system does is provide much more automatic and faster process planning with fewer people. Because it is more consistent, changes can be made faster and the whole production control system is more efficient. It speeds up the manufacturing process and has the beneficial side effect of reducing inventories of parts and tools. The special benefit of the work measurement system in this scheme is that standards are readily available and contribute to the rapid development of precise methods in the production process.

Beginning with design, this system can take a part specification and run it through the classification system to identify it according to part, function, shape or any other characteristic desired. The computer searches the database and retrieves similar parts that have been designed in the past. It also pulls out information on how the part was made, including materials, condition of materials, fabrication equipment and what machines or tools were used.

At this point, work measurement is added to the formula. If a new calculation of a labor standard is needed, it can be performed on the computer. Or an existing labor standard can be retrieved from the database and combined with the process information.

Advantages of such a system are immediately apparent. Consistency is a major one. Given the basic information needed to process a part, five planners will come up with five different process plans. Each of these plans will vary in efficiency, depending on the experience and skill of the planner involved. The computer, however, calculates the fastest and most cost-effective way to produce the part every time.

According to a recent article in AMERICAN MACHINIST, the greatest portion of nonproductive time on the shop floor results from idle time that occurs because the planners are unable to write a process with enough detail to keep production moving in a continuous flow. In an average workshop, a part spends only five percent of its time on a machine tool. And, of that five percent, less than 30 percent involves actual metal removal. We can have the most modern and efficient machine and cutting tools in our plants, we can have robots and NC machines row upon row, but if they are not operating, the whole investment is lost. The computer eliminates idle time because it calculates the most efficient route for a part to take through the manufacturing process.

STARTING POINT

The final question, then, is where to begin? How can a manager take the first step down the road to total automation and what is the best first step? We believe that establishing good computerized standards is probably the first goal. There are several reasons for this approach. First, standards are going to be absolutely essential to the effective operations of the other parts of the system. Second, standards require a discipline of methods which is a step toward automation. Third, standards are now available on software programs that are easy to install. And fourth, standards can provide immediate productivity improvements that make it easier to convince management and the workers that further automation of the work process is achievable.

Along with the benefits of the automated factory, there have been some criticisms of such a highly computerized operation. And there is no doubt that it is certainly going to generate changes in the workplace. Progress always exacts its price in one form or another. Workers are going to have to learn new skills, especially those employees who are replaced by machines. But this has always been the case in every great industrial advance.

The problem is not if, but when and how we are going to make these advances. In our presentation today, we have tried to indicate the direction this new technological development is taking us and the pace at which we are traveling. The automated factory is certainly not a reality today, but we can already see the beginnings of such a system in place. Very few companies, if any, have been able to do the whole job. But such a conversion must be made if we are going to realize substantial productivity gains in the near and long term future. We are literally standing on the brink of another Industrial Revolution today. As in earlier revolutions, those companies that are able to adapt to the new climate will grow and prosper. Those companies that do not adopt new methods will wither and disappear from the industrial scene.

Flexible Manufacturing Systems: Combining Elements To Lower Costs, Add Flexibility

By H. Thomas Klahorst
Kearney & Trecker Corp.

Flexible manufacturing systems (FMS) can be defined as a group of machines and related equipment brought together to completely process a group or family of parts.

Their primary components (see Figure 1) include:
1.) Machine tools.
2.) A material handling system.
3.) A supervisory computer control network.

The secondary elements which must be added to complete an FMS environment include:
1.) Numerical control (NC) process technology.
2.) Spindle tooling.
3.) Workholding fixtures.
4.) Operations management.

The major task in FMS application is to insure the smooth combination of these elements. The application of FMS to a variety of production problems has been increasing steadily. This trend is primarily a function of the benefits being derived from such systems.

The anticipated benefits which manufacturing people expect upon implementing a flexible manufacturing system can be categorized in three major groups:
1.) Increase production flexibility related to product design or production volume.
2.) Reduce operating costs or initial investment.
3.) Improve ability to respond to the market and its unpredictable demand characteristics.

In the systems that have been implemented, the benefit relationships are:
☐ 55% related to cost reduction programs.
☐ 30% related to market response improvement.
☐ 15% related to flexibility in production.

The types of problems which FMS can be applied to vary greatly. In order to meet these varying demands, modules representing machine tools, material handling and computer control can be defined and combined in innumerable ways to create the proper systems design.

Machine tools—Generally, the FMS concept will be applied to a production environment that can be defined as mid-volume, mid-variety. Therefore, most of the machine tools will feature NC technology and elements of flexibility which allow them to address numerous part types with no changeover.

Material handling—Almost every benefit of a manufacturing system is affected directly by the selection of the material handling system. The basic characteristics a material handling system must have include:
☐ Lack of floor obstructions.
☐ Low cost per unit transport.
☐ Low cost per foot transport.
☐ Design configuration flexibility.
☐ Proven reliability.
☐ Quietness.
☐ Expandability.
☐ In-process dynamics.

Table 1 evaluates some of the forms of material handling and how they relate to these objectives.

Horizontal forms of material handling must be selected based upon the application's task—no one type will solve all problems. In systems where random delivery and high delivery flexibility are required, three system types seem best suited:
1.) Tow-line.
2.) Guided vehicles.
3.) Power and free.

When part size or higher levels of production are desired, two different forms of material handling lend themselves:
1.) Roller or rail-type.
2.) Shuttle car.

In all applications, the selected method of material handling is a function of values placed on the different objectives of a material handling system. The "objective weights" in the chart are variables which must be assigned by the builder/user design team.

Figure 1: Primary Components of a Flexible Manufacturing System

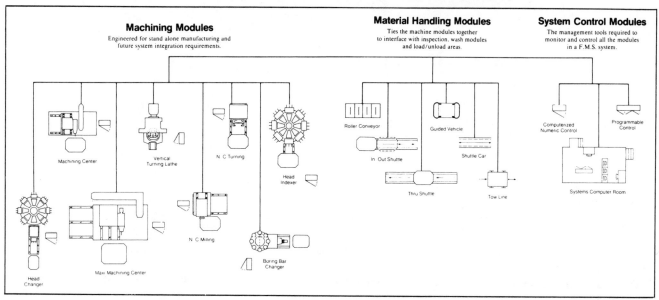

Computers—The control systems for FMS now utilize a series of control systems. Each level of control performs to a limited decision level:

☐ *Level 1:* The most basic units of control, including programmable controllers and computerized numerical control (CNC) units of machinery. This level of control is "logic" oriented and is heavily dependent upon the physical equipment in the system for its structure. Equipment can be operated in a manual mode at this level of control.

☐ *Level 2:* The parts flow and production process are controlled at this level. It includes programmable logic plus minicomputers, and it interfaces with the Level 1 elements. A semi-automatic mode of operation is possible at this level of control.

☐ *Level 3:* Management information, real-time routing, decisions and manpower direction exist at this level of control. It is software oriented and does not change because of system components. At this level, full FMS automatic operation is achieved.

Aspects of integration

Integrating an FMS can cause several business issues to be raised, including:

☐ *Who* should do it?
☐ *When* should it be done?
☐ *What* are my responsibilities as the final user?

Let's consider these issues—one at a time.

1.) *Who should do it?*

The manufacturing system is comprised of a number of elements:

Element	% of Project Value
Machine	50%
Material handling	10%
Controls	8%
Tools and fixtures	25%
Services	7%
Total	100%

Table 1: Material Handling System Evaluation

Objective Weight:	Open Floor 15	Low Unit Cost 10	Low Foot Cost 10	Design Flexibility 15	Reliable 20	Quiet 5	Expandable 10	Process Buffer 10	Speed 5	Score (100 = Perfect)	Comments
1. Tow-Line	.9	.9	.6	1.0	.8	1.0	.9	.8	.2	82.5	Proven
2. Guided Vehicle	.9	.3	.9	1.0	.5	1.0	.9	.5	.7	73.0	New potential
3. Shuttle Car	.2	.2	.5	.6	.8	.9	.6	.2	.7	51.0	Proven—some problems
4. Roller/Rail	.1	.5	.1	.6	.9	.9	.4	.2	.3	46.5	Obstructive
5. Power & Free	.4	.7	.8	.9	.6	.6	.6	.8	.4	65.5	Not tested—some problems
6. Stacker Crane	.1	.1	.1	.4	.6	.7	.8	.2	.8	39.0	Restrictive configurations
7. Elevated Track	.1	.6	.3	.6	.6	.8	.9	.8	.7	56.0	New technology

Rating: 1.0 = Objective completely met.
 .5 = Fair compliance to objective.
 .1 = Objective minimally satisfied.

This breakdown of a typical system, based on Kearney & Trecker's experience with manufacturing projects, indicates that the machine tool builder—and the industrial engineer who works with him or her—are directly responsible for 50% and oversee another 25% of any given system's project.

2.) *When is an FMS applied?*

A few guidelines can be defined as to when a manufacturing system should be applied:

☐ When part size and mass exceed "jib crane" standards.

☐ When production volume is in excess of two parts per hour.

☐ When processing requires more than two machine types to complete a workpiece.

☐ When more than five machines are required.

☐ When phased implementation is planned so that material handling provisions can be incorporated in the initial phases and bad habits can be avoided from the start.

3.) *What are the responsibilities of the user?*

The tendency to assume that the system's builders will solve all problems must be avoided. Compared to the life of a system, the builder is involved for a very short time. Responsibility must be with the user. Let's review some of them:

☐ Installation planning involvement: piping and wiring methods.

☐ Installation manpower supplied during physical installation of the equipment.

☐ Maintenance training: hands-on training before product shipment and on-site training during the installation and test activity.

☐ Management awareness: The user's various levels of management should take part in training classes and simulation exercises to acquaint them with the nature of the production system purchased. Failure of management to understand the system's capabilities is a danger to be avoided.

☐ *Problem definition:* Forming the foundation for all work is the definition of the task, established by defining three parameters. These three parameters outline the solution which will be implemented.

1. Part mix: What is the total universe of part types the FMS will address? Certainly, current parts must be defined, but, of equal importance, future parts should also be defined. Pretend that five years have passed—how is the part mix different? If it is not different, go out ten years.

2. Production volume: What is the production per year, per part? If you don't know, guess or estimate. Once again, look out five years: How has time affected production volumes?

3. Business manufacturing philosophy: There are always a few basic business operating philosophies which will have a major impact on system design; these should be defined at the start of a project. Examples include:

☐ Batch or random processing.

☐ Part design change frequency.

☐ Stability of market demand.

Summary

No one flexible manufacturing system design is capable of satisfying all the varying elements in the range of manufacturing problems. Each application must establish its own criteria and objectives. From this the FMS design can be selected.

Regardless of design, some conclusions can be drawn from the operating experience gained from manufacturing systems:

☐ Their application frequency and scope will increase for the next ten years.

☐ The use of computer technology will increase.

☐ Greater automation will occur in systems, including robot load/unload functions, automated tool delivery and tool setting facilities and integrated storage facilities.

These changes are being initiated due to the results automated manufacturing systems yield:

☐ Direct labor content in the workpiece is reduced to one-fourth of previous levels.

☐ Indirect and factory burden costs are reduced to one-eighth of previous values.

☐ Floor area is 60% for the same productivity.

☐ Lead time is reduced five or six times.

☐ New product designs are accepted with minimum cost.

The combined technologies of machines, material handling and computers are providing major advancements in productivity, and this trend is increasing.

H. Thomas Klahorst has been manager of special product sales for Kearney & Trecker Corp. since 1975. Klahorst, who joined Kearney & Trecker in 1968, has held several other positions within the company, including sales trainee, product application engineer, project manager, sales engineer and supervisor of systems sales. He holds both bachelor's and master's degrees in business administration from the University of Wisconsin.

Management Looking At CIMS Must Deal Effectively With These 'Issues And Realities'

By Robert L. French
R.L. French & Co.

A number of previous articles in this series have touched on the subject of management commitment and involvement. The intent of this article is to focus on the problems and opportunities provided by computer-integrated manufacturing systems (CIMS) from the perspective of operating management.

Although some significant progress has been made during the last 30 years, by and large existing manufacturing control systems range from moderately successful to disastrous failures.

As of the summer of 1984 *we have not made* the significant breakthrough that is necessary to allow manufacturing control systems to achieve their true potential.

The most common explanation of this failure is that we have lacked "management commitment," and if only those dullards who are making the decisions would listen to those of us who are enlightened, we would be on the way to great leaps forward in productivity, and the "factory of the future" would become the "factory of the present."

This series has established that the concept of CIMS encompasses an integrated data base that flows through the entire process from design (CAD/CAM) to financial control, with all intermediate transactions—such as production scheduling, process automation, process control, material handling and storage, and maintenance scheduling and control—being driven from the same data base.

Let's look at 15 "issues and realities" that face the operating management considering a CIMS:

1.) *The present system of accounting and control is not adequate to measure or support CIMS.*

Without being critical of our colleagues in the accounting profession, manufacturing accounting has not changed significantly during the last 30 years.

Since the end of World War II, the American system of management has focused primarily upon the effective use of direct labor. Both the industrial engineering and accounting professions have concentrated almost exclusively upon the efficient use of the direct labor hour.

A great deal less attention has been given to an accurate allocation of overhead costs. Even today, a great many plants use plant-wide rather than work or load center burden rates.

This results in serious distortions of cost between simple bench operations and those performed on CNC machines, for example. As the shift from a labor intensive to a capital intensive manufacturing base has accelerated, accounting and control systems have not been reevaluated accordingly.

As an illustration, American manufacturing and control practice has essentially ignored setup time for years. With the advent of CIMS the Japanese have recognized the critical importance of setup and have given setup time the same type of engineering effort as a direct production operation.

This is merely the tip of the iceberg—the entire area of cost effectiveness and system justification is crying out for fresh approaches and concepts.

2.) *Very few companies can afford the capital investment necessary to make all existing plant machinery compatible with CIMS.*

While there are a great many CNC machine tools now in use, a very high percentage of the work being performed is still being done on previous generations of equipment. Almost every plant is a mixture of new and old. Very few have the luxury of all new equipment.

3.) *Product design and manufacturing processes have a very short life cycle.*

Technology is exploding in arenas other than computerization. The decision to commit extensive resources to CIMS to completely automate a process carries the assumption that the process itself is secure and assured of a long life cycle. There are very few products or processes that are not either already under pressure from alternate materials or subject to complete redesign for the future.

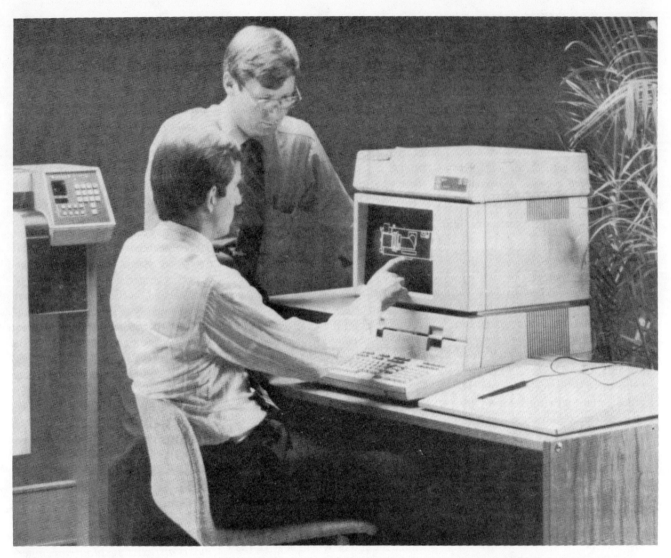

Managers implementing a CIM system face the challenge of retaining still useful portions of the present process while introducing new tools that will increase operational effectiveness. (Photo courtesy of Carrier)

4.) *Manufacturing strategy may be compromised in favor of marketing strategy.*

It is very easy to add products to those offered by a company, but it is extremely difficult to take them away. To maintain credibility in the market, a company frequently must continue to manufacture products to support the customer even though volume and facilities are not compatible with CIMS.

This may appear to be a communications problem or even one of relative power status between the manufacturing and marketing sections of the business, but in reality, the marketing strategy of retaining "customer support" types of relatively obsolete products may be essential to the survival of the business.

5.) *It is difficult to justify the cost of data base development and maintenance.*

One of the biggest obstacles to successful inventory management and shop floor control systems has been the failure to maintain bill of material structures and shop routings. During periods of intense cost pressure, such as the recent recession, the temptation is very great to eliminate some of the tedious detail that is necessary to keep the manufacturing data base fully viable. It is very difficult to quantify the value of accurate data base records.

6.) *A great deal of hostility toward computerization continues to exist at all levels within the typical manufacturing firm.*

Although the massive invasion of personal computers into classrooms, homes, retail stores, etc., has greatly changed the level of the general public's understanding of computerization, it has only driven underground the hostility of those who will see their jobs change drastically or disappear altogether due to CIMS. At this juncture, middle management is threatened. Hostility now takes more subtle forms.

7.) *Technical skills in CIMS remain in short supply.*

While there has been a massive shift in job training, and a great many people are learning to be programmers, there is not an even distribution between specific availability and need. While there may be an abundance of entry level COBOL programmers, there are critical shortages of skilled computer maintenance technicians, CAD/CAM engineers, etc.

Also, since those with experience specialize in the operating system and software of a given manufacturer, those skills do not readily transfer to another manufacturer or software system.

These inbalances create continual turnover among those with high-demand skills, and therefore threaten the success of a CIMS program after management has undertaken the project.

8.) *Program languages are not compatible among various CIMS elements.* This phenomenon is true on two levels:

☐ Incompatibility between various stages of CIMS. For example, at this stage, the programs that drive CNC machines are not compatible with shop floor control and MRP systems.

☐ Incompatibility between systems from various hardware manufacturers for a given segment. For example, accounting software packages for one set of hardware do not interface with shop floor control systems from another manufacturer.

9.) *The management generation gap:*

The managers in senior level positions in most manufacturing companies today have managed during the punched card era with all of the rosy promises and false starts of the last 20 years. These people have seldom been "hands on," and have only experienced the disappointment of unrealistic expectations.

At the other end of the spectrum, the young engineers and managers just entering the work force have been very much "hands on." It is very difficult to these two groups to communicate effectively.

10.) *CIMS is trying to hit a moving target.*

American manufacturing techniques and philosophy are in the midst of very dramatic change. The success of Japanese manufacturing methods, the move to worker participation in the decision making process, etc., will have a dramatic impact on CIMS systems in the future.

Traditional manufacturing reporting systems are designed around the "structured" organization. Managers are accountable for performance and lead their people toward the achievement of goals through direct supervision and control. The impact of some of the newer concepts of the behavioral sciences upon CIMS is not altogether clear.

11.) *Many manufacturing plants contain a mixture of batch and flow-through production.*

While it is relatively easy to design systems for either "batch" or "flow-through," the mixture of both in the same system makes the control process more difficult. For example, such finishing operations as tumbling, painting, plating, degreasing, washing and heat treating are typically batch operations and require a grouping of flow-through products. These processes therefore resist the total integration of computerized control.

12.) *Volume and mix is dynamic.*

A given manufacturing plan must inevitably make a choice about the assignment of resources. The use of a flexible manufacturing system may be designed with a given set of parameters.

The realities of the market may change those parameters drastically. A given product may "take off" on relatively short notice, which will necessitate dedicated production facilities and therefore alteration of the CIMS design.

13.) *Traditional forms of functional organization may be entirely outdated.*

One of the first questions that always arises with the installation of a data base system is, "Who is responsible for data base mainte-

Problems

- Inadequate measurement system.
- Partially obsolete facilities.
- Short product life cycles.
- Obsolete products.
- Inadequate data base.
- User hostility.
- Shortage of technical skills.
- Incompatibility between systems.
- Management generation gap.
- Changes in management philosophy.
- Facilities with mixed processing.
- Dynamic volume and mix.
- Outdated organization.
- Varieties of process options.
- Loss of superior/subordinate support.

nance?" Should product engineering have complete control of the parts master and bill of materials, or should manufacturing?

Considering that the data base really now serves the entire organization, who is knowledgeable and powerful enough to effectively manage it to meet all functional needs?

A strong case can be made that a new general manager level position is needed to manage the data base. The person with the qualifications to perform the task should have functioned as the operations general manager and gained a broad understanding of the total business before assuming this position.

With the establishment of such a position, the actual ownership of all data base information would be transferred to the data base general manager. Data base decisions would therefore be made based upon the best interests of the entire company rather than the partisan interests of any one department.

14.) *In many manufacturing facilities the variety of process options is exponential.*

Many manufacturing software packages make it possible to designate "alternate process routings" which indicate authorized alternate methods for use if the need arises.

The need to use an alternate process can be created by a number of conditions, such as a breakdown of the primary resources or a volume and mix which overload the primary. In certain environments this condition can become extremely complex.

> **Approach**
> - Realistic assessment.
> - Functional impact.
> - Company-wide impact definition.
> - Realistic plan.
> - Commitment to strategy, not details.
> - Feedback and participation by function.

15.) *The shock of change can destroy credibility and support by superiors and/or subordinates.*

It takes very strong leadership skills to successfully implement a program with the massive impact of a CIMS. For an operating manager to be successful in this venture, the need for broad-based support and realistic time expectations is absolute.

Support may wear thin or waver from both above and below. The board of directors may be willing to commit adequate resources, but the working level staff may become disillusioned by the extensive changes required in the details of practically every job in the company.

The operating manager contemplating CIMS must recognize these realities and be prepared to effectively deal with the problems that will develop.

Prescription for success

To be successful in implementing CIMS, management must do an extraordinary amount of homework. The installation of a computerized system that is totally integrated literally goes to the very core of the management process itself. In every organization a culture and "style" exists that may be threatened by the disciplines of the integrated system.

The effective operating manager thoroughly understands the environment in which the management process takes place. The challenge is to retain those portions of the present process that have made the organization successful while introducing a new set of tools that will greatly increase the effectiveness of the process.

The following are general guidelines, but there is no "pat" formula. The leader of the effort must carefully think through the strategy that will be most effective in his or her particular situation.

1.) Assess as realistically as possible the probable impact upon each functional area of the business, identifying both short term problems and penalties and long term benefits *to that sector of the organization.*

2.) In a one-on-one or small group situation for the leadership of each function, share both the concerns and opportunities and obtain a consensus as to why the effort should be undertaken even from the relatively narrow view of a given function.

3.) Consolidate the concerns and opportunities from each functional area into a company-wide list and add additional concerns and opportunities that are company-wide in scope.

4.) Taking into account the types of issues and realities previously enumerated, establish a realistic plan for the CIMS project effort, including funds, key manpower and time requirements. Make reasonable al-

> **Technique**
> - Dialogue.
> - Design.
> - Documentation.
> - Dedication.
> - Discipline.

lowances for unexpected delays and errors.

5.) Obtain approval at the board of directors level for the strategy and the concept—not necessarily specific project fund authorizations (these should be handled one at a time as specific hardware and/or software is secured).

6.) Feed back to each functional area, again on a reasonably personalized basis, the plan of action and the participation required.

7.) Initiate the process employing the five "D's" necessary for the successful detailed implementation of a CIMS as described in the following section.

The five "D's"

Dialogue—The dictionary defines dialogue very simply as "conversation." Conversation, in turn, is defined as the "exchange of thoughts by talking informally together."

Training, on the other hand, is defined as "practical education in some profession"—or development of knowledge by training or teaching.

What is needed in the CIMS installation process is much more emphasis on the former technique and much less emphasis on the latter. We need less lecturing about techniques such as "pegging" or "phantoms" or "lot sizing" and much more exchanging of ideas concerning alternate manufacturing routings, labor reporting problems, material substitutions, etc.

The technical aspects of system design and data base organization are not the primary concern of the user. The dialogue must take place in the *user's frame of reference*. It must deal with the realities of alternate process routings, for example, rather than vague systems jargon.

The implementation process must be handled as a two-way dialogue rather than as a classroom or audio visual exercise. The purpose of the dialogue is to coordinate the *capabilities* of computerized systems (without burdening the user with mechanical techniques) with the *needs* of the user, which are often already set forth in some type of manually prepared record or report.

The height of absurdity is to ask what the user "wants" in the computerized system without discussing what is available on a routine basis, as well as what can be done in the future with varying degrees of cost and difficulty.

Design—For a manufacturing control system to be successful, one of the most critical elements is the design of the basic record files. If the part master, routing, tooling, and shop work order files have structured data fields for every variable of the manufacturing process, the system can be designed to provide the user with almost any conceivable service. However, a great deal of dialogue is necessary to be certain that these basic files are properly structured.

Another critical element of design is the provision of adequate code fields. If enough codes are provided, the computer programs can respond to all of the variable realities that exist in the actual manufacturing environment.

Documentation—The third "d" stands for *user* documentation. During the dialogue about system design, the latest thinking needs to be continually documented in the form of a user's system manual.

The primary purpose of such procedural documents is to provide a continuing track record tracing the progress of the system that will facilitate more dialogue. The user documentation also becomes the primary reference resource for the programmers who are converting the design to program code.

Obviously, if the user system is well documented, the basis for training data input people is also provided. Rather than being limited to vague generalities, however, this training is "function specific," illustrating exactly how a particular document is to be processed, the codes to be used, the time considerations if the system is batch entry, etc.

Dedication—A great deal of dedication is required by the entire organization. Disappointments and delays will inevitably occur. Gathering, organizing and loading all the basic file data into the system are very difficult tasks.

It is to be expected that some of the needed data will be in very poor condition in the existing records. Mistakes will be made, and work will have to be redone. Therefore, the total cost of setting up the system may exceed even the best informed "up front" estimate.

Discipline—This is a harsh word, and may frighten some companies away from the entire idea. After all, in manufacturing plants there is as much art in the entire process as there is science. The idea of hard and fast rules is repugnant to many.

If the system is properly designed, however, it will be responsive to the "real world," and the only discipline required of the users will be that they cooperate with the system and recognize the need for providing adequate information and data.

In some instances, shop people have the "bravado" attitude that "we are the doers, and therefore, we are entitled to ignore the system if we choose." Obviously, this attitude must be reformed.

If a great deal of effort has gone into making the system as "real world" as possible, the users of the system must recognize their obligation to apply the disciplines necessary to allow the system to function properly.

Summary

It is not practical to arrive at all the answers "up front" before starting the process. The development of a CIMS is an evolutionary process that begins from a charter or concept, but is not set in concrete from the beginning.

The following are some new patterns that may develop:

☐ Entrepreneurial common sense may be substituted for elaborate justification formulas.
☐ The investment in data base development and maintenance may be recognized as an unavoidable cost of doing business.
☐ Facilities not compatible with CIMS or those retained temporarily for marketing reasons may be isolated into another administrative unit.
☐ Dialogue will greatly reduce hostility.
☐ Opportunities in skill growth will be provided for those displaced by CIMS. A great deal more effort may be devoted to retraining.
☐ Senior management will gain in understanding of CIMS, thus reducing the generation gap.
☐ Policy decisions regarding future product strategies and management methods will be forced into focus.
☐ Fresh organizational patterns will develop.
☐ Renewed confidence in the quality of leadership will overcome the shock of change. Everyone takes pride in a successful effort.

Putting all the above into perspective, CIMS in one form or another will be installed in practically every manufacturing company in America over the next few years. Exciting opportunities for dramatic increases in productivity are immediately ahead. Effective leadership during this challenging period will be at a premium, and those companies that effectively make the transition will have a bright and secure future. IE

This article was developed through the cooperation of CIMS series editor Randall P. Sadowski and CIMS committee member Ed Fisher.

Robert L. French spent the first 20 years of his career as an industrial engineer and plant manager with various manufacturing firms. In 1974 he formed R. L. French & Co. Inc., management consultants, of South Bend, IN. He is also president of On-Line Data Inc., a computer service firm, and is a senior member of IIE.

IV. Planning and Applications

Systems and Concepts

Computer Integrated Manufacturing Systems for Small Firms

Case Studies

EDITORIAL OVERVIEW of Section IV

As can be seen from the preceding sections, we are on the brink of achieving computer integrated manufacturing systems. Much of the requisite computer storage, communication technology, and subsystems are available; the task facing us now is to transform them into workable systems. While no fully integrated manufacturing systems exist, substantial progress has been made toward this goal.

The four papers in the first subsection identify the key planning elements and concepts necessary for CIMS implementation. The authors of these papers discuss the importance of building the system in modules so that the necessary capital and training can be spread out over a period of time. The necessity of developing and educating all workers involved in the manufacturing system is stressed by Manchuk. Hardware and software considerations are discussed by Lee, Mosier, and Harkrider.

Since small, as well as large, firms can profit from CIMS concepts, two articles on how to implement CIMS in small firms are included. Lee Hales provides a good overview of CIMS and the long-term planning required from the perspective of the small firm, while Salomon and Biegel discuss the economics of CIMS in small firms.

This section concludes with five case studies of CIMS development and implementation. Frank Curtin discusses the cellular approach taken by General Electric in Louisville, Kentucky and Erie, Pennsylvania. The Louisville plant makes consumer products (dishwashers) while the Erie plant makes diesel locomotives. These two plants support the premise that the CIMS concept applies to large _and_ small projects. John Huber then discusses the history and development of integrated systems at Grumman Aerospace under the Tech Mod Program—a program that looks at the entire plant when developing the systems. This leads to Pratt and Whitney's "Factory of the Future," in which jet engine components are manufactured. John Fargher provides an in-depth look at the Naval Air Rework Facility at Cherry Hill, North Carolina. He focuses upon the planning and control system, and provides details on the data and material management. In our last case study, Gary Redmond discusses the development of a flexible machining system at FMC Corporation.

Reprinted from *Industrial Engineering*, September 1983.

How To Define And Plan A System For Integrated Closed Loop Manufacturing Control

By Edward J. Anstead
Sperry Corp. Computer Systems

The need for effective manufacturing control systems has never been greater than it is today. Profits are down in many industries, and productivity is lagging. Every firm wishes to maximize customer service with minimum inventory investment and efficient plant operation.

But since each goal is guarded by a different department within the company, conflicts arise. If each department establishes a control system to optimize its own position, these conflicts can escalate. In response, some manufacturers have turned to integrated manufacturing control systems to help run their businesses.

The purpose of this article is to define the requirements of an effective computerized manufacturing control system, to show the importance and interrelationship of each requirement and to tie the overall requirements together into a closed loop system. The term "closed loop" manufacturing control system (MCS) means that the planning functions which feed the execution functions also receive feedback so that replanning can occur.

Figure 1 shows a schematic plan of a closed loop system that addresses such questions as:

☐ Which products should be produced in the next year? (Forecasting)
☐ How many should be produced? (Forecasting/master scheduling)
☐ How should each product be made? (Engineering design)
☐ Was it made to specification? (Quality control)
☐ Is capacity adequate at the plant or corporate level? (Rough cut capacity)
☐ Is capacity adequate at the work center level? (Capacity requirements planning, production scheduling)
☐ When are parts or subassemblies needed? (Material requirements planning, inventory control)
☐ Are work orders on schedule? (Shop floor control)
☐ What price should be charged? (Marketing)
☐ How much should it and did it cost? (Cost accounting)
☐ Are needed parts on order? (Purchasing)

Building the system

Information which feeds into the closed loop manufacturing control system is shown in Figure 2.

Although effective control systems have specific differences, they have

An integrated MCS (above) touches every department in an organization. Its real time updating capability is crucial in making timely decisions in the distribution area (left).

the following common attributes:

☐ *Shared data requirement:* The ability to access data supplied by other departments is a critical requirement of a manufacturing control system. The circular format in Figure 2 was chosen to emphasize this concept.

If the production control department wishes to release a rush work order to production, for example, it must first determine whether all parts are available for fabrication and assembly. This is accomplished by reviewing the bill-of-material file to determine what is needed to build the product. Then a check of the stock status file is made to see if all parts are in stock.

After a "what if" interrogation of the shop floor priority schedule to see how other work in progress will be affected, the work order is released to the floor.

In the above example, acting on incomplete information could result in releasing orders lacking necessary parts into an already overloaded work center, which would require avoidable expediting. An effective manufacturing control system must have shared data capability.

☐ *Real time access:* The only constant in a manufacturing environment is change. In the previous example, production control called up the stock status file to determine whether parts were available before releasing the rush work order to the floor.

Receipts and disbursements in a storeroom are dynamic. In a batch environment, records are typically updated at the end of each day. The stock status screen might show that all items were available, but certain parts might have been removed, and the actual item count might be zero.

Without real time updating, users cannot rely on batch reports, and will physically count the parts on hand before releasing a work order.

☐ *Data accuracy:* The need for accurate data is critical in manufacturing. Hundreds of decisions are made and actions taken each day. If the data do not reflect reality, users will abandon the system because they can't trust what it tells them. Management may point to a complete and formal MCS that in reality is not being used.

Inaccurate data in the system cause problems which can lead to the conclusion that the MCS is to blame. In fact, poor data integrity is the cause of many control system failures.

☐ *Company-wide support and education:* The decision to undertake a manufacturing control system touches every department in the organization. If the data processing department begins such a program alone, it is doomed to failure.

If one or two more departments are involved, it may survive. But successful installation depends on companywide support of the MCS. From executives to daily users, each

Figure 1: Schematic Plan of Closed Loop System

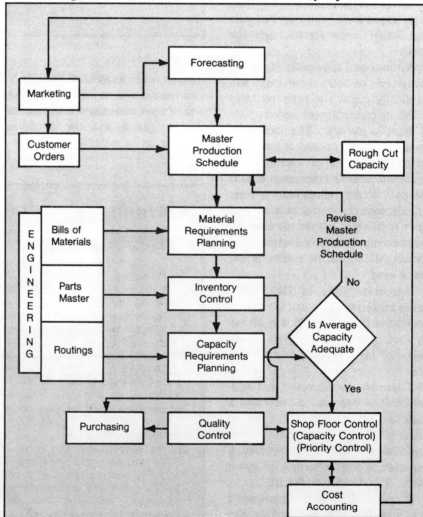

person must understand the demands of the system and the discipline which must be maintained.

Supportive statements are not enough. Each person must learn why the system should be supported. When personal benefits are recognized, employees will endorse the system and enforce its discipline. A broad-based education program will help assure company-wide support and a strengthening of the MCS.

Getting started

The installation of a computerized MCS is a major undertaking. Obviously, a computer must drive the system because of the on-line shared data requirements and the large transaction volumes. Three options are suggested:

☐ Hire a reputable consultant who will identify where you are now, help in the selection of computer software and hardware, educate your employees and oversee the implementation to completion.

☐ Evaluate software and hardware vendors yourself, including their dedication to manufacturing, their successes in other installations and both pre- and post-implementation support.

☐ Develop your own in-house application software and educate your own employees. This is the longest term solution, because thousands of man-hours are required to write, debug and document a large MCS. The risk of a non-integrated data base is also present.

For further reading:

Orlicky, J., *Material Requirements Planning,* McGraw-Hill, New York, NY, 1976.

Peterson, R. and E. A. Silver, *Decision Systems for Inventory Management and Production Planning,* John Wiley and Sons, New York, NY, 1979.

Plossl, G. W. *Manufacturing Control: The Last Frontier for Profits,* Reston Books, Reston, VA, 1974.

Wight, O. W., *Production and Inventory Control in the Computer Age,* Canners Books, Boston, MA, 1974.

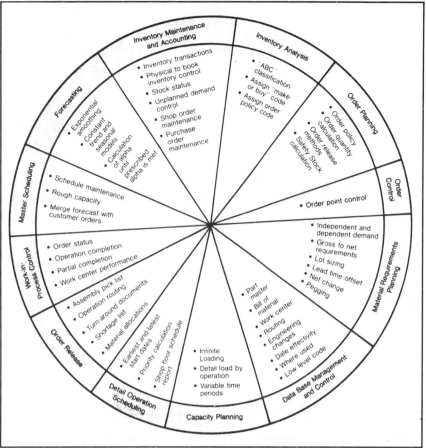

Figure 2: Information Feeding Into Closed Loop MCS

Benefits of an Integrated On-Line Closed Loop Control System

☐ *Inventory reduction:* Through material requirements planning and tightened inventory controls, the inventory investment will decrease.

☐ *Labor cost reduction:* Working on the right items at the right time, balancing shop load, eliminating bottlenecks and shifting priorities are all features of the MCS which will reduce overtime costs. Tracking actual versus standard labor costs will point management toward corrective action.

☐ *Increased customer service:* When customer order entry, manufacturing and shipping are all on-line, a customer's inquiry can be immediately answered. This enhances the firm's credibility and service image. Furthermore, shipping complete orders on time builds additional good will. Customer responsiveness increases sales.

☐ *Improved morale:* A successful MCS gives employees a positive feeling about themselves and the company. Information they need to do their job is available. Departments pull together instead of in conflict. Employee productivity is increased.

☐ ***If a system is properly installed, users can expect an active rather than a reactive environment.*** The closed loop feedback feature allows managers to re-evaluate plans quickly when actual results deviate from plan.

Edward J. Anstead is a manufacturing consultant with Sperry Corp. Computer Systems. He holds BSIE and MBA degrees from Pennsylvania State University. Anstead is a senior member of IIE and president-elect of the Harrisburg, PA, chapter.

CIM Project May Be Doomed To Failure If Key Building Blocks Are Missing

By Stan Manchuk
McDonnell Douglas-Tulsa

Computer-integrated manufacturing systems (CIMS) are a pervasive management strategy, and their development represents a conscious long-term implementation of strategic, operational and tactical plans to achieve high productivity and efficiency in plant operations.

Many factors can hinder the successful development of computer-integrated manufacturing systems. Most are related to incomplete development of some of the subsystems, interfaces and support structures and/or to high capital requirements.

As CIMS receive more attention and evolve into the fully integrated systems that competitive industry needs, it is well to ask ourselves what elements are critical to successful CIMS implementation and if there are key building blocks, the absence of which dooms the CIMS attempt to failure.

This article explores the various components of CIMS, their purpose and their relationship to the other components, and attempts to identify the elements that are missing or inadequately addressed in many CIMS developments.

The missing and/or often inadequately addressed elements can be classified under one of three categories:
☐ Managing and organizing for CIMS.
☐ Developing/refining subsystems.
☐ Integrating subsystems.

Before any substantial effort is made to improve any particular operating situation by installing CIMS, an understanding is needed by management that the existing operating situation is not satisfactory for the following kinds of reasons:
☐ Competition has a technological and/or quality edge.
☐ Management controls are not satisfactory.
☐ The level of technology or automation employed is not adequate to compete.
☐ Obsolete plant.

These are the types of problems that CIMS can solve. Problems such as inadequate knowledge base, declining (total) market or obsolete products *cannot* be solved by installing a CIM system.

CIMS development, which provides the opportunity for solving technological coordination and control problems, involves considerable financial and technological risk. Unless management has a thorough understanding of the cost, benefits and risks associated with CIMS development and has committed the resources necessary for its achievement, the CIMS will fail.

A good method for developing this requisite managerial understanding is by developing a model of the needed CIMS, assessing the current status of existing programs and planning the steps that will carry the company toward realization of the CIMS. It is only by doing productivity studies in detail, with committed personnel and top management funding and support, that any significant progress toward authorization to proceed can be made. These plans and studies will be the road map for the project and will help those charged with its implementation to prudently select the next project from the array of options available.

As CIMS are pervasive, all groups and functions that will be affected must be informed and their commitment and understanding secured. All groups must understand that the structure of the organization will be impacted, that various procedures will be changed, and that fewer support and direct personnel will be needed in the factory of the future.

The realization that with changed plant design goes changed organization structure is often missing. The inability to adapt the organization structure may doom a technologically sound CIMS to failure. The choice of doing little or nothing is not available, and continuing improvements are, and will continue to be, the order of the day. This course requires a substantial front-end investment and commitment in time, resources and, particularly, people.

Adequate planning is in too many instances a missing element in CIMS. A long-term funded and dedicated effort is in order here, starting small with one or two people and gradually building into a larger effort involving more functional organizations.

Substantial debugging and rethinking will be needed and provided for during the building of the CIMS, and detailed planning and review can help avoid costly problems caused by software, hardware, organizational and personnel incompatibilities. To

The Canada-Tulsa Technology Transfer Story

From a recent McDonnell Douglas Canada-Tulsa technology transfer, it can readily be seen that there are known processes and approaches that we all may employ to construct our own versions of CIMS. Not included in the discussion here, but a vital part of the continuing effort on the factory of the future concept at McDonnell Douglas-Tulsa, are a variety of company proprietary technical/scientific technology transfers that have been carried out or are still taking place, including DGS (distributed graphics system), DNC (direct numerical control), and MDC APT (automatic programmed tool).

This technology transfer, a first within the corporation, was a large-scale transfer of business systems applications for the manufacturing, materials and fiscal divisions from Toronto, Canada, to Tulsa, OK. Included for manufacturing support was a network of systems applications supporting work order release, shop planning and scheduling, work-in-process inventories, changes and shortages in the orders, bill of materials generation for various programs, parts tracking, open door master (Tulsa developed) and reporting.

Included for purchasing and material was a network of systems applications supporting purchasing, pricing, vendor masters, statistical inventory, MRP-type inventory and bill of materials generation (explosions and implosions). Included for financial management were systems applications supporting materials accounting, accounts payable, and time and attendance and labor reporting (the ATLAS system being processed at Douglas Aircraft in Long Beach).

The planning for this transfer employed a two-tier approach. First, a goals-oriented overview was developed in sufficient detail to identify the major projects and tasks to be undertaken. This was to provide an understanding of all the resources required, costs involved and benefits to be achieved.

Second, as changes and improvements were identified, task forces were assigned. Such task forces must be knowledgeable regarding the processes involved to design and detail a program that will achieve the desired results, including scheduling and coordinating of all resources.

For both levels, coordination and monitoring of milestone events is an absolute necessity and has to be agreed on by all parties concerned. The transfer thus was a complete strategic and tactical study, followed by implementation. The entire planning exercise and the justification and authority to proceed were based largely on economic cost displacement.

At its peak, the project involved approximately 130 people—including clerical, user, management, systems analyst and programmer personnel from both Tulsa and Toronto and a small contingent of sub-contract programmers who were retained to convert a small proportion (approximately 20-30%) of existing applications on Tulsa's UNIVAC computers to IBM computers.

All these people were given detailed task assignments on specific project teams, and some people worked on a number of projects and tasks; the management and coordination of all the resources involved was a complex task in itself.

Considering the many variables involved and the fact that the MDCAN systems adapted and adopted for Tulsa had to perform in Toronto prior to being transferred to Tulsa, the only practical course was to maintain an identical environment and convert Tulsa from UNIVAC to IBM. In the time available, it was simply not possible or practical to approach this technology transfer in any other way.

From the start of the study to actual implementation, the project took less than two years and was accomplished on time and within budget. A good part of this success was due to focusing on the projects and tasks that were necessary to achieve the intended results, and thus keeping things on track. We are currently realizing a better than anticipated payback, even though implementation was complicated by the loss of two major contracts and a 14-week strike.

Reviewing the conversion of approximately 20 to 30% of existing applications from Univac 9480 and 90/70 computer equipment to dual IBM 4341 computers involved education and training of all EDP personnel

---

obtain best results, quality assurance (quality engineering) needs to be an integral part of the above planning process in order to avoid incurring additional manufacturing costs of nonqualified processes, including equipment and tools.

Effective automation presupposes this integrated effort. If we expect to obtain automated parts production, various elements must exist or be created:
- Group technology.
- Knowledge of process capabilities.
- Tool standardization.
- Process sequencing/flow material handling.
- Jigs and fixtures, set-up requirements.
- In-process sensors, gauging (quality engineering).
- Inspection procedures.
- Scheduling.
- Inventory goals.

From the above it is readily apparent that CIMS development is not a wish list kind of thing. It has to be based on adequate studies that demonstrate that the plant, equipment, processes and tooling can provide the necessary capability and capacity for a given level of production.

A requisite for success is selection of the proper individual to head the CIMS development team. He or she should possess a through knowledge of the type of manufacturing operations performed by the firm, the current information systems and the state of the art in computer technology. The team leader also needs the interpersonal skills to handle opposing groups within the organization and obtain compromises that each group will support without reducing system effectiveness. This individual must be far-sighted and be placed high enough in the organization to enforce decisions. Adequate support staff must be provided, or the leader will be so immersed in the details of the system that he or she may lose sight of the overall CIMS goal.

Integrating systems and data

A method for determining the best

and was supported by MCAUTO (McDonnell Douglas Automation Co.). Acquisition of equipment and system operating software was also through the auspices of MCAUTO.

Education and training of the Tulsa functional users was provided primarily by the MDC (McDonnell Douglas Corp.) functional user community, assisted by IRM (information resources management) personnel. It consisted largely of on-the-job training by the MDCAN functional user personnel, including management people dedicated to this task.

Detailed overviews and step-by-step explanations, coupled with hands-on participation, were the main approaches used. It is fair to say that the functional user communities became throughly immersed in their projects, and the knowledge they acquired and put to use was a large part of the reason for our success.

For the EDP personnel, the requisite training courses were determined in conjunction with, and supplied through the auspices of, MCAUTO. A phased schedule for all personnel training evolved and was coordinated with detailed assigned involvement with specific system applications products to obtain and maintain close cooperation of all personnel. Information resources management (IRM) personnel and the functional users were brought together to work in teams.

Planning and implementation was largely a joint effort of all parties concerned, including headquarters information resource management. A considerable amount of front-end effort was required, but it was very important to the successful completion of the technology transfer. If you don't know what you want to do, you shouldn't start.

The original study, although a very good strategic guideline in many respects, did not allow for the tactical plan of detailed assignments of tasks and personnel necessary to the implementation of each particular project and task within project. Implementation order was dependent on the interrelationships of the various systems applications.

This sequence considered applications within a functional area first and then in relation to other functions. In working with interrelated systems, the order of precedence is vital, as you are trying not only to convert from a manual or inadequate system, but also to maintain integrity of records and data throughout—a formidable task that requires a dedicated, controlled management effort.

Building the tactical plan consumed much time and effort and involved considerable travel to Toronto by Tulsa personnel. Subsequently, Toronto personnel traveled to Tulsa to ensure successful system application transfer and implementation.

Coordination, cooperation and progress were assured by senior management personnel at both components. The fact that when the exercise started, E.A. Reece was vice president and general manager of both facilities, and both he and the MIS director (the author) were transferred to Tulsa, was helpful in this regard. Contractual arrangements for support were negotiated with MDCAN and handled as normal intercomponent work.

Management support from all involved functions and from top management is critical, as people must be made available and dedicated to assigned tasks, and other normal work assignments must be interrupted. The level of support from MDCAN was negotiated as part of the original study plan and was further refined and expanded from the results of the detailed tactical plans. There was consistency throughout, and the differences that occurred were due to changes in approach dictated by a mutually agreed upon tactical plan.

Of course, there was an impact on both components, although considerable effort was expended to monitor and contain it while keeping to the schedules agreed on. As the work progressed, it became obvious that a new computer operations area would be needed at Tulsa, and this was implemented as planned in time for the arrival of the second 4341 IBM computer.

This effort was also coordinated by the plant engineering people. Originally it was hoped that the existing computer operations area could be rehabilitated, but because of various concerns, available space, power requirements, etc., a new site at the Tulsa plant was decided on.

---

manufacturing processes to use in manufacturing a particular design is the first link in integrating CAM and CAD. Automation of process planning is a must; support of the manufacturing design function rests on a necessary foundation of group technology, classification and coding.

A departure must be made from the use of manual efforts to create new planning processes for parts, especially those similar to existing parts, to reduce the substantial cost and effort involved. Computer technology is available—i.e., CAPP (computer-aided process planning)—and its feasibility must be determined. It might be economically justifiable to use CAPP to restudy current production methods, or it might be applied only to future production, because the cost of incorporating it for existing production may be unnecessarily high.

Computer-automated process planning has proven to be very difficult to implement. Like any other complex task, it must be decomposed into smaller tasks that can be adequately defined for particular production processes—e.g., sheet metal, machining, subassembly and assembly, and those for particular products.

It is the knowledge of the people actually working with the processes and products that is important here (a kind of expert system). We must record this knowledge before the craft skills on which it is based become extinct as the older generations leave industry.

Development of this process planning data base, stressing its knowledge content and assuring collection of this knowledge for later coding, is vital to the acceptable functioning of the system. Reliance on a commercially developed system which has not been modified to capture the special expertise that has developed through years of working with a particular product mix may seriously weaken, if not destroy, the quality of your process planning.

Assuming that design is computer assisted, it is logical that the geometric definition will be supplied in a digital format. These geometric data can be combined with the process planning data and processed by a computer-assisted numerical control programming system to produce the part program.

This program can be further conditioned (compiled and post-processed on a computer) for a particular combination of machine tools and machine tool controllers to allow for alternate routings to meet scheduling constraints within capacity limits. With computer numerical controls, the additional capability to enter in the part program changes at the machine tool controller level is obtained.

In CIMS, the part geometry, routing and tooling suggested by this automated planning process can be used with the various expert systems to achieve plans that are consistently a step closer to optimum and/or less costly than manually generated plans. The linking of CAD, CAPP and automated parts programming using expert systems is under intensive study, but has not achieved commercial application and must be considered a missing element at the present time.

Product forming operations

The next subsystem encountered in part production using CIMS would be the actual product forming operations (chip making operations in discrete part production). Direct numerical control enables part programs to be downloaded from a host computer to a supervisory computer on demand; further downloading to a specific machine tool is also possible.

In addition, simultaneous data collection can provide management information on part/process progress for an individual or combination of parts and machine tools. This enables closed-loop, computer-based production control for discrete manufacturing processes, which is a must if a truly integrated manufacturing and control system is to be developed.

This capability exists in part for certain production processes. One such process is tube bending at McDonnell Douglas using a CADD (computer-aided design and drafting) system to design, perform design analysis and determine structure and tube routing requirements, which are then transferred to a manufacturing data base, converted to manufacturing bend data and downloaded to computer-controlled bending machines to produce the parts.

By integrating a DNC (direct numerical control) system with its feedback mechanisms, you can provide shop managers with more effective control of their environment while reducing the cost of operation.

Looking to the future, plans to provide facilities engineering with systems application tools and linkages to manufacturing systems and with DNC to continuously monitor the machine tools and condition should be incorporated. Also, a serious look must be taken at appropriate management control systems that will assist managers in their work as a by-product of scheduling and production control systems, specifically by giving them such items as cost of production and order status.

To provide this information, it is important that all these functional groups be included in the review, analysis and implementation phases so that their needs can be incorporated in the final design. An agreement on a material, part and order identification system that allows proper accrual of all pertinent costs is essential if manufacturing, finance and marketing are to be able to adequately perform their functions in an integrated CIMS.

The integration of CAD and CAM into a CIMS is an ongoing evolutionary process. As areas of economic benefit are identified, they should be targeted and plans made for their realization.

Robotics can be treated in much the same manner, particularly if we treat the robot as a computer peripheral and use the computer as a simulation tool. A robot can be made an integral part of a DNC manufacturing cell and/or an FMS (flexible manufacturing system), with its array of business application support systems, to achieve further economic cost benefits.

While this process may, at the current state of the art, supply only a rough order of magnitude level of control (production and labor leveling), this is an important starting point. You must have a computer-based master scheduling, work order release and shop loading system with some kind of data collection requisite (feedback mechanism). Otherwise, the background data and procedures to enable realization of CIMS potential benefits are missing.

In order to realize the CIMS potential, an efficient scheduling system must be used to coordinate the production facilities. A good master scheduling system is vital, and this, of course, must be computer based, as it is simply not possible to manage the information volume in real time and at the same time to handle the "what if" queries satisfactorily any other way.

There are many variables that the system must take into account, including machine capacity, capability and availability, and labor profiles (loads). Data, of course, are dependent on a manufacturing/material requirement planning process, which in turn is dependent on a bill of materials approach.

Detail (shop floor) scheduling is a further scheduling refinement, but it should not be attempted except in isolated instances until the gross forms of control are installed and working.

If efficient master scheduling is missing from either your current system or your planned first steps toward CIMS development, your plans are in jeopardy. Decisions to install computer-aided design, computer-aided process planning and direct numerical control equipment hinge on volume considerations, and keeping them loaded once installed. If you cannot do this, it is unlikely that your investment will pay off.

Work order release, shop loading and detail scheduling are among applications systems that collectively reflect shop floor control. Others are time and attendance recording, labor reporting and progress reporting (parts tracking).

These tools can provide a visible picture of the size and utilization of the work force, the amount of work in the shop and the labor hours expended on scheduled production. Such systems are available commercially and, assuming that the requisite data can be gathered, should not be missing from any CIMS.

Organizing for CIMS

Development of CIMS is a long-term effort which requires coordination of many functional groups. The kind of coordination and participation that are needed do not happen by chance—they must be planned for and receive top management support.

The study team approach, coupled with continuing systems applications, planning and implementation groups, can make CIMS a reality. The important thing is to get people working together, forgetting the separation of worker, manager and professional and concentrating instead upon the actions required to make the CIMS effective.

The various needs of differing functional groups must be met, and this is best achieved through the formation of strategic and technical planning groups representing all disciplines. These groups should meet on a regular, continuing basis, as frequently as necessary, and with the functional authority to commit to particular courses of action.

Perhaps the most difficult chore will be integrating the process and support functions, because there is no clearcut approach for achieving this, but a first step can be bringing the business and technical sides of information resources management under the central direction of one senior manager as discussed previously. It's

> **" All functional areas must be included in the process by which the master plan is modified. It is simply not possible for each area to 'do its own thing' without referencing and coordinating with all other impacted functions. "**

far better to have more coordination and less confusion than little or none and disaster as an ultimate outcome.

Since CIMS are computer-based technologies, they are burdened with all the related communications processing and storage considerations. Additionally, considerable variation in the type and sophistication of available hardware and software exists, and micros, minis, mainframes and terminals must function as an integrated whole.

With any continuing and evolving CIMS, interfacing problems will be ongoing. They can be minimized to a great extent by insistence on a standard computer architecture (computers, storage and communications) and strict adherence to the master plan for CIMS development. The executive in charge of development and operations must enforce adherence to the plan, but must allow newer technologies to be introduced as appropriate.

All functional areas must be included in the process by which the master plan is modified. With CIMS, it is simply not possible for each area to "do its own thing" without referencing and coordinating with all other impacted functions.

Central to a CIMS is a data base, simply because a multitude of separate file structures for each specific is unwieldy when inquiry and manipulation of information are pertinent to more than one functional group. However, it seems that data base development is going in two separate unrelated directions—business and scientific/technical—with precious little coordination between the two.

Whether or not one data base approach will be adequate to meet the disparate needs of these fundamentally opposed groups is immaterial. The data base(s) must be complete. Missing data will impair CIMS effectiveness as surely as, if more subtly than missing communication among the manufacturing, accounting and marketing data bases.

It is perhaps easiest to evolve a data base from functional requirements for a particular application and correlate it with purchasing, inventory control, materials accounting, etc., to arrive at common data elements. Proceeding in this fashion from the inside to the outside will provide the greatest opportunity for evolving a successful data base.

A data base administrator in the information resources management group who will provide the necessary support and coordination is a must, but the planning team must determine the structure. Due to the cost of CIMS development efforts, it is worthwhile to try to transfer systems and subsystems that others have developed and are using to the developing system.

The transport of systems applications from one environment to another is fraught with danger. This is a technology transfer in a technical or scientific sense, and although it is not easy, it is easier than a technology transfer involving business, industry, company, plant, etc.

A careful, documented study is required to ascertain for any technology transfer that the right conditions prevail for successful transport. Prerequisites are top management support, adequate level of funding, educated and trained personnel, functional user support, participation and leadership (mandatory), and a willingness to adopt, adapt and integrate the new technology into the business operations.

Considerable flexibility and transportability are available, but much effort must be expended on telecommunications, computer hardware and software, and networking and standardization to achieve the desired results. How much effort is required in each transfer is a function of the number of systems applications, their sophistication and their interdependencies.

All of the above must be covered by an agreed-upon company policy regarding who does what processing where, and be coordinated and controlled to maintain the effectiveness of a CIMS. The site of the CIMS must conform to all needs for site protection, processing security and physical security consistent with an "uninterruptible" computer processing environment. Early and continuing attention to site planning is essential to the ongoing success of a CIMS, because as a company becomes increasingly dependent on the continuing functioning of computer systems, it cannot afford long-term interruption of computer facilities. Often missing in CIMS development are safeguards that will allow the system to continue to operate when some system degradation has occurred.

Modification of the F4 phantom fighter was among Tulsa projects that were facilitated by CIMS development.

The computers upon which the factory of the future is dependent must be reliable, available and maintainable. For processes and industries that are computer dependent, the loss of these computer facilities can ultimately lead to business failure. Computer disaster/recovery planning then becomes essential to survival.

This planning requires application-by-application review with the users to determine critical needs and backup requirements with reference to a partial, as well as a complete, loss of facilities.

Project planning is responsible for the provision of in-flight recovery for on-line applications, back-up recovery for batch applications and sufficient computer resources to allow the switching of critical applications from failed units to operating units. Long-term, long-range planning is essential, and the proper management of computer facilities is the vital ingredient in proper computer operations support.

At some point, re-design will be necessary to allow incorporation of new technologies. You must retain the ability to make continuing improvements in order to meet changing operating conditions.

For the well being of any CIMS, it is important that all future computer-aided design efforts support the digital representation of the design. It is only in this way that adequate downstream communication, from design engineering to manufacturing engineering to the production shop floor and back to the business systems using the expert systems that have been set up, will be obtained. Incorporated in this production design database are all the support elements (other support data bases) that are needed to enable the production of parts from the original design.

Computer-aided design, coupled with computer-aided process planning that also incorporates the necessary cost elements, allows more accurate cost estimating for bid proposals. If the master scheduling system allows "what if" questions to be entered, a process exists that allows for a quick response to marketing inquiries. Obviously, this is an ideal condition, but even if it exists only partially, there can be considerable savings in manpower and cost.

A continuing R&D emphasis is necessary to improve these processes, and one must pay continuing attention to the transfer of information from one discipline to another. This requires long-term coordination, cooperation and understanding.

Refining subsystems

Few general recommendations can be made concerning the development and refinement of the various subsystems with respect to the plan for CIMS development except that the developers:
☐ Evaluate the planned subsystem with respect to the plan for CIMS development. Be sure it is logically the next element to be incorporated (i.e., the subsystem necessary to realize the planned benefits).
☐ Coordinate and communicate with all concerned functional groups regarding their reactions to designs, support that they must supply and suggestions.

Typical subsystem/functions that must be incorporated into a CIMS are:
☐ Communication protocol.
☐ Data base management.
☐ Documentation standards.
☐ Management controls.
☐ Work order release.
☐ Shop scheduling.
☐ Inventory control.
☐ Order control.
☐ Bill of materials generation/explosion.
☐ Parts tracking.
☐ Labor reporting.
☐ Purchased parts control.
☐ Vendor masters.
☐ MRP.
☐ Accounting materials and labor.
☐ Accounts payable.

The expert type systems relating to design of the part, routing, and tooling must be incorporated as well. The missing system elements for CIMS depend on the current level of plant operations—what is well established in one firm may be missing from another. An internal audit of capabilities must be carried out and coupled with CIMS goals to form realistic plans for development.

Summary

CIMS development is no different from any other industrial engineering project. Careful planning, development of economically justified plans and astute management will lead to success. Although some elements are not yet fully developed, many of the elements do exist, and with proper attention to the specific capabilities, expertise and problems of the firm, CIMS can enable firms to reduce costs and gain competitive advantage. **IE**

This article was developed through the cooperation of CIMS series editor Randall P. Sadowski and CIMS committee member John Nazemetz.

Stanley Manchuk is director of information resource management at McDonnell Douglas Tulsa. He transferred to Tulsa from McDonnell Douglas Canada during the technology transfer project described in the article. Prior to joining McDonnell Douglas he was employed by Control Data Canada, Cybershare Ltd. and Westinghouse Canada Ltd. Manchuk has published articles on plant management in production/inventory control and preventive maintenance.

The Development of a Computer Integrated Manufacturing System: A Modular Approach

Scott E. Lee
Master of Science
Industrial Management
Clarkson University
Potsdam, New York

and

Charles T. Mosier, Ph.D.
Assistant Professor
Industrial Management
Clarkson University
Potsdam, New York

ABSTRACT

This paper discusses the design, development and implementation of a Computer Integrated Manufacturing System (CIMS) from a modular point-of-view. The advantages and disadvantages of a modular approach are discussed in the context of database development and hardware requirements.

The Distributed Data Base (DDB) approach is explained as an effective CIMS structure. Several examples of modular implementations are cited as well as a conceptually complete modular system.

INTRODUCTION

In an effort to address an area of apparent interest to today's manager involved in the manufacturing environment, this paper will discuss the implementation of a Computer Integrated Manufacturing System (CIMS) from a modular point-of-view. The topic of modular implementation is of particular interest to the manufacturer who presently has a great deal of capital resources tied up in existing computer-based information systems or simply does not have the capital resources necessary to fully integrate his manufacturing operations by means of one, large-scale implementation project.

The concept of a modular implementation procedure allows an evolutionary development of a computer integrated system. This has obvious benefits to the user including limited capital outlay and limited personnel commitment. The decision to use a modular implementation approach is easy for the organization to make. What's not so easy is determining how to best perform this implementation and end up with a fully integrated information system free of excessive redundancy.

The goal of this paper is to address this question. The following format will be used to examine this problem and present solutions to it.

Part 1: The Concept of Modularity

Part 2: Database Considerations

Part 3: Hardware Considerations

Part 4: A Model of Modular Implementation

The first section, the concept of modularity, will focus on the costs and benefits of a modular implementation. The trends leading to modular development in information processing will also be discussed. Part two, database considerations, will address the urgency for total integration of information in a modular configuration. Means of achieving this goal are discussed. Part three, hardware considerations, deals with the problem of hardware obsolesence and redundancy. Modular development requires well laid plans for hardware acquisition and integration. Finally, Part four will present a somewhat theoretical model of a modular implementation. Published implementation experiences will be referenced.

PART ONE: THE CONCEPT OF MODULARITY

A large majority of today's CAD, CAM, and CAD/CAM systems in use throughout the world may be classified under two general catagories [5]. These include:

1. The small-scale specialized turnkey system

2. The large-scale mainframe based system

As the name implies, the turnkey systems are generally designed for specific applications. They are self-contained systems, providing the required hardware and software necessary for the ready application of the system to a particular need. On the other hand, the mainframe-based system is an all-encompassing system with a fully integrated data base providing shared information to the many application users.

Conceptually, the differences between these two systems may be characterized by a centralized versus decentralized organization [1]. In one case, a computer application system is developed. This system is a set of computer programs which, taken together, input, store, process, and output information required to support specific management situations. The traditional data processing network observed in most organizations evolves through the continual addition of application systems. Along the way, manufacturing firms are implementing the elements of CAD and CAM technology in a piecemeal approach. And as a result, when one of the isolated systems completes its function, it is usually unable to pass the results on to another computerized or automated function [4].

On the other end of the spectrum, a centralized organization via data base management gained recognition as a means of counteracting the problems generated by computer applications systems. The data base approach attempts to standardize processes used to perform functions common to all applications. This organization provides a central information base used by all members of the organization. The data base management system successfully integrates all aspects of the organization, although the addition of applications to the data base over time becomes quite troublesome. The initial applications usually run smoothly, but when several applications are piled up, the problems begin to emerge in the form of excessive response times or alarming operational costs [8].

The concept of modularity is best illustrated by point C in Figure 1. Point A represents the decentralized, applications management structure whereas point B represents the centralized, data base management structure. Somewhere between the two lies the modular management structure. This organization is characterized by individually implemented application systems set-up in a network of interconnected computers rather than one powerful mainframe. Overall system operation is controlled by a dedicated central CPU that stores the main data base [6].

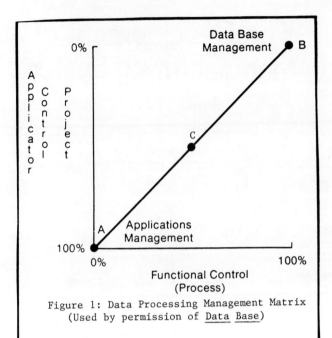

Figure 1: Data Processing Management Matrix
(Used by permission of Data Base)

This structure of system organization is best described by the Distributed Data Base (DDB) concept. The DDB organization is a hybrid concept, taking the best from the centralists and the best from the decentralists and welding them into one data processing management strategy [1]. The key to this strategy is to integrate as many functions as possible in a systems-like approach. This allows information from one application area to be readily accessed by other applications.

Modular development of an integrated system presents many benefits to the organization. Several of these benefits include [5]:

1. Since data for each application does not have to be re-entered, costly redefinition and reformatting are eliminated.

2. Transcription errors are alleviated as well as lead times.

3. Since the information is more accurate, product quality and reliability are enhanced.

4. Integration improves the communication between manufacturing and engineering organizations.

The benefits of a modular implementation are extensive, but they come at a cost. The strategic planning, development and implementation of such a system is critical in order to assure hardware compatibility and software integration as more and more applications are implemented.

PART TWO: DATA BASE CONSIDERATIONS

As described in Part One, the data base organization for a modular implementation is best described as a Distributed Data Base (DDB). This data base organization networks multiple CPUs together to provide a more powerful computing resource and also provides the ability to interface with any corporate computer environment [2]. Connected to the central CPU are one or more remote minicomputers that drive clusters of intelligent terminals. Also connected to the central CPU are intelligent workstations that combine an intelligent terminal with local mass storage and additional software (See Figure 2).

These workstations and remote minicomputer clusters have sufficient computing power to operate as stand-alone units, calling upon the central CPU only for large computing tasks such as finite element analysis [6].

The key to CAD/CAM's ability to integrate a variety of the manufacturing firm's functions and processes is the common engineering and manufacturing data base. This reservoir of information allows CAD/CAM systems to support a variety of manufacturing and engineering applications in a multi-application environment. A data base captures all relevant information about a company's products and all information needed to put the product into production. When a product is complex and composed of various subassemblies and component parts, each individual product element will have detailed manufacturing and engineering information associated with it [5]. Figure 3 on the following page gives a detailed illustration of the development structure of a CAD/CAM data base.

The heart of the CAD/CAM data base is the product model. This is a geometric description of the product that is input to the CAD/CAM system by the engineering design function. The product model is used for other engineering functions such as layout, detail design, engineering analysis and technical illustration. As additional work is done by these functional groups, these engineering applications supplement and reuse the product model information.

With the product model data in place, a host of manufacturing applications may make use of this information and continue to add to the data base. In a Distributed Data Base environment, however, problems exist which are critical to the development and use of the DDB. Several system characteristics must be maintained in order to avoid the undesirable consequences of these problems. These include

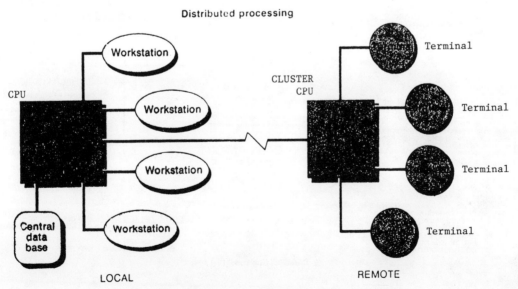

Figure 2: Distributed Processing Configuration
(Used by permission of Machine Design)

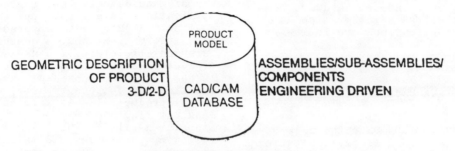

CAD/CAM data base: product model core.

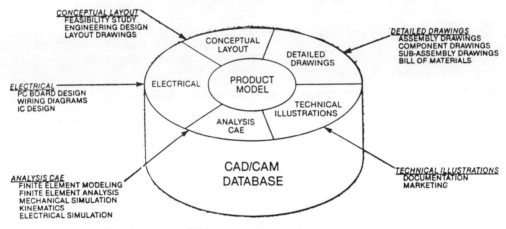

CAD/CAM data base: engineering applications.

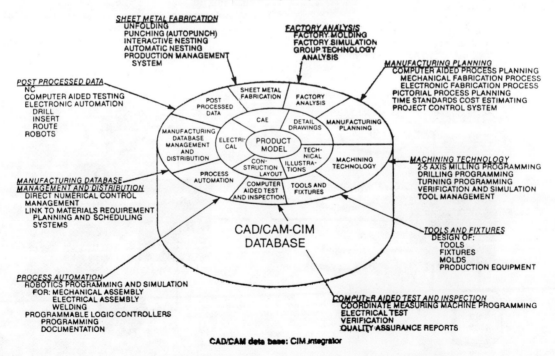

CAD/CAM data base: CIM integrator.

Figure 3: Development structure of a CAD/CAM Data Base
(Used by permission of Design News)

the following [7]:

1. Prevention of nonplanned data redundancy and of possible data inconsistencies when developing new applications.

2. Statement of standards for data definition and utilization.

3. Efficient query management.

4. Efficient use of the computer networks.

Adherance to these system characteristics requires well managed, systematic design and development of each application from the start.

PART THREE: HARDWARE CONSIDERATIONS

In a time when computer technolgy is changing at such a rapid pace, planning future hardware requirements can become a nightmare for the manager. This can hold particularly true in the modular Computer Integrated Manufacturing (CIM) development project.

Hardware obsolescence certainly has posed its fair share of problems for the manager over the past 20 years. For example, the process industries have gone through seven generations of control computers since 1960 [9]. To stay with your existing computing facilities in the face of enhanced capabilities of new hardware can put you in a threatened competitive position in a short period of time. O'Donnell suggests that the impact of obsolescence may be minimized by ensuring that installed equipment remains serviceable and by facilitating migration to new systems when appropriate.

Modular system development allows integration of discrete application subsystems. These subsystems are generally in the form of traditional turnkey systems designed to perform specialized tasks. These turnkey systems provide mini-computer capability to perform each respective function by means of its own CPU. The integration of the common data base is of the greatest significance to the system user. This allows users at different nodes in the distributed network access to any relevant product or planning data located in the network.

A prime example of a modular implementation system on the market today is Sperry Univac's Unis CAD system [10]. Unis CAD's main attribute is its use of a mainframe to maintain a single central database that as many as 256 interactive graphics terminals can share. Univac has devised interfaces for Megatek and Adage graphics terminals which themselves perform computing in the process of manipulating images on their screens. This leaves the mainframe to handle a common parts data base, other Unis modules, and whatever additional tasks the user needs to perform.

Univac has provided interfaces for Unis CAD to work in conjunction with other Unis modules so that an integrated system is available to whatever degree the user desires. Unis CAD can be used on a stand-alone basis or with as many Unis modules as are needed. Present modules available with the system include basic design, drafting, numerical control facilities, hybrid geometric modeling, and engineering analysis. Modules on the drawing board include macro capability, and bill-of-material input to other Unis modules.

Another modular implementation, of particular interest to the organization with several application systems already in place, was performed by John Deere and Works [4]. Deere Tractor Works consists of seven separate buildings. Nine minicomputers operate ten systems to control the manufacturing processes and movement of materials throughout the facility.

The two significant characteristics of Deere's system that catagorize it as a modular operation include:

1. The minicomputers are also data processing and business management computers, linked with a host computer at John Deere's remote computer center. The minicomputers essentially drive the entire factory in response to instructions received from the host.

2. The communication between the minicomputers takes the form of queries to determine availability of parts and subassemblies, to ensure that a tractor does not start down the final assembly line unless all major components required for its completion are available.

PART FOUR: A MODEL OF MODULAR IMPLEMENTATION

To best illustrate the development of a modular Computer Integrated Manufacturing

System, it is helpful to describe the theoretically "complete" system. For the purpose of this paper, General Electric's "Factory With a Future" will be presented as a model of the completely integrated manufacturing system. Descriptions of various module developments will be based on actual implementation experiences by various firms.

Figure 4 on the following page presents a modular illustration of GE's "Factory With a Future." As the illustration shows, three levels of data communication exist. At the top level is the data management and communications system which provides all internal networking services. As subsets to this system are the computer aided engineering, computer aided manufacturing, intelligent warehouse and operations management systems. Associated subsystems to each of the subsets compose the third level of communication. Implementation of system modules is likely to take place at the subset level.

The subset level is generally within the capability of today's minicomputers. For example, in the case of the computer aided engineering subset, a minicomputer can typically support the associated subsystems. These subsystems include CAD, analysis, NC programming, mold/tool design, QC planning and process planning. Of course, it is possible to obtain traditional turnkey systems capable of providing the individual subsystem applications. The problem with developing the modular system from the subsystem level is the difficulty in successfully networking an efficient information flow as additional applications are implemented.

Modular implementation offers great benefits to the organization, but requires taking a hard look at all of the possibilities and then developing a flexible, long-range, step-by-step plan that is uniquely suited to your particular operation. This is particularly important in terms of designing and developing the corporate data management and communications database.

Referring to GE's "Factory With a Future," the following modular development is possible:

1. Introduction of Calma, GE's computer-aided design sytem can be implemented as an individual module. This module creates a data base that serves as a common denominator for design and engineering.

2. The addition of CAM software allows machining time and process planning information to be created at the same time as design. Because CAD and CAM share a common data base, every design change automatically results in appropriate modifications to the data base for programming your numerically controlled machine tools.

3. Computer-aided engineering software may be implemented as the next module. It allows the design and test of the product with the existing data base while, at the same time, an integrated CAM system is creating pre-production manufacturing data.

4. Introduction of the manufacturing resources planning system enables the integration of many of the key aspects of the firms operations into one central data base. Engineering changes can be entered quickly that directly access the data base information already established by the CAD/CAM system.

Referring back to Sperry Univac's Unis CAD system, the benefits of a modular system development in a distributed data processing environment are obvious. Downloading a lot of the CAD/CAM operations to the minicomputer frees the mainframe to run other business programs. The mainframe CPU is only called on to perform large-scale system work.

Hewlett-Packard has made use of the modular system concept with distributed data processing at several of their facilities [11]. At their integrated circuit facility, the parametric device test and engineer quality analysis system are configured in a distributed computer hierarchy. The supervisory computer provides test programs and patterns to be downloaded to the test system's computers, as well as format conversion and packing to ship the data to an engineering data base on the manufacturing planning level.

At their metal/plastic parts fabrication facility, the direct numerical control/work order grouping system is set up in a distributed heirarchy with a supervisory computer for programming NC sheet metal punches.

The model modular Computer Integrated Manufacturing System (CIMS) will successfully tie together all data communication in the organization. From the office - MRP, NC Part Programming, Energy Management, Group Technology(GT), Computer-aided Process Planning(CAPP), and Computer-aided Design(CAD). From control requirements - Process Control, Adaptive Control, Programmable controllers, NC, CNC, DNC, Minicomputers, Microprocessors, Vision systems, Sensors, and Diagnostics. From

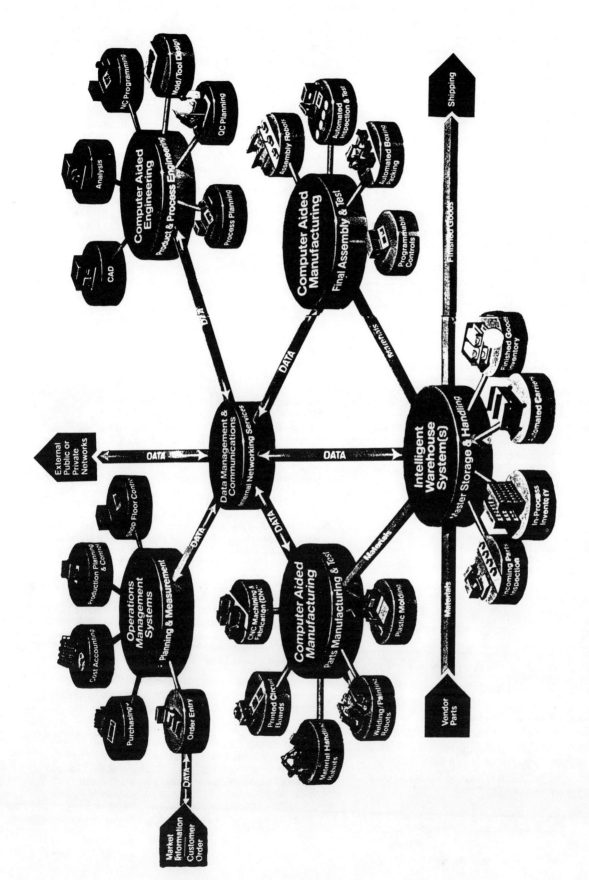

Figure 4: General Electric's "Factory With a Future". A conceptual model of the complete modular Computer Integrated Manufacturing System.

handling requirements - AS/AR systems, Robot Handling, and AGV. And finally, from processing requirements - Machine tools, Automated Inspection, Work performing Robots, and Automatic Assembly.

These are the components of today's manufacturing facility that must be brought together via fast, efficient data communication.

CONCLUSION

Modular Computer Integrated Manufacturing System development appears to be the most viable alternative of the future. Today's manufacturer is looking for a means of introducing high technology into their operations in a way that is profitable to the organization in the short-run and strategically sound for the organization's long-run goals. Modular system development offers the manager this very capability.

A strategic long-range plan must be developed for such a system development. This well laid plan will allow efficient distributed data base development as system modules are progressively implemented. The end result of such an implementation stategy is a successful information network between all aspects of the organization, free of data redundancy and hardware obselescence.

REFERENCES

[1] Appleton, D.S., "DDP Management Strategies: Keys to Success or Failure," Data Base, Summer 1978

[2] Greene, A.M., "Modular Approach to Low Cost CAD/CAM," Chilton's Iron Age, Vol.225, June 16, 1982

[3] Greene, A.M., "Will Turnkey CAD/CAM Systems Remain Viable?," Chilton's Iron Age, Vol.225, Nov 19, 1982

[4] Hegland, D.E., "Putting the Automated Factory Together," Production Engineering, Vol.29, No.6, June 1982

[5] Lerro, J.P., "CAD/CAM System: foundation of manufacturing automation," Design News, March 1, 1982

[6] Marshall, P.S., "Do It Yourself CAD/CAM," Machine Design, Vol.24, No.10, May 6, 1982

[7] Martella, G. and Schreiber, F.A., "Creating a Conceptual Model of a Data Dictionary for Distributed Data Bases," Data Base, Summer 1979

[8] Molina, F.W., "A Practical Data Base Design Method," Data Base, Summer 1979

[9] O'Donnell, M.M., "Planning for Obsolescence: Innovation vs Maturity," InTech, Vol.29, No.9, Sept.1982

[10] Verity, J.W., "CAD Meets CAM," Datamation, Vol.28, June 1982

[11] "CAD/CAM - The Factory Integrator," Production Engineering, Vol.30, No.4, April 1983

BIOGRAPHICAL SKETCH

Scott E. Lee is a Masters degree candidate from Clarkson University in Potsdam, NY. His area of specialization is Manufacturing Systems focusing on the integration of information systems with the factory floor. He has served as a Graduate Assistant while pursuing his Masters degree and has performed consulting work for IBM in Endicott, NY related to his research in the area of visual inspection tasks. Prior to graduate school, Mr. Lee was employed as an Industrial Engineer for Welch Allyn Medical Products Division where he was involved in the development and implementation of a Materials Requirements Planning system. He was employed by the Bendix Corporation Electrical Components Division as a summer intern while pursuing his B.S. in Industrial Distribution from Clarkson. Mr. Lee recently accepted a position with the General Electric Company in Schenectady, NY as a Systems Engineer for the Industrial Management Systems Operation. He is an Associate Member of IIE.

Charles T. Mosier is Assistant Professor of Industrial Management at Clarkson University in Potsdam, New York. He received his Ph.D. in Operations Management from the University of North Carolina at Chapel Hill. His current research focuses on Group Technology and Computer Integrated Manufacturing implementations. His articles have been accepted by the International Journal of Production Research, Management Science, and various AIDS Proceedings.

Reprinted from the 1984 Fall Industrial Engineering Conference Proceedings.

Implementing Flexible Automation

F. E. Harkrider
Vought Aero Products Division
LTV Aerospace and Defense Company

ABSTRACT

With a sharp focus on aerospace industry manufacturing technology trends and product design developments, the Vought Aero Products Division (VAPD) of the LTV Aerospace and Defense Company has sought to meet the needs of its current and future production programs through implementing automated systems that offer the flexibility required for varied and dynamic product lines. The most pressing technological need was found in the area of detail part machining.

The initial development, design and implementation of a Flexible Machining System (FMS) was successfully undertaken, and the system was put into operation on 2 July 1984. The coordinated efforts of the Division's Industrial Modernization (IMOD) organization (in development, design and implementation) and the Production organizations (in final implementation and operation) involved significant innovations in management, control and factory integration. The pitfalls and successes in the design, implementation and production phases of the FMS effort are presented in the discussion that follows. The management tools and operating methods used in integrating the computer-oriented system into a conventional human-oriented factory environment are described along with the resulting impacts on existing organizations and environments.

INTRODUCTION

The VAPD FMS was the end result of extensive inhouse investigation of alternatives to the purchase of additional conventional machines or vendor offloading to meet current B-1B detailed part machining capacity needs. Studies verified that the implementation of a computer-integrated automated machining system offered a 3-to-1 productivity increase over the conventional machining alternatives. The business opportunities and technological advances, that were offered through the implementation of an automated system comprised of proven technology elements, provided VAPD with an attractive, low-risk alternative to maintaining the status quo in meeting current and future program needs.

Systems, industrial and computer engineers in VAPD's IMOD organization, developed the FMS concept by matching a requirement with technology applications. The IMOD team designed the system, justified the costs and negotiated the issuance of the purchase order to the selected vendor. After extensive internal development and design efforts, Cincinnati Milacron, Inc. (CMI) was selected as primary vendor for the FMS. CMI was chosen on the basis of the firm's cost, design, and commitment to massive software development and demanding schedule requirements.

Throughout the implementation phase of the FMS project, a close working relationship was developed and maintained out of necessity between CMI and IMOD engineers. IMOD coordinated the effort with VAPD functional manufacturing units during the design phase of the effort. During implementation, a VAPD Manufacturing Operations manager was appointed to interface the system into production. This interface included the coordination of Manufacturing Engineering, Quality, Material Control and Facilities tasks in order to prepare the system for production on 2 July 1984. Meanwhile, IMOD retained the responsibility for design trade-offs, delivery and final implementation of the system. The methods used to accomplish the tasks involved in design and implementation of the FMS, along with the problems encountered and the solutions employed, are described in the paragraphs that follow. The process detailed in this discussion resulted in the successful implementation of the FMS, which VAPD believes is the most advanced manufacturing system in the world.

FMS REQUIREMENT IDENTIFICATION

In mid-1982, VAPD was faced with a machining workload that exceeded the inhouse capacity. We had competitively sought and won sufficient work to overload our present equipment and had planned to subcontract work that we could not do internally. The primary reason for the overload was the activation of the United States Air Force (USAF) B-1B Program in which VAPD serves as aft and aft intermediate fuselage (AIF) subcontractor to Rockwell International (RI).

During this same time period, the IMOD organization was formed within VAPD. Under the direction of G.E. Ennis, IMOD sought to capitalize on the opportunity presented by the B-1B workload and to simultaneously modernize the manufacturing capabilities of VAPD.

In pursuing the opportunity to apply new technologies to modernize the VAPD factory and improve productivity and cost-effectiveness on the B-1B Program, IMOD engineers performed feasibility studies in order to cost justify the FMS and to document its potential benefits. IMOD then used these results to sell the FMS concept; first to Division and Corporate management and then to RI and the USAF.

VAPD implemented the system because it proved to be an attractive business proposition, not just for the sake of technology. The return on investment was calculated and analyzed and found to meet Corporate requirements. The LTV Aerospace and Defense Company was successful in securing USAF indemnification for the system as part of the B-1B subcontract.

The resultant business arrangement enabled VAPD to implement the first system in its overall plan for a computer-aided Multiproduct Factory of the Future (FOF) with minimal risk, while the U.S. Air Force realized substantial program savings. This opportunity was truly unique in that it was an everybody win situation. The U.S. Air Force could realize millions of dollars in savings, while VAPD used their portion of cost savings to modernize outdated, labor-intensive manufacturing capabilities.

SYSTEM DESIGN

System Design Parameters

Since the manufacturing concept was based on current and clearly established program needs, designing the FMS was a rather well-defined, but complicated task. The system had to be sized and designed to accommodate the B-1B workload. Unlike some other systems that were initially designed and then applied in production, the VAPD system had to be designed to fit the work and to be integratable into the Division's manufacturing environment.

Compliance with Current and Future Needs. A part of that integration, however, required that the system not just comply with current needs and operations, but with those involving manufacturing operations planned or projected for five to ten years in the future. Such system compliance with future requirements was necessary in order to assure that the FMS would provide an integratable initial link in the VAPD plans for a Multiproduct FOF.

Key Integrating Link of Information. This integration requirement demanded the inclusion of an informational software link with existing computer systems, in order to obtain and plan tasks and to interface with the majority of factory areas that are human oriented and dependent. To serve the informational needs of an FOF computer-oriented environment, the FMS had to be designed for integration into a Computer-Aided Factory Integration (CAFI) system.

This CAFI will be comprised of a network of interconnected, hierarchically structured computers and information systems that will essentially comprise the brain or control mechanism of the automated factory, and will serve to unify and coordinate the systems within the VAPD Multiproduct FOF. FMS informational software links had to be so designed to allow their use as the foundation for the VAPD FOF functional hierarchy of computer control that will eventually link every element in the factory.

Design Phase Scenario

Design Effort Participants. The FMS design effort was performed by an IMOD team comprised of the IMOD director and specialized systems, industrial and computer engineers who were familiar with the hardware and operating software systems that currently control Corporate and Division operations.

Workload and System Sizing. The IMOD design team surveyed the B-1B detail machined part task and identified a candidate workload comprised of 541 part numbers requiring a four-axis machining capability. These parts were included in those originally planned for outside vending in order to meet required production quotas. To allow maximum productivity and the manufacturing flexibility to respond to the needs of a low-rate (four-ship-set-per-month) program, the team concluded that the workload should be processed through the system in minimum lot size quantities.

This minimum-batch sizing was ideally set at single-ship-set quantities at a time. This serial (one at a time) manufacturing capability demanded the flexibility in order to meet B-1B machined part needs. The serial approach also minimized costs incurred in stored materials and work-in-process inventory.

With the workload sized and quantified, the IMOD engineers determined that a system made up of eight machining centers, with a significantly sophisticated and responsive software system, could accommodate the B-1B detail machined part requirement and provide adequate support for projected machined part needs.

Hardware/Software Element Scoping. With the workload and overall system sizing determined, elements were defined that would enable the FMS to perform the desired functions. The IMOD team's attention in this effort was focused on the definition and capabilities of the significantly sophisticated and responsive software system that was the enabling element in the workload and system sizing effort. The software system was the primary control element required to keep the cell productive and truly flexible on a 24-hour, six-day-per-week schedule.

Hardware element scoping efforts resulted in the determination that the FMS should include queueing modules that would allow for interfacing of manned loading/unloading of the workload interface, as well as for queueing enough work for an unmanned third shift. The hardware scoping isolated specific needs for an inspection capability, a wash station to remove machining chips prior to inspection, and an integrated chip/coolant collection system. To allow maximum machining center spindle utilization, the FMS required the automatic dispatch of the parts to various process/operation stations within the system. Because of the significant cost of the system, the designers saw a need for a complete module redundancy so that discrete failures in the system would not cause significant loss of production.

Design Simulation/Verification

After the IMOD team sized and scoped the workload and system, the generally defined work content and system software and hardware elements were used in a computer simulation model. The modeling activity was conducted in order to verify the design and its functionality in accomplishing the desired production goals.

Once the simulation was completed, IMOD formed a team comprised of representatives from the manufacturing functional areas to review, critique, and offer suggestions concerning the operational affect of FMS in the respective production functional areas.

FMS Design Documentation

The FMS workload and elements, as identified in the design phase effort, were validated through the extensive modeling, simulation and analysis activities that were undertaken by IMOD systems engineers. The confirmed design elements were then described in the Systems Requirement Document (SRD). This SRD was held as the final authority in all contract negotiations, in the final award, and in acceptance testing.

SYSTEM DESCRIPTION

Major Elements

The VAPD FMS, as designed by the IMOD team, is shown in Figure 1. The system basically consists of:

- Eight four-axis computer numerically controlled (CNC) machining centers equipped with 90-cutter tool magazines and pallet delivery/discharge systems
- Automatic wash station
- Two coordinate measuring machines (CMM)
- Two ten-position carrousel queue modules, each with two load/unload stations
- Automated storage and retrieval system (AS/RS) for cutters
- Two fixture buildup stations
- Dual-flume chip removal/coolant distribution system to support both aluminum and nonaluminum material machining and segregate chips under computer control
- Four-vehicle, computer-controlled wire-guided cart (robocarrier) system
- Forty machine work pallets
- System control computer network including direct numerical control (DNC) capability and a computer-driven software system that directs system operations.

Functional Operation Scenario

The eight machining centers are individually equipped with controllers that communicate with the system computer. The machining center controller interfaces with the system computer to determine its specific task. Up to 16 parts can be loaded on a pallet for machining at one time. The controller drives the machining center to:

- Shuttle a pallet load of parts to be machined into the input queue
- Cut the designated parts from the pallet load of blanks in a conventional NC mode
- Rotate the pallet (if required) to machine the remaining blanks on the other pallet riser face(s)
- Shuttle the completed pallet load of parts to its output queue for robocarrier pick up.

The controller also performs diagnostic functions to determine machining center status and assess malfunctions. Sufficient memory capacity exists within the machining center controller to store NC programs for up to 48 parts to be machined whenever the system computer communication is not available.

In addition to the worktable surface directly adjacent to the spindle on each machining center, the module features two input and output queues. The input and output stations were included within each center in order to meet the design goal of maximizing spindle cutting time by providing accessible queue space.

After a completed work pallet is shuttled to the output queue, a robocarrier is dispatched to the center to transfer the pallet to the wash station. The wash station removes the chips from the machined parts on the pallet riser and then dries them to prepare the parts for inspection.

The robocarrier retrieves the pallet from the output queue of the wash station and subsequently transfers it to one of the two CMMs for inspection. The CMM inspects all of the machined parts

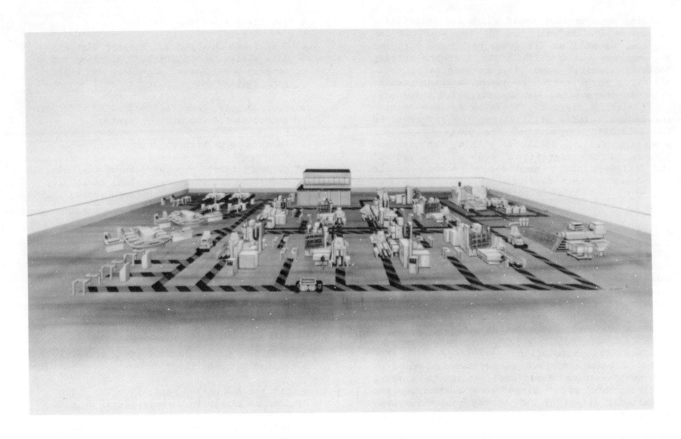

Figure 1 VAPD Flexible Machining System Layout

for proper tolerances under NC control and returns the results to the system computer for disposition. When the inspection operation is completed, the CMM shuttles the pallet to one of its output queues and the computer dispatches a robocarrier to deliver the completed/inspected parts to one of the carrousels for unloading.

Informational Operational Scenario

The functional operation of the FMS may seem simple until the complexity of the information flow is examined. The elements addressed in the FMS information flow include:

o Required quantities and need dates of parts

o Number of passes required for each part to machine opposite faces

o Numbers and types of pallet riser fixtures required to hold the blanks and partially machined parts during machining

o Occupation maintenance and statusing of module queues

o Resource (blanks and NC programs requirements)

o Cutting tool management

o System scheduling.

These information elements are managed and controlled by the FMS computer system network. The successful use, interchange and manipulation of the system's information elements via the computer system truly make this Flexible Machining System unique and unparalleled in its function and capabilities.

Computer System

The FMS computer system is based on the Digital Equipment Corporation (DEC) PDP computer. The system is centered around a PDP 11/44, which serves as the primary hardware driver for the FMS computer network. The FMS software, however, is the system element that enables the unmanned control, operation and monitoring of the machining function. Software is the element that makes the system.

Software Description. The unique architecture of the FMS software scheme, as shown in Figure 2, is the cornerstone of the system's operational success. The software is the commodity that organizes and applies 18 pieces of complicated machinery, a host of computers and controllers, man-years of design and thought -- to specific, tangible production goals aimed at maximized productivity and minimized cost.

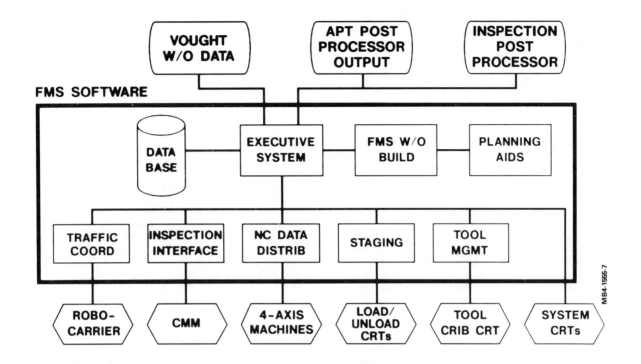

Figure 2 FMS Software Architecture

The FMS software is the key that makes the VAPD system more capable and flexible than any other. It provides for a smooth operation, while allowing for a calculated recovery in the event of a malfunction. The software interfaces with existing systems with minimum programming effort to determine assigned tasks and to report successes.

In the design effort, the IMOD team determined that the goal for the software system was to assure that any human operator assigned to the FMS would be bored. The team further stipulated that such an operator would only be informed of operating decisions, instead of making one. This software design criteria was considered essential for establishment of an FOF computer-oriented system as opposed to a modernized version of the human-oriented current environment.

Team members reasoned that if the operator assigned within the FMS were to make decisions, then the quality of the product and the productivity of the operation would be limited to the cognitive skill and capability of that person. Further limitation would be placed on such an operation since the same operator is incapable of working three shifts per day and would therefore be unable to make responsive, specific decisions to optimize throughput and minimize work-in-process inventory on a continual basis. IMOD team members believed that if the FMS software made the decisions, then they could be optimized to provide the maximum throughput for all shifts. Only in this controlled manner could parameters that manage the system be changed and the results on productivity measured.

Based on the clearly defined software design criteria, the IMOD team structured a software system that alleviates the human operator of all discretionary input. The validity of this design goal is now being put to the test in the recently implemented FMS. The system, in full operation, should act as a machine, commanding the tasks assigned in a timely manner, but without uncontrolled perturbations.

Computer System Operation. The computer system software includes the executive system that enables the FMS to become a productive operational reality. As shown in the figure above, the executive system retains the status of the elements of the system in order to interrogate the data base, and thus maintains the system in an operational mode.

The executive system controls the output to the carrousels, the machining centers, the chip/coolant system, the wash station, the CMMs, and the cart system, while it manages the queues inherent in such a complicated system. To accomplish this task, it communicates with four other DEC computers, eight CMI machining center controllers (which challenge the PDP 11/44 for intelligence), and 16 other controllers within the system.

Additionally, in order to efficiently produce high-quality detail machined parts, the PDP 11/44 periodically downloads work orders, due dates, material availability statusing, NC machining programs, NC inspection programs, and recently effected changes thereto from the Corporate open order system.

When part machining is completed, the computer system then informs the appropriate Corporate information systems of this fact, and reports detailed accounts of time utilization and inspection results. This two-way communication capability with Corporate systems allows FMS to function, even in the conventional factory environment, as a totally integrated system.

HARDWARE/SOFTWARE IMPLEMENTATION

Responsibilities, Tasks and Transitions

The design effort that resulted in a viable FMS concept required a significant commitment and risk. However, despite the unique challenges that the design and subsequent procurement efforts had produced, the implementation of the actual FMS hardware and software posed issues and tasks that collectively appeared to be quantum leaps beyond the former activities in complexity and demand.

The implementation tasks broached the responsibilities and areas of expertise of a variety of organized disciplines within VAPD. These tasks included:

o Facilities modification and factory rearrangement

o Machining center and CMM NC program development

o Computer interface program design and development to link FMS and Corporate system data bases

o Tool, fixture and cutter design, building and/or purchasing

o FMS production manager selection

o Software design, programming and checkout to prepare the FMS for production.

IMOD also had to coordinate tasks with CMI to finalize the design details, provide the interface between CMI and the user groups, and resolve unanticipated issues and questions. One of the initial implementation tasks was to develop a PERT chart of the functions, identify the specific tasks, and establish budgets for the tasks.

Through the design and vendor selection phases and into the preliminary implementation effort, the IMOD team retained and exercised primary control and responsibility within VAPD for the FMS project. However, as implementation progressed to the point that FMS integration into existing factory operations was an issue, responsibility for the final implementation tasks was conveyed to a Production organization team led by the FMS production manager and comprised of representatives from manufacturing units that would ultimately be affected by the FMS operation.

After the tasks were initiated, the FMS production manager was named. He assumed the responsibility for coordinating the efforts of the manufacturing units for completion of their tasks. He had a vested interest in the successful completion of these tasks, since he was to be responsible for the operation of the FMS on 2 July 1984.

Management Chart Room and Weekly Meetings

One of the most useful management tools employed to assure that the 2 July 1984 production date would be met was the establishment of a management chart room. The IMOD team initiated the charting of every identified task involved in the implementation effort. A sample chart for NC tool tape statusing is provided in Figure 3. The charts included weekly goals and major milestones. As the implementation tasks progressed, separate charts were prepared to follow the progress on critical action items and problem areas. These task charts were all posted in the management chart room.

Once a week, a meeting was held in the chart room. All participants in the FMS implementation effort were represented in these conferences along with a management group comprised of vice presidents and directors. At each meeting, IMOD team representatives would present the charts prepared for the implementation tasks. The FMS production manager would then present status charts from the functional manufacturing units. These presentations included discussions of the progress and problems associated with the hardware and software implementation. After the charts were presented, the performance-to-budget status for each unit was shown and additional discussion ensued.

The establishment and use of the management chart room proved to be an exceptionally beneficial management tool. The charting and posting of task statusing allowed the team to follow overall and individual activity progress from week to week. Comparisons and schedule departures were evident to all participants and management. Posted task charts provided subtle incentives to those responsible for discrepancies to pursue a speedy resolution. The charting and simple routine scheduling of a weekly meeting encouraged the speedy resolution of issues and accelerated the decision-making process prior to the meeting.

The production manager also held a weekly working meeting to provide communication at the level at which the actual work was being accomplished. In the weekly gathering, problems were addressed and in many instances resolved, responsibilities were assigned, and fires ignited under deserving participants. These working meetings forced assigned personnel to face up to their problems, to realize the impact on the schedule and other units, and to take steps to

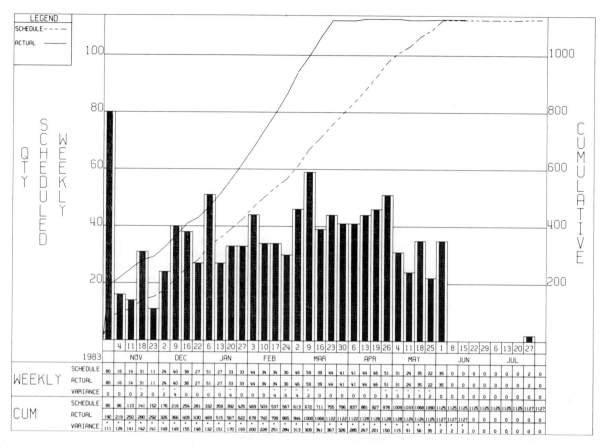

Figure 3   FMS NC Tool Tape Status Chart

correct the situation. IMOD team representatives were active in these meetings, because they had to insure that the design of the system was compatible with the planned implementation.

<u>IMOD Implementation Issues</u>

Many of the problems that occurred in the implementation process could not have been anticipated in a project of this size. IMOD team tasks, which addressed both hardware and software elements, seemed to attract problems. Some of these issues were even humorous (in retrospect), while some should have been more aggressively guarded against.

<u>Unforeseeable Issues</u>. One of the more retrospectively humorous occurrences involved the laborious testing of the FMS computer room fire extinguishing system. All personnel were fully instructed on the various conditions or actions that could set off the system. They were further briefed on the method for prohibiting the release of the fire-extinguishing halon gas in the event that a false alarm was sounded. All personnel were so instructed, that is, except for the repairman who came to replace a bad compressor in the air conditioning system for the computer facility -- the fire extinguishing system worked!

<u>Human Intervention Issues</u>. However, despite the best attempts at problem trouble-shooting and prevention, the FMS implementation effort did involve certain issues inherent when human beings are involved in any activity. Some individuals have a better idea on how the system should work than the designers. Many of the ideas suggested by manufacturing area technical personnel were good and needed to be incorporated into the system. Conversely, some of their well-intentioned suggestions were based on their knowledge of the existing, human-oriented factory environment, which the designers of the FMS were fervently trying to change. An excellent example of such a situation occurred in the selection of parts that could be mounted on the same pallet for machining (i.e. pallet set groupings).

The IMOD team developed computer programs for pallet set grouping selection that sorted candidate part numbers with common due dates, cutting tools, proportions, sizes and etc. However, some FMS implementation participants selected parts for pallet set groupings based on the commonality principle instead of using the available computer listings.

The dichotomy of pallet set groupings was realized when the due dates of specific parts included in a grouping were found to be weeks apart. In some of these instances, parts that had not been scheduled for stock preparation passes through the FMS were already required for a second pass.

Hindsight on such conflicting technical approaches resulted in the conclusion that personnel involved in such activities must be channeled to follow the specified design and analytical methods, and not the methods they find most familiar or convenient. This disciplining must be managed in such a way, however, that the experience and initiative offered by the assigned technical personnel is not stifled. The expertise and technical prowess of the personnel involved in the implementation of automated technology systems is essential in order to fine tune the design so that it may become a production reality.

Communication Issues. The IMOD organization was further identified in the FMS implementation effort in the role of communicator. The project was a very large undertaking with a very tight schedule that could not slip. At the outset, IMOD management assigned a project manager who had a counterpart at CMI. These two individuals were the official communicators between the two companies. IMOD conducted bimonthly design meetings with CMI to discuss design changes for updates on implementation, schedules, progress and problems in general.

In the relationship between the two companies and their managers, a log of issues was established for design and problem areas. This log was reviewed monthly and served well to document and track the issues that are traditionally lost in the mountain of letters and telegrams between participants.

However, the communications were not without drawbacks. Responsibilities for hardware and software implementation tasks were divided among the IMOD groups that were organized to accommodate the FMS workload. Communication among the IMOD groups and individual members was not an issue since the personnel were all physically located in the same facility area. However, the official channel for communication was not adequate due to the dynamic nature of the FMS design and implementation effort and the time lag created by the requirement of using official channels of communication.

The need for sometimes daily contact between CMI technical personnel and those at VAPD was impacted due to the timeliness and interpretation inherent in communicating through project managers. Ultimately, the project managers were used to convey formal transmissions and to relay factors that would affect price or schedule. The end result was that the leaders of the VAPD and CMI software groups were able to discuss their problems in a timely manner without interpretation.

Software Issues. As the implementation phase progressed, it was apparent to all FMS participants that CMI had underestimated the effort required for the software development. This factor seriously endangered the 2 July 1984 startup date. The software development lag had a significant impact since the FMS was designed to be run without and cannot accommodate manpower. This design requirement meant that the computer system had to have software capable of transmitting NC programs to the machining centers, running the robocarrier cart system, and managing the schedule.

As a workaround to the software task scoping shortfall, CMI broke the software into two phases, offering VAPD the software essentials by 2 July 1984 and proposing delivery of complete software to meet the specified system requirements in 1985. However, despite the efforts of both VAPD and CMI, this compromise schedule was also unattainable. By 2 July 1984, CMI software developers provided only the bare essentials of the sophisticated software system, delivering a skeleton operating system for production startup.

Hardware Issues. Hardware problems were also causing serious schedule compliance difficulties. While most of the hardware was essentially off-the-shelf, virtually every module had to be engineered and modified for the additional requirements. The resulting delayed equipment shipping dates were seriously impacting NC program proofing schedules that assumed the installation and operation of some of the FMS equipment.

CMI addressed their responsibilities by committing additional resources to the project. They had already provided VAPD with one, and then a second machine tool to aid in proofing the geometry in NC part programs. VAPD also had the use of a consignment CMM and the vendor allowed us to send a team to their facility to conduct additional proofing of the NC inspection programs. CMI committed to building a permanent test cell for continuing software development. Prior to this time, they had a test cell built in their Cincinnati, Ohio, facilities consisting of one each of the hardware elements that were to be delivered to VAPD. During the last few months before the FMS was activated for production, CMI conducted testing on a 24-hour, seven-day-per-week basis in both VAPD and CMI facilities.

Resolutions and Results. During the month of June, 1984, VAPD and the IMOD team had anticipated that by this point in time, program proofing would be completed and the system would be undergoing testing in the production mode. However, the month of June was spent completing installation tasks and in testing the hardware and software. The final skeleton version of software was installed less than a week before cell startup. The stage was then set for a rude awakening -- real data is not the same as test data.

PRODUCTION IMPLEMENTATION

On 2 July 1984, the Flexible Machining System was turned over to Production as planned. The system had moments of greatness and moments of disappointment. At the time of production startup, CMI had delivered all hardware with the exception of one machining center. One CMM was semi-operational and the other was being installed. Software elements that were noticeably absent when startup occurred included such features as stability, recovery from malfunctions, diagnostics, and adequate scheduling routines to support a more manual system than anticipated at this stage.

VAPD demonstrated some delinquencies in providing the resources required for production operations, as well. There was a shortage of proven NC programs because of inadequate availability of the machining centers for proofing. All of the material for machining was not available. The result was that only a limited number of pallet sets were available for machining. Unfortunately, all of these available pallet sets needed the same types of pallets to do the first operation (stock preparation). The result was a difficulty in keeping the system active when it was operational.

In light of all of these factors, the FMS production manager made the decision to use three of the machining centers to complete part proofing. Additionally, the CMMs were not included in the initial system operation because of a shortage of proven NC programs and because the software to integrate the machines into the FMS was not complete. One machine was used for proofing until the second was operational and then it, too, was used for proofing.

The first week of production was acceptable in light of all of the problems and the abbreviated scenario described. However, room for improvement was evident. CMI was operating the computer system because the software was not yet in a stable state and there was no documentation or recovery aids. Each week the system became more stable. In an attempt to load the FMS, a concerted effort was conducted to identify work, which could be run now, and the work that could be run in near time with the minimum manual effort. The combined efforts of IMOD, the functional manufacturing units of VAPD and the additional work that CMI has devoted are paying off. The moments of greatness are more productive and are lasting longer. The moments of disappointment are shorter.

CONCLUSION

The implementation of the FMS into the VAPD factory has been a qualified success. The system workload was planned to be light while the problems were resolved and the problems rose to the occasion to challenge the planned workload. The true test will occur when the workload escalates to capacity and the system operates for 24 hours per day, six days per week.

Present loading problems indicate that the system was designed correctly. The main indications are that human operators cannot make the decisions in a timely manner to run the system in even a semiautomatic mode. The number of parts, pallets, cutting tools, passes per part, etc. are just too enormous for a person to comprehend. Already, evidence indicates that human intervention greatly reduces the flexibility and the productivity of the FMS.

BIOGRAPHICAL SKETCH

F. E. Harkrider is the Manager of Information Systems within the Industrial Modernization organization of the Vought Aero Products Division, LTV Aerospace and Defense Company. His responsibilities include the design of software systems to modernize manufacturing capabilities at VAPD. His prior experience included the managing of an organization which designed and developed software for engineering and manufacturing systems. Mr. Harkrider holds a BSEE degree from The University of Texas at Austin and an MBA from the University of Dallas.

Reprinted from Industrial Engineering, June 1984.

How Small Firms Can Approach, Benefit From Computer-Integrated Manufacturing Systems

By H. Lee Hales Management Consultant

Computer-integrated manufacturing (CIM) is a way of doing business, not just a specific system or set of applications. The focus of CIM is on the sharing of information by engineering, production and various support groups. In its most comprehensive sense, CIM automates the flow of information between the operations and activities of the manufacturing firm.

Obviously, CIM encompasses a large number of computer and automated systems, many of which are listed in Figure 1. The chart shows dozens of potential projects for even the smallest manufacturing firm. In fact, with today's microcomputers, the small batch manufacturer can obtain many capabilities whose use has been restricted in the past to large, process-oriented concerns.

Clearly, we need a plan to work our way through the many opportunities ahead. And while each plan will be different, there are some common steps that are necessary for success.

1.) *Set meaningful objectives*. The objectives of CIM are many, but generally include shorter development cycles for new products and designs; improved quality; reduced inventories; flexibility; and automation of small batch, discrete parts production.

The relative importance of these objectives will vary from time to time and from plant to plant. As the emphasis shifts, so too should the priority and sequencing of CIM applications (see Figure 2). The IE can play a valuable role in communicating this relationship to management and lobbying for continuity in directions and objectives.

2.) *Audit current conditions*. Once the objectives and their importance have been established, the next step is a systematic audit of current conditions. The purpose of this audit is twofold: to assess the status of current computer and automated systems and to determine the relative importance of future CIM applications.

The audit needs to consider all stages of the manufacturing operation and all levels of activity. It must give equal attention to the internal needs of engineering, production and various support groups.

The CIM audit should also reflect the external influences on the busi-

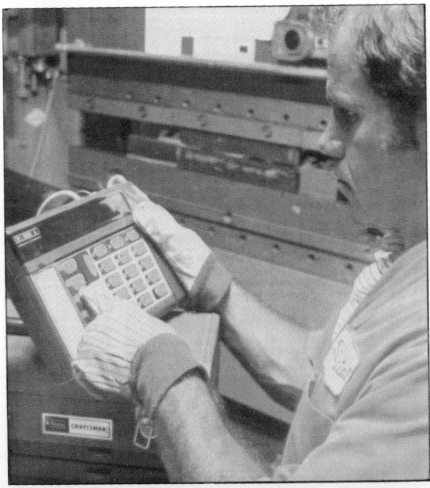

The compact terminal being used here to enter work order information enables non-computer experts to communicate with a host system. Its keyboard was designed to be used while wearing gloves. (Photo courtesy of Hewlett-Packard)

Figure 1: CIM Overview Chart

STAGE OF OPERATIONS		ACTIVITY LEVELS		
		BUSINESS/PLANT	DEPARTMENT/CELL	WORKSTATION/MACHINE
ENGINEERING	DESIGN	Program development. Design standards. Group technology coding. Central parts database. Computer system interfaces.	Project management. Design procedures. Group technology coding. Local parts database. Computer system interfaces. Training & instruction.	Computer-aided design, incl. molds, dies, tools & fixtures. Group technology part retrieval. Logic & circuit design. Schematics. User interfaces.
	ANALYSIS	Research & development. Laboratory management. Design standards. Central engineering library. Computer system interfaces.	Laboratory management. Design & testing procedures. Local engineering library. Expert systems. Interfaces. Training & instruction.	Computer-aided engineering. Geometric modeling. Finite element modeling. Kinematics. Performance simulation. Logic & fault simulation. Producability analyses. Cost estimating.
	DOCUMENTATION	Documentation standards. Document & information transmission & retrieval. Archives. Data management. Technical publications. Computer system interfaces.	Documentation procedures. Document & drawing input. Micro/reprographics. Printing. Plotting. Scanning. Character recognition. Photocomposition. Interfaces. Training & instruction.	Computer-aided drafting. Word processing. Bill of materials. Process instructions. File & tape generation.
PLANNING	PART PLANNING	Group technology coding. Process planning standards. Central libraries—standard times & costs; tooling & materials. Decision support systems. Expert systems. Interfaces.	Group technology part retrieval. Process planning procedures. Local libraries of standard data, tooling & materials. Decision support systems. Expert systems. Interfaces. Training & instruction.	Computer-aided process planning. Cost estimating. Producability analyses. Process instructions. File & tape generation. Process verification. Methods analyses.
	PRODUCTION & INVENTORY PLANNING	Logistics. Operations research. Manufacturing resource planning. Forecasting. Capacity planning. Master schedules. Lot-sizing rules. Inventory planning. Make/buy analyses. Decision support systems. Interfaces.	Line balancing. Machine loading & scheduling. Process simulation. Lot-sizing. Methods analyses. Decision support systems.	Time standards & allowances. Yield data. Feed & speed data. Job sequence. Methods analyses.
CONTROL	PROCESS, SHOP & MATERIALS CONTROL	Performance standards & measures. Reporting policies. Control system interfaces.	Monitoring systems—machine, process, manpower, materials. Data transmission & analysis. Displays. Reporting systems & procedures. Statistical process control.	Data collection systems & devices—microprocessors, programmable controllers, gauges, valves, alarms, meters, switches, sensing, encoding, scanning, labeling, machine vision, character recognition, keyboards, terminals.
	INSPECTION, TESTING & QUALITY	Quality assurance goals & standards. Program management. Field data collection. Service/warranty database.	Inspection & testing procedures & standards. Monitoring & reporting systems. Statistical process control.	Automatic testing equipment. Computer-aided testing & inspection. Diagnostics. Verification, detection & measurement systems & devices—gauges, meters, switches, sensing, scanning, machine vision, character recognition.
PRODUCTION	PART & TOOL PRODUCTION	Research & development. Testing. Program & project management. Operations management.	Transfer lines. Special systems. Flexible manufacturing systems. Cellular manufacturing systems. Machining/fabrication centers. Robotics.	Standard machines & process equipment. Drives, motors, control devices & systems—numerical control, direct numerical control, computer numerical control. Adaptive control. Robotics.
	ASSEMBLY	Research & development. Testing. Program & project management. Operations management.	Automatic assembly systems. Flexible assembly systems. Assembly conveyors. Robotics. Interfaces to handling, inspection & control.	Automatic assembly machines. Semi-automatic equipment. Robotics. Interfaces to handling, inspection & control.
	HANDLING & STORAGE	Interplant handling systems. Physical distribution. Packaging & container standards. Bar coding standards. Dispatching systems. Fleet management.	Conveyors & transporters. Automated guided vehicle systems. Automated storage/retrieval systems. Sortation systems. Carousels. Order picking & dispatching systems.	Workplace handling devices. Robotics. Containers, pallets, totes. Controls, motors, drives. Data collection systems & devices—sensing, encoding, labeling, scanning, character recognition, machine vision.
SUPPORT	FACILITIES ENGINEERING & MAINTENANCE	Program & project management. Facility planning & design. Construction. Preventive maintenance. Operations management—power & utility systems.	Project management. Plant layout, materials handling analyses. Work order systems. Status, monitoring & security systems. Diagnostic systems & equipment.	Workstation layout & equipment. Support machinery & equipment. Drives, motors, & controls. Data collection systems & devices—gauges, valves, alarms, switches, meters, testers, recorders, sensing, scanning, vision.
	PERSONNEL MANAGEMENT	Organizational planning, policies, & procedures. Manpower planning. Collective bargaining. Incentive systems. Benefit program design. Performance standards & measures. Safety & health goals.	Work rules. Reporting systems & structures. Training & instruction. Safety & health protection.	Work rules. Reporting systems. Training & instruction. Ergonomic analyses. Methods analyses. Time standards & allowances. Performance rating. Safety & health equipment & procedures.
	DATA PROCESSING	Program & project management. System design & development. System architecture & hierarchies. Protocols, standards, interfaces, access. Capacity planning. Data processing & management. Telecommunications.	Project management. System design & development. Networks & gateway configurations. Applications development. Distributed processing. Office automation.	User interfaces. Hardware configurations. Distributed processing. Personal computing. User support. Office automation.
	ORDER PROCESSING	Sales & booking policies & procedures. Forecasting procedures & systems. Central customer database.	Scheduling & releasing. Interfaces to accounting & production systems. Local customer database.	Order entry. Status checks & queries. Office automation—word processing, scanning, telecommunications, keyboard & terminal devices.
	ACCOUNTING	Accounting & reporting procedures. Audit policies & procedures. Budgeting, planning & appropriations. Financial reporting. Physical inventory policies. Computer system interfaces & design.	Transaction processing systems—general ledger, accounts payable, accounts receivable, invoicing, funds transfer, cash management, payroll. Job & project costing. Sales & inventory analysis. Fixed asset management.	Transaction reporting. Data entry & retrieval. Status checking. Office automation—word processing, character recognition, micrographics, keyboard & terminal devices, telecommunication.

ness—foreign competition, labor relations, tax incentives—in addition to such internal factors as existing systems, technical skills and organizational structure. As one who works at all levels, with all affected groups, the IE should be a key contributor to any CIM audit and an important reviewer of its findings.

3.) *Define specific projects.* Each desired application will lead to one or more development projects. Proper definition of these projects requires a good working knowledge of current technologies and commercially available systems.

It is vital to keep abreast of what is possible, what has been accomplished

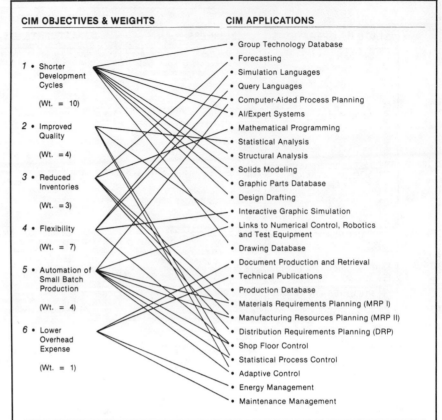

Figure 2: Sequencing of CIM Applications

Figure 3: CIM Audit

Vowel	Priority & Meaning
A	*Abnormally high*—supports several critical objectives; required for competitive reasons; current systems inadequate.
E	*Especially important*—supports a critical objective; required for competitive reasons; current systems need improvement.
I	*Important*—supports a critical objective; required for competitive reasons; improvements already under way.
O	*Ordinary*—supports long-term objective; current systems OK.
U	*Unimportant*—does not support important objectives; current systems OK. (Not shown on chart.)

| | STAGE OF OPERATIONS | ACTIVITY LEVELS | | |
		BUSINESS/PLANT	DEPARTMENT/CELL	WORKSTATION/MACHINE
		OBJECTIVES / PRIORITY		
ENGINEERING	DESIGN	1 E	1 I	1 E
ENGINEERING	ANALYSIS	1 E		1 A
ENGINEERING	DOCUMENTATION			1 E
PLANNING	PART PLANNING	1,5 I		1,5 E
PLANNING	PRODUCTION & INVENTORY PLANNING	2,3 A	2,3 A	
CONTROL	PROCESS, SHOP & MATERIALS CONTROL	2,3,4,5 A	2,3,4,5 E	
CONTROL	INSPECTION, TESTING & QUALITY	2 A	2 I	2 I
PRODUCTION	PART & TOOL PRODUCTION		5 O	5 O
PRODUCTION	ASSEMBLY		5 O	5 O
PRODUCTION	HANDLING & STORAGE			

The CIM audit sets priorities for each stage of operations and level of activity. Priorities are assigned using the vowel-letter code at left. In general, high priorities are placed in applications areas which support multiple and highest weighted objectives. In this example, the objectives refer to the list in Figure 2.

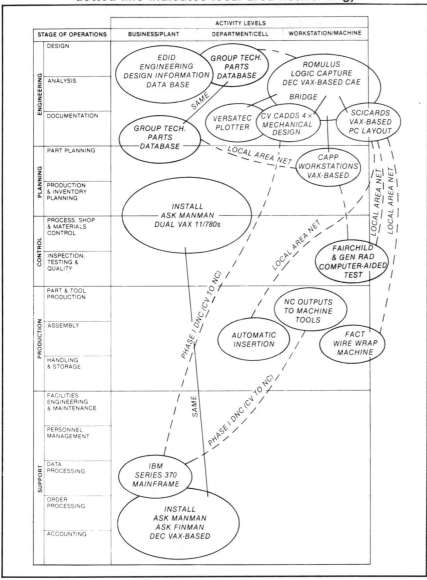

Figure 4: CIM Projects (solid line means direct link; dotted line indicates local area networking)

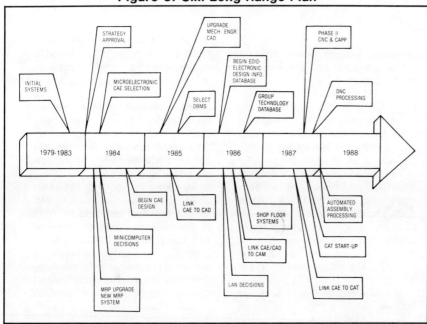

Figure 5: CIM Long-Range Plan

by others and what remains to be developed and proven. Large industrial companies are dedicating sizable staff groups to this purpose.

For those with computer and automated systems in place (and who doesn't have some?), the definition of specific projects will center on the links between these systems, or the lack thereof. Shared databases, communications, controls, network design, replacements and upgrades of existing systems—these are typical "first projects" for those who already have computer-aided design, numerical control and automated handling in place.

4.) *Follow a logical sequence.* By its very nature, CIM ties together different functions and organizational groups. Projects cannot be pursued in isolation when integration is the goal.

Consideration must also be given to available funding and to the capacity of key people to plan and implement while still performing their important day-to-day responsibilities.

The pace of technological advance is another factor whenever computer systems are planned. While it is generally unwise to chase the state-of-the-art, timing of product reviews and major purchases should consider impending advances in price and performance.

A good project sequence will incorporate all of these factors into a digestable, results-oriented schedule.

5.) *Assign the proper team.* Key people make it happen. Questions of ownership, authority and control are central to the success of CIM. Communication, cooperation and compromise are essential between engineering, production and data processing groups.

The lead group will vary from project to project and time to time, but the needs of each group must always be considered. A steering

CIM planning in action

Figure 3 (page 46) gives results of a systematic CIM audit. The priorities are those of a medium-sized manufacturer of computer systems and machinery. The company's products range from standard minicomputers and terminals to large, specialized systems.

Annual product sales are $120 million. Production is centralized in one facility that houses 500 employees. Of these, 350 are direct labor and 150 are indirect. The industrial and manufacturing engineering functions employ 10 to 15 people, depending upon the level of new product introductions.

Manufacturing cycles range from 30 days on standard minicomputers to nearly a year on large systems. In both cases, roughly 30% of the cycle is spent in final assembly and test. While the majority of orders are for the standard, shorter cycle products, the larger systems provide most of the revenue and consume most of the manpower.

The company purchases many electrical components, but fabricates many mechanical and sheet metal parts in-house. Typical release quantities range from five to 15 pieces, and these represent a 90-day supply of the parts involved.

Because the change in computer technology is so rapid, the company's most important CIM objective is the shortening of development cycles from design concept to production. As a result, engineering applications show a high priority.

The goal is to shorten the time required for circuit design, analysis and layout. Better systems are sought for schematic entry, logic simulation, timing analysis, fault grading, etc. Longer-term needs include a central database for electronic design information. This would aid the conceptual phase of circuit design and analysis.

In manufacturing, lead times are long and promised ship dates are critical. New government orders will require component lot controls and some tracking of individual parts. There is a consensus throughout the firm that current production planning and shop floor control systems must be replaced, including order entry.

Note that even though this company produces thousands of mechanical and sheet metal parts, the electronic side of engineering is currently the top priority. In another firm or another industry, or with different production volumes, the priorities might be quite different.

The CIM projects identified in Figure 4 follow directly from the priorities established in the CIM audit. At this point, we can begin to visualize the necessary links between specific systems and equipment.

The opportunity chart, of course, is merely an overview—an idea generator and communications device. Each project will need a well developed and fully documented plan. At this point, formal system development methodologies should be applied (see "For further reading").

The time line in Figure 5 shows the sequence for those projects identified in the preceding steps. The time line milestones are coded to confirm that their sequence reflects the priorities established in the CIM audit. The precise timing of events must wait until the development methodologies have been applied and converted to project schedules. However, a rough cut can be made very early for budgeting and staff planning.

As shown in Figure 6, the same chart used for CIM audits and project definition can also be used to structure the steering committee and assign project teams. In some cases, it may prove desirable to realign the organization for the duration of major CIM projects. The chart can identify where such changes should occur.

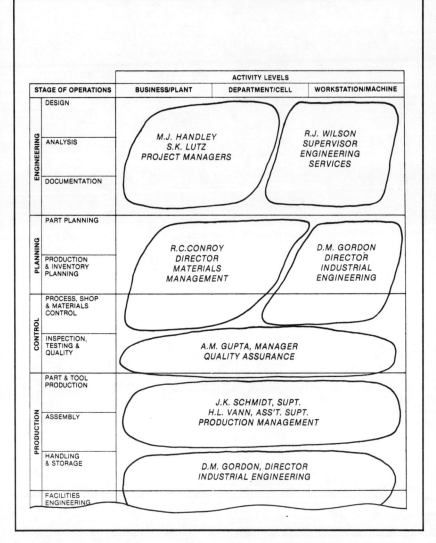

Figure 6: CIM Responsibilities

committee may be useful, in which each member is ultimately responsible for several specific projects and applications, yet all members participate in the design and selection process.

The industrial engineer is both a user and an important source of shared information. In the course of a day's activities, the IE may routinely move among engineering, process planning, quality control, production and facilities support. Following the steps we have defined, the IE can use these contacts to provide the big picture, give direction and help plan for the success of computer-integrated manufacturing.

For further reading:

Bravoco, Ralph R., "Information Systems Engineering—An Overview," Society of Manufacturing Engineers, Technical Paper MS84-183, 1984.

Le Clair, Steven R., and Thomas W. Hill, Jr., "Functional Planning Approach Maps Out Connections Between People and Systems," *Industrial Engineering,* April 1984.

H. Lee Hales is a Houston-based consultant who specializes in facilities planning and factory automation. He works with engineers and data processing on a variety of computer and automated systems and advises software and systems suppliers on the needs of industrial managers and planners. Hales is the author of a forthcoming book, *Computer-Aided Facilities Planning* (September 1984), and co-author with Richard Muther of the two-volume set *Systematic Planning of Industrial Facilities.* He is editor of *CIM Strategies,* a newsletter on computer-integrated manufacturing, and a contributor to Auerbach's *CAD/CAM Series* and to McGraw-Hill's *Management Handbook for Plant Engineers.* A senior member of IIE, he also belongs to CASA-SME, the Institute of Management Consultants and the Independent Computer Consultants Association. Hales holds BA and MA degrees from the University of Kansas and an MS from the Massachusetts Institute of Technology.

Assessing Economic Attractiveness Of FMS Applications In Small Batch Manufacturing

By Daniel P. Salomon
and John E. Biegel, P.E.
University of Central Florida

Evolutionary developments in manufacturing technology and rapid growth in the capabilities of digital computers have given rise to the cellular manufacturing system and its highly automated form: the flexible manufacturing system (FMS). This type of production system is capable of producing high quality products in small lots at a low cost.

A successful FMS requires that similar parts be identified and grouped together into part families to take advantage of similarities between geometric shapes, sizes and processing steps.

Any workpiece in an applicable part family can be randomly introduced in the FMS by clamping it in place on a fixture that is specifically designed to hold the different parts within a family. The fixture is then attached to a pallet and/or cart that will transport the part from station to station.

Workpiece setup is accomplished in the load/unload area and therefore is external to the FMS. The pallets and/or carts form part of a material handling system that is controlled by the central computer to achieve independent movement of the palletized workpieces as well as temporary storage or banking of the parts.

Tool setup is done off-line and the preset tools required for a job are loaded into the tool drum at the station.

Since setup time and its related fixed cost are eliminated or drastically reduced, it becomes as economical to produce parts in small lots as it is to produce them in large lots (see Black, "For further reading"). This situation has made economic order and production quantities concepts, as well as conventional methods of economic justification, obsolete.

This article presents some important considerations that can help the analyst assess the economic attractiveness of an FMS for a specific manufacturing application.

Direct cost reductions

Flexible manufacturing systems call for considerably less direct labor than conventional manufacturing systems. However, direct labor cost does not increase in proportion to direct labor content, because the workers' degree of specialization increases, and this increases hourly compensation.

On the other hand, reducing direct labor diminishes operator errors. This in turn drastically reduces rework and virtually eliminates scrap. Because less raw material is used to produce the same number of pieces, there is a potential for material cost reductions also.

In 1973, Hutchinson and Wynne (see "For further reading") simulated the operation of an FMS configuration that incorporated the features of a system that was already in commercial production. Advantageous sets of good operating decision rules were found. They also concluded that "parts made on an FMS as compared with stand-alone NC or DNC production offer unit part cost reductions of 30 to 70%."

Cook (see "For further reading") used available data in 1975 and found that an FMS needs only between 10 and 30% as much direct labor as a standard job shop of similar capacity.

Klahorst reported in a 1981 article that through the use of an FMS, direct labor content in the workpiece is reduced to one-fourth of the previous levels (see "For further reading").

Our empirical analysis in this area shows that direct cost reductions can vary widely (from 3% to 66%) depending on variables such as the proportion of the direct cost that is contributed by labor and by materials, current and expected

FMS can reduce labor content in a workpiece by significant percentages. (Photo courtesy of Xerox)

scrap levels and the percentage increase of unit labor cost due to a higher degree of worker specialization.

Utilization and lead time

Cost reductions are also obtainable by improving machine utilization and reducing manufacturing lead time. To analyze these cost effects we first need to define the time parameters related to how the machines are utilized and how the parts spend their time in the shop and then determine how they could be affected by FMS applications.

☐ *Manufacturing lead time* (T_{ml}): Total time to process the part through the plant.

$$T_{ml} = n_m(T_{su} + QT_o + T_{no})$$

n_m: Total number of machines through which the part must be routed.

Q: Batch size.

☐ *Machine setup time* (T_{su}): The period in which the machine is being prepared, parts are waiting and there is no part loaded in the machine.

☐ *Non-operation time* (T_{no}): The time that an individual part spends during transportation, delays and inspections. An automated material handling system, controlled by the computer that contains the manufacturing data base, permits an FMS to move parts efficiently and reduces transportation times and delays.

☐ *Operation time* (T_o): The time an individual part spends on a machine.

$$T_o = T_m + T_h + T_{th}$$

☐ *Machine time* (T_m): Time in which the part is actually being worked on. The advantages of combined and/or simultaneous operations can result in shorter machining times in an FMS.

☐ *Workpart handling time* (T_h): Time the workpart is being handled on the machine. Since in an FMS the palletized parts are registered automatically at each work station, T_h is considerably reduced,

☐ *Tool handling time* (T_{th}): Tool handling time per workpart is the average time taken to change and adjust the tooling while the part is on the machine. In an FMS, tools are pre-set off-line by loading them into the tool carrier. Automatic tool changing devices provide for storing, selecting and changing of tools and permit substantial reductions of T_{th}.

All the non-productive times (T_{su}, T_{no}, T_{th}, T_h and T_{mw}) could potentially be eliminated or reduced with an FMS. Determining how much the reduction would be requires a thorough analysis of the specific FMS operation.

Simulation studies help generate data on system performance that can be used to produce tables such as Table 1, which utilizes the optimistic-pessimistic format.

Using the information given in Table 1, we can calculate how much of the time the machine is being productive. Figure 1 depicts how the machine is utilized for the most likely performance of an FMS compared to a conventional system. Figure 2 shows how an average part spends its time while in the shop for the most likely performance of an FMS compared to a conventional system.

Because every part passes through

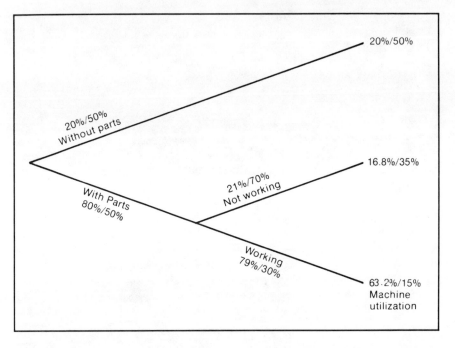

Figure 1: Distribution of Machine Time for the Most Likely Performance of an FMS Compared to a Conventional System (FMS%/CS%)

the system much faster, at any given time fewer parts remain unfinished, waiting for a machine to work on them or waiting for other parts to be assembled.

The inventory carrying cost reductions that can be expected are approximately equal to the percentage reduction of manufacturing lead time. Our study showed that reductions in the neighborhood of 81% can be obtained from the most likely performance of an FMS.

Increased machine utilization results in the reduction of the capital recovery cost allocated to the product. The magnitude of this reduction depends on the performance, the economic life and the investment required for the FMS, as well as on the cost of capital and the method of cost allotment used by the firm. A 28% reduction in this parameter was obtained on our study.

Typically, a comparison of an FMS and a conventional system (CS) is a comparison between a high investment, low operating cost system and a low investment, high operating cost alternative. Among the factors that significantly affect operating costs are maintenance, energy and indirect labor costs.

Maintenance costs: As a result of greater complexity and increased machine utilization, preventive maintenance costs can be expected to be higher for an FMS than for a CS. On the other hand, the FMS is continuously monitoring itself, and a machine breakdown causes the parts to be routed through other machines. In such a case the FMS would not operate at the same level of efficiency, but nevertheless would continue to operate.

Corrective maintenance costs are contributed mainly by parts and labor, with a low contribution due to production time loss. Corrective maintenance costs associated with machine breakdown can be expected to be lower for an FMS than for a CS.

Energy costs: With the application of adaptive control technology, the machine tools and the material handling system of an FMS make efficient use of energy. However, energy costs would more likely be higher for an FMS than for a CS because of the energy consumed by the automated material handling system and the central computer.

Indirect labor costs: Full integration of computer-aided design and manufacturing (CAD/CAM) is a major prerequisite of an effective FMS. Such integration offers the following advantages:

☐ Data are originated in machine language as a by-product of the design phase.
☐ Once in machine language, data are never re-entered, but may be added to, deleted from or modified.
☐ A single database is used throughout the entire planning and control process.

As a result, fewer personnel are required to handle and communicate the information. Indirect labor costs would more likely be lower in an FMS than in a CS.

Facing uncertainty

Estimates of the parameters we have discussed should be used to compute the total operating cost of the system (TOC). However the estimates are obtained, conditions of uncertainty exist for the TOC parameter.

Another parameter for which uncertainty exists is the economic life of the FMS. Increased machine utilization can result in faster deterioration and a shorter economic life than expected. There is uncertainty associated with high technology obsolescence that can also result in a shorter economic life than expected.

Sensitivity analysis can be performed to analyze how sensitive an economic measure of effectiveness is to changes in any of the uncertain parameters and to determine under which conditions the FMS can be expected to be a better investment alternative than a CS, and vice versa.

In this phase, the implications of taxes should be considered. We performed a sensitivity analysis on the after-tax equivalent uniform annual cost (ATEUAC) for a CS and an FMS, in compliance with the accelerated cost recovery system (ACRS), and observed that:

□ The FMS alternative gets more attractive than the CS as the economic life of the system gets longer (see Figure 3). For very short economic lives, the savings in operating costs do not occur during a sufficient period of time to offset the capital recovery costs of the higher FMS investment. For long economic lives, the ATEUAC of an FMS and a CS are not sensitive to this parameter, and the difference between the ATEUAC of the FMS and that of the CS does not increase significantly.

□ The ATEUAC of a conventional system is less sensitive to changes in the cost of capital than the ATEUAC of an FMS. The FMS alternative gets more attractive than the CS as the cost of capital decreases.

We also analyzed the effects of the following ratios on ATEUAC:
□ TOCR: Ratio of total operating cost of FMS to TOC of CS.
□ IR: Ratio of investment in FMS to investment in CS.

It can be seen in Figure 4 that the FMS alternative gets more attractive

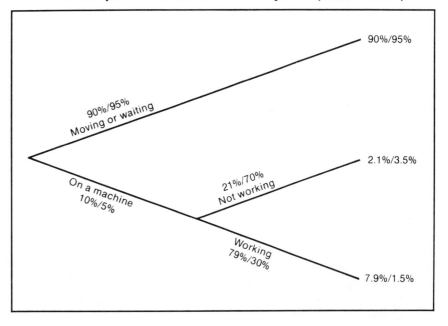

Figure 2: Distribution of Manufacturing Lead Time for an Average Part for the Most Likely Performance of an FMS Compared to a Conventional System (FMS%/CS%)

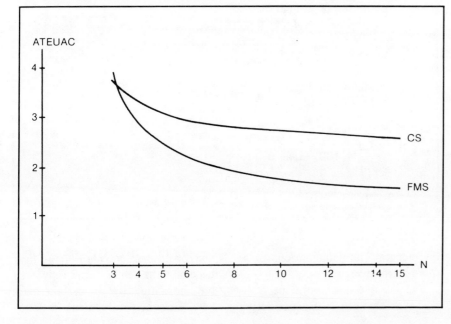

Figure 3: After-Tax Equivalent Uniform Annual Cost (ATEUAC) of FMS and Conventional System as a Function of the Economic Life of the System with Investment Ratio (IR) = 2, Total Operating Cost Ratio (TOCR) = 1/3 and i = 10%

as IR and TOCR decrease.

Incremental implementation

Another factor that could have an impact on replacement studies of this kind comes as a result of one of the characteristics that sets the FMS apart from alternative concepts: its suitability for incremental implementation. A firm can start with a single standalone machining center and build the system around that machine through the addition of other elements in planned steps.

This ability to expand as production dictates reduces the risks associated with market forecast errors and also means that substantial reductions in capital recovery costs are obtainable. The magnitude of these reductions depends on the cost of capital, inflation, costs incurred due to incremental implementation, how often the FMS is able to expand to meet production requirements and how fast these requirements increase.

Our empirical analysis in this area showed that in the current economic and financial climate, a firm can save approximately 12% in capital recovery costs if an FMS is implemented on an incremental basis as compared to total implementation. In analyzing the sensitivity of capital recovery costs to changes in parameters, we found out that higher cost-of-capital/inflation ratios produce an increase in the percentage of the savings in capital recovery costs.

If inflation is equal to the cost of capital, there are savings to be realized. This is due to the deferment of costs incurred by the incremental investment program.

High costs related to the incremental implementation plan considerably reduce the savings and may even offset them, especially when inflation is higher than the cost of capital. If the FMS is able to expand faster, the possibility of shorter intervals between expansions makes incremental implementation plans more attractive.

The FMS offers enormous flexibility for changes in product design and/or production volumes and may become the only alternative to remain competitive in certain markets. Design, development, installa-

Table 1: Comparison of How Machines and Workparts Spend Their Time in the Shop of a Conventional System and in an FMS using the Optimistic-Pessimistic Format

Line No.	Parameter	Conventional System Performance*	FMS Performance** Pessimistic	Most likely	Optimistic
1	Percent of machine time the machine spends without parts (T_{mw})	50%	35%	20%	5%
2	Percent of machine time that there is a part on the machine ($T_{mn} = 100 - T_{mw}$)	50%	65%	80%	95%
3	Percent of time that the part is not being worked on while on the machine $\frac{(T_t \cup T_{th})}{T_o}$	70%	35%	21%	7%
4	Percent of time that the part is being worked on while on the machine $\frac{(T_m)}{T_o} = 100 -$ Line 3	30%	65%	79%	93%
5	Percent of manufacturing lead time that the part spends either moving or waiting $\frac{(T_{no} \cup T_{su})}{T_{ml}}$	95%	92.5%	90%	85%
6	Percent of manufacturing lead time the part spends on a machine $\frac{(T_o)}{T_{ml}} = 100 -$ Line 5	5%	7.5%	10%	15%

*Source: Merchant 1977.
**Exact values depend on the specific FMS.

tion and testing of an FMS require management's full commitment and a large investment that must be justified. The criteria and information presented here are intended to aid the analyst in the economic justification of an FMS and facilitate communication with top management for effective decision-making.

For further reading:

Black, J T., "Cellular Manufacturing Systems Reduce Setup Time, Make Small Lot Production Economical," *Industrial Engineering,* November 1983, pp. 36-48.

Canada, John R., and White, John A., *Capital Investment Decision Analysis for Management and Engineering,* Englewood Cliffs, NJ: Prentice-Hall, 1980.

Cook, Nathan H., "Computer Managed Parts Manufacture," *Scientific American,* February 1975, pp. 22-29.

Groover, Mikell P., *Automation, Production Systems, and Computer-Aided Manufacturing,* Englewood Cliffs, N.J.: Prentice-Hall, 1980.

Groover, Mikell P., and Hughes, John E., "Job Shop Automation Strategy can Add Efficiency to Small Operation Flexibility," *Industrial Engineering,* November 1981, pp. 67-76.

Hegland, Donald E., "Flexible Manufacturing-Your Balance Between Productivity and Adaptability," *Production Engineering,* May 1981, pp. 39-43.

Hutchinson, George K., and Wynne, Bayard E., "A Flexible Manufacturing System," *Industrial Engineering,* December, 1973, pp. 10-17.

Klahorst, Thomas H., "Flexible Manufacturing Systems: Combining Elements to Lower Costs, Add Flexibility," *Industrial Engineering,* November 1981, pp. 112-117.

Merchant, Eugene M., "The Inexorable Push for Automated Production," *Production Engineering,* January 1977, pp. 44-49.

Ostwald, Phillip F., *Cost Estimating for Engineering and Management,* Englewood Cliffs, NJ: Prentice-Hall, 1974.

Pappas, James L.; Brigham, Eugene F.; and Hirschey, Mark, *Managerial Economics,* 4th ed., Hinsdale, IL: Dryden Press, 1983.

Park, William R., *Cost Engineering Analysis,* New York: John Wiley & Sons, 1973.

Salomon, Daniel P., "Models for Economic Evaluation of Flexible Manufacturing Systems," Masters' Research Report, University of Central Florida, 1983.

Weston, Fred J., and Brigham, Eugene F., *Managerial Finance,* 7th ed., Hinsdale, IL: Dryden Press, 1981.

Willis, Roger G., and Sullivan, Kevin H., "CIMS In Perspective: Costs, Benefits, Timing, Payback Periods are Outlined," *Industrial Engineering,* February 1984, pp. 28-36. **IE**

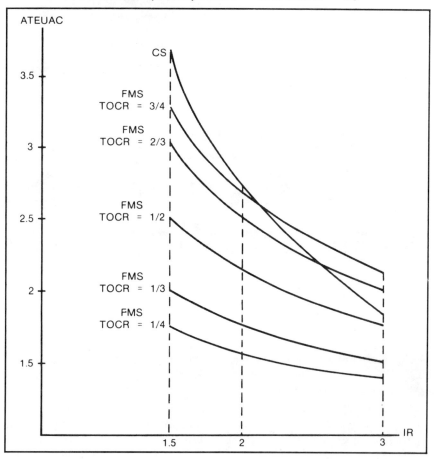

Figure 4: After-Tax Equivalent Uniform Annual Cost (ATEUAC) of FMS and Conventional System as a Function of Investment Ratio (IR) and Total Operating Cost Ratio (TOCR) with i = 10% and N = 10

Daniel P. Salomon recently received his MS in engineering administration from the University of Central Florida. He recently accepted a position with Corporate Investments of Florida as an investment analyst. He obtained the BSIE degree from the Universidad Iberoamericana in Mexico City. Salomon has worked as manager of informatics and systems services in the government agency responsible for strategic planning for central Mexico, where he was mainly involved with transit network analysis and transportation systems evaluation. His experience also includes small business financial management consulting and teaching statistical methods. Current interests are in the areas of manufacturing management systems, economics of advanced technologies and cost reduction programs.

John E. Biegel, P.E., is a professor of engineering at the University of Central Florida. He holds degrees from Montana State University, Stanford University and Syracuse University, and is a registered professional engineer in Florida and New Mexico. He has taught at Kansas State University, Syracuse University and the University of Arkansas, and has worked in industry with the Ford Motor Co. and Sandia Corp. His current interests are in CAM, AI and expert systems.

Reprinted from Industrial Engineering, September 1983.

Automating Existing Facilities: GE Modernizes Dishwasher, Transportation Equipment Plants

By Frank T. Curtin
General Electric Co.

Factory automation comes in all sizes, but the shape and the way it gets put into place don't vary that much. Whether it's a relatively modest project, a complete rework of an existing factory or the construction of an entirely new facility, the same basics apply.

However, that's not to say it's easy. The requirements are far too complex to be reduced to a formula. Moreover, the process calls for expertise in a wide range of disciplines, including long-term planning, product design, process and information flow, purchasing and equipment installation.

Making it all come together requires a willingness to give up old habits and accept new concepts. But perhaps even more important, it requires exceptional coordination of expertise you have in-house with expertise you hire from consultants or expect from suppliers.

Most automation projects do not provide the luxury of a whole new facility, but rather involve modernization of existing facilities and working around production schedules with the least possible interruption of ongoing manufacturing requirements.

In this article, I'll try to illustrate the complex nature of automation planning and implementation by describing two different factory modernization projects within General Electric.

This "tale of two cities" involves a $38 million automation project at GE's dishwasher plant in Louisville, KY, and a multi-year $316 million project at our transportation equipment plant in Erie, PA.

These modernization projects have been enormously helpful in identifying and honing the skills needed for factory automation planning and implementation as well as in developing the hardware and software technology needed for future automation projects.

They have also given GE showcase factories that demonstrate the value of flexible automation in creating competitive capacity.

Cellular approach

Both projects have taken the cellular approach; that is, they consist of islands of automation which are very nearly or completely "closed-loop." Taking automation in steps allows a framework for finite planning, yet permits flexibility regarding the more distant future. As projects continue to evolve, islands of automation can eventually be tied together with increasingly sophisticated materials and information systems.

In Louisville

Integrating product and process design called for extensive preparation long before the first modification was evident on the factory floor. GE conducted or commissioned some 35 separate market research studies to be sure the end product dishwasher would meet market expectations well into the future.

At the same time, GE involved its suppliers in the product/process integration. Automated equipment is the most disciplined inspector you can have. It has little ability to allow for out-of-tolerance parts. Therefore, critical to quality (CTQ) standards, which allowed variances of one in 1,000 parts, and very critical to quality (VCTQ) standards of one in 10,000 were set.

Focusing the plant

Focusing a plant is largely a matter of determining what is and is not economically automatable, what should be moved and where, and what would be more efficiently purchased than made. This discipline is also very effective in reducing the numbers and volumes of parts to be handled.

In the Louisville plant, for example, the more than 4,000 parts and assemblies coming in and going out of the facility will eventually be reduced to about 800. And the amount of cash tied up in inventory will be reduced by two-thirds.

Injection molding

At the heart of the entire automation project was the new Permatuf tub and door. The Permatuf tub, which was already in production, was redesigned with computer-aided design (CAD) assistance for more productive integration into the manufacturing process.

The tubs are molded in 22-ton molds that run fully automatic on six 3,000-ton injection molding machines controlled by six GE series six programmable controllers. These are closed loop systems that monitor 32 different temperature points, 10 velocities, five gates, seven pressure points and a selection of counters and self-diagnostics.

Before the application of the programmable controllers, the process was monitored by a worker and took an average of 120 seconds. The current cycle rate has been reduced almost to the theoretical limit. Inner doors are molded on three 2,000-ton injection molding machines similar to the ones which mold the tub.

Tub assembly

Tubs are delivered by a new conveyor system to a highly sophisticated non-synchronous assembly system with no direct labor operators. The line is loaded and unloaded by Cincinnati Milacron robot and employs a number of procedures new to dishwasher manufacturing, such as ultrasonic upsetting, flame abrading, metal fastitch joining, pierce riveting and vibration welding.

The programmable controllers (PCs) control the sequence of operation at each station and track any component faults through the system. All fault data are transmitted by the PCs to the plant quality information system. For certain critical relationships, the system even transmits quantitative data such as carrier number, tub cavity number, screw torques and dimensional relationships.

Conveyor system

The various islands of automation—injection molding, tub assembly, inner door assembly, and the final assembly and testing areas—are tied together mechanically by a 2½-mile conveyor transfer and storage system which replaced about nine miles of conventional monorail conveyor. This system is capable of delivering the tubs and doors to their respective assembly carousels on a first-in/first-out (FIFO) basis and returning the completed assemblies to a storage bank before final assembly.

Before the new conveyor system was installed, the tubs were handled 27 times by workers. Now, they are handled once.

Six programmable controllers with more than 3,000 inputs and 2,500 outputs control the conveyor system and are, in turn, controlled by the computer director.

Final assembly and testing

Tub and door assemblies are loaded by robot onto an assembly carrier

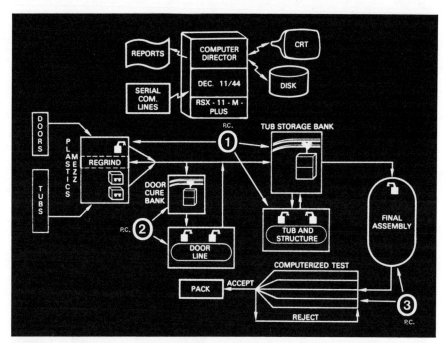

Diagram shows layout of the Louisville dishwasher plant conveyor system.

which can be swiveled horizontally and vertically to facilitate the work of the assembly line operators. The system allows operators to work on stationary units rather than moving ones.

Operators pull a green handle to release units to their next station, but only when they are certain a quality job has been done. If the assembly cannot be completed properly within a reasonable time, the operator releases the unit with a yellow handle, which automatically alerts the system to a problem.

The unit then completes its route through assembly, but is automatically diverted to a repair bay, where the fault is corrected *before* the unit goes into testing.

Completed dishwashers are taken into the test area in the same carrier that took them through final assembly. A laser reader identifies each carrier for the series six PC controlling the testing.

During normal testing, the electrical signature, functional components, noise level and water usage are checked. Test and repair data flow, real time, into the plant quality information system, from which they can be accessed by shop operations personnel on their own CRTs. This immediate feedback allows "root cause" fixes to be initiated as soon as the fault is first identified.

In Erie, PA

In Erie, General Electric chose to automate a 70-year-old factory building rather than move its manufacturing operation.

The total automation effort will involve some 1,000 separate projects in Erie and at a satellite plant in nearby Grove City. Diesel engine production will move entirely to the Grove City plant, and the various islands of automation involved in diesel engine construction are being tied together. That, in turn, will allow the Erie plant to focus more sharply on locomotive assembly.

To date, about 30% of the modernization project has been accomplished. Major projects completed so far include a flexible machining system, an automated steel plate burning/nesting operation, two automated warehouses, and a learning and communication center.

The $16 million computer-controlled flexible machining system (FMS) is the first of its kind in GE and one of the relatively few flexible machining systems in the world.

The system currently has the capacity of producing more than 5,500 motor frames annually, but ultimately can be expanded by 60%. It consists of:
☐ Nine heavy-duty machine tools with GE Mark Century 1050 computerized numerical controls and automatic or robotic tool changers. There are two vertical milling machines, three horizontal machining centers, three heavy-horizontal boring mills and one medium horizontal boring mill.
☐ A custom-built fixture set-up station.
☐ A 212-ft rail-tracked, chain-driven automated transporter which moves frames among 21 load/unload stations. The transporter is controlled by a GE series six model 6000 programmable controller interfaced with the executive computer.
☐ A DEC PDP 11/44 executive computer which directs and tracks the activities of all systems.
☐ An automatic chip removal and coolant recirculating system.

Dishwasher tubs are molded in fully automatic injection molding machines run by programmable controllers at the GE Louisville plant.

The computer, acting on production requirements entered by the system manager or the computerized material requirements planning (MRP) system, displays the sequence of parts to be loaded or unloaded for maximum system productivity. After each frame is loaded and the computer has been signalled that it is ready, the computer selects the work station that can perform the first machining operation and will be available in the shortest time.

The computer then directs the material transporter to deliver the part to the proper machine tool, downloads the machining program to the tool's NC control and—after the workpiece is in place—initiates the program. An average of eight machinings are required.

If a machine tool is down for repairs or maintenance, it is automatically isolated from the system, and production is rescheduled to the remaining stations.

A management information module in the system allows direct interrogation and provides logs of production performance, machine and tooling usage, failure data and so on. The machines can be operated in an independent, local mode in case of system failure.

More than 500 cutting tools are stored within the nine machines. Those tools critical to smooth operation are automatically monitored and replaced by backups based on usage or broken tool detection.

The machining sequence, which once took 16 days and involved 70 persons, now is done in just 16 hours with just two people per shift. Bottom line on this automation investment has been a 240% increase in employee productivity, improved part consistency and quality, and a potential 100% increase in capacity in 25% less floor space.

Nesting system

The steel plate burning/nesting operation provides one of the first glimpses of the "paperless factory" that has been so long envisioned. It is the first system in the world to link all elements from part design through plate burning, including material handling and information processing.

Nesting is a totally integrated

closed-loop system. Production tools needed from material requirement planning are reviewed and updated daily. The engineering CAD system provides the parts configuration for electronic transmission to an auto-nest system.

This batch-run system nests the parts and generates images on the Calma Graphics Systems screen for review and "fine tuning" by a human operator. Source output from the Calma system is then transmitted to the host computer for post processing and distribution to the CNC burning machines.

The plasma-arc unit can cut two plates simultaneously with four cutting torches at speeds up to 150 in. per minute. A second unit is planned later in the program. Together, they will handle two-thirds of the total burning load. This represents a 100% labor productivity gain and a potential savings of more than $400,000 a year in material.

Automated warehouse

A $4 million high-rise Auto/Stack-M system by Harnischfeger was the first of three computer-controlled storage and retrieval systems now being implemented.

The system makes more efficient use of space and provides real-time inventory control. It includes three separately controlled cranes and two input/output staging areas. It has 8,100 openings with pallet weight capacities up to 4,000 lb. More than 50,000 different in-process storage parts will be handled each year, which provides in-process storage for a sheet metal fabrication facility.

This system is managed by a DEC 11/70 computer which is part of the business inventory management system. The storage systems will interface through a DEC 11/23 computer, and a GE series six model 6000 programmable controller will direct the input/output conveyors and palletizers.

In any modernization and automation, it is of key importance to train personnel to use the new equipment. The most efficient system in the world is useless without shop floor support.

Rather than create a facility strictly for training GE people to use and maintain factory automation equipment and industrial communications systems, however, the scope of the project was expanded to include training of customers in the operation and maintenance of the increasingly sophisticated products we are producing.

Learning center

The Learning and Communications Center has five basic areas:
☐ Classrooms that allow flexible teaching formats for large or small groups.
☐ A high technology area that provides instruction in the latest state-of-the-art technologies, such as microprocessors and NC controls. Training techniques include use of computers, simulations and advanced audio-visual equipment for individualized programmed training. In addition, costly failures can be simulated to increase the likelihood of avoiding them in the field.
☐ Heavy apparatus laboratories that permit hands-on instruction in overhauling and troubleshooting of equipment ranging from motorized wheels and diesel engine components to an entire locomotive.

☐ A visitors wing for meetings and small conferences.
☐ A communications center with a 300-seat auditorium for employee information conferences as well as multi-use meeting rooms for customers, employees and visitors.

Summary

Alleviating people's fears of automation is a prime consideration with regard to planning an automation project. In many industries, automation is an absolute necessity to maintain a competitive position. It will cause dislocations and require retraining from top management on down, but if management is sufficiently dedicated and solicits the help and cooperation of its work force in a planned and sincere way, resistance can be minimized.

It is difficult to overemphasize the value of input from hourly shop floor workers in planning and implementing automation systems. No one knows the nitty-gritty of a job better than the person who does it.

In short, the keys to automation are planning, coordination and integration at all levels. Reach out for all the experience you can find. Look to your hourly employees, to your own design and production engineers, to your upper management, to suppliers and to consultants. Don't spend for the sake of spending, and do not try to shortcut the procedure. In industrial automation, mistakes can be costly and long-lived.

Frank Curtin is general manager of the factory automation systems department of General Electric Co. His responsibilities include planning, developing, marketing and implementing automated systems. His component provides system integration products and services for advanced applications of factory automation technology both inside GE and outside on a commercial basis. This includes systems engineering, information systems and flexible manufacturing cells.

Technology Requirements to Support Aerospace Plant Modernization

John Huber
Grumman Aerospace Corporation
Bethpage, New York 11714

ABSTRACT

Plant modernization is most effectively accomplished by a proper blending of capital investment in new equipment with technology development. Grumman Aerospace Corporation, in conjunction with the Naval Air Systems Command, has embarked on a program which integrates manufacturing technology developments with a corporate plant modernization program. The program is a three-phase, five-year effort in which a sixteen million dollar technology effort has been coordinated with a one hundred and fifty million dollar capital investment program to reduce manufacturing costs. Discrete technical projects addressing eight principal aerospace manufacturing areas have been identified. The key elements being addressed in the program are the new technologies required for effective use of new manufacturing processes on a complex, high quality product, at low production rates.

BACKGROUND

For over 50 years the Grumman Aerospace Corporation has been a major producer of tactical aircraft for the United States Navy. At the present time Grumman is a prime contractor for NAVAIR on five aircraft: (a) F-14A Fighter Aircraft; (b) E-2C Early Warning Aircraft; (c) A-6E Attack Bomber; (d) EA-6B Electronic Countermeasures Aircraft; and (e) C-2A Cargo Aircraft, shown in Fig. (1). In addition, Grumman produces the USAF EF-111 tactical jamming system, shuttle orbiter wings and a variety of other aircraft structures under both military and commercial subcontracts.

Over the past few decades several significant changes have taken place in the aerospace industry. The aircraft structure has changed from sheet metal to a heavy dependence on complex machined parts with materials changing from aluminum to titanium, and more recently, to composites. As a result of the changes, the new aircraft have a longer life span and are more versatile, but they are also more complex and costly. Consequently, the resulting production rates are significantly lower with the costs significantly higher, see Fig. (2 & 3). During this time period, the key requirement of manufacturing technology was to develop the processes needed to produce these new aircraft rather than processes to reduce production cost. Restrictive Department of Defense procurement policies and uncertain production rate projections only

Fig. 1. Grumman Production Aircraft

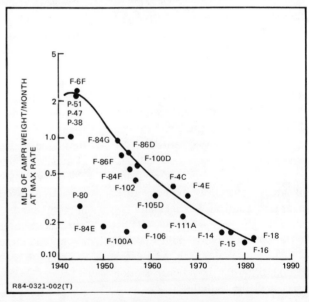

Fig. 2. U.S. Fighter Production 1944 – 1984

fueled the drivers for rising cost by diluting and/or diverting the contractor's incentive to make the investments needed to address the new military aircraft potential procurements. In recent years the Department of Defense has recognized these problems and has aggressively pursued a series of innovative programs designed to motivate the industry to invest

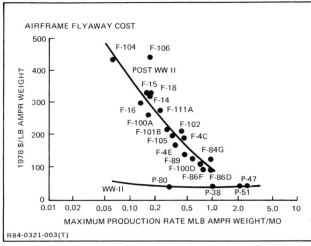

Fig. 3 U.S. Fighter Production Costs

in modernization to reduce costs. The first of these programs was sponsored by the Air Force at General Dynamics in 1979 where manufacturing technology was coupled with corporate capital investments to successfully reduce costs on the F-16 aircraft.

The Navy has sponsored a similar program for its contractors. Grumman is one of the first NAVAIR contractors to participate in the program. The program is directed at achieving cost reductions on existing contracts, consequently the technology requirements are biased toward the product mix in production. At Grumman, all five aircraft programs are mature; that is, they are on the flat part of the learning curve. These complex aircraft, fabricated at low production rates, will be required into the 1990s. For the future, the program must provide a capability to manufacture new designs currently being passed down to manufacturing via a CAD/CAM link.

The NAVAIR Program at Grumman, started in 1982, has three objectives:

- Improve in-house and supplier productivity

- Reduce aircraft unit costs

- Enhance industrial surge ability.

The program is divided into three phases, see Fig. (4).

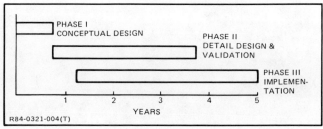

Fig. 4. Program Schedule

- *Phase I — Conceptual Design* — Investigation and development of manufacturing technology modernization plans leading to Navy/Grumman cost reduction targets and incentive agreements

- *Phase II — Detail Design and Validation* — Detail design of the improvement concepts identified in Phase I, validation of their development and definition of corresponding preliminary implementation plans

- *Phase III* — Implementation in production of Phase II results.

The work done in Phase I focused on the Grumman five-year capital plan and its interaction with those specific primary production operations that promised to yield significant economic benefits. An analysis of our overall production structure indicated that Grumman manufacturing operations could be divided into eight principal functions: Sheet Metal Detail Parts, Machined Detailed Parts, Assembly, Electrical, Composites and Bonding, Management Information Systems, Quality and Material Handling, see Fig. (5). The first five areas are discrete manufacturing operations and in general represent plant and/or responsibility center divisions. The last three functions run across all of the discrete functions and provide a necessary integrating force. A "top down" factory analysis studied manpower, floor space, existing capital equipment in each manufacturing area and the five-year capital investment plans for the area. It determined the current cost levels for these areas and evaluated new concepts to improve their productivity. The interactions of the capital investment plans with the technology needs to allow systems to operate more effectively, were identified and evaluated to establish risk, costs and development planning requirements.

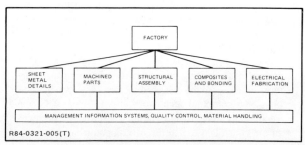

Fig. 5. Key Manufacturing Centers

As a result of the Phase I analysis, Grumman prepared a balanced five-year, twenty-five project effort in concert with Corporate and Government Owned and Contractor Operated (GOCO) capital commitments. The projects were identified on the basis of the following Phase I conclusions:

- Maximizing benefits accruing from capital investments in detail parts fabrication requires establishment of specific productivity improvement goals. To achieve these goals, plans must cover:

- Factory flow and layout
- Information systems driving the factory
- Techniques for modifying tools to accommodate the new equipment
- The relationship between the new equipment and the stocking arrangements for raw material and parts
- Modification of the organizational structure to function efficiently with the new equipment and systems, consistent with the goals

- Assembly labor expenditures approach 50% of the total product labor cost. To obtain significant productivity improvements via intensified use of capital equipment, we must overcome the effect of low production volume by introducing more flexible assembly equipment, capable of processing families of assemblies possessing common structural forms

- For Quality Operations to keep pace with improvements yielding increased part throughput, without requiring increased quality labor expenditures and without lowering quality awareness, development of an automated, easily accessible part quality history data base is required. Effective use of this base will lead to reduced Quality costs via earlier identification and resolution of specific part problems. Properly implemented, it will lower production labor costs by serving as a focal point for future Manufacturing and Engineering-directed improvements

- Automation of storage and handling techniques for raw material, tools and in-process parts offers excellent potential for significant indirect labor cost reductions. Current plans, while a major step forward, must be expanded to ensure coordination of these techniques with factory operations

- The importance of computerized management information and factory operating systems points up the value of quickly completing the integration of current and near-term systems into the enhanced factory. The growing emphasis on computer technology in the fields of Group Technology and Process Planning requires a commensurate controlled expansion of computer capability

- Automated digitizing techniques are needed to ensure cost effective implementation of numerically controlled equipment on current production programs. It must be coordinated with Grumman Engineering CAD/CAM systems data base to ensure maximum benefits to present and future procurements.

- A training program must be developed which will provide the technique for achieving the proper skill mix to obtain the maximum benefit from reduced staffing levels reflecting productivity improvements.

TECHNOLOGY PROJECTS

In general, all of the technologies identified to support plant modernization are related to the impact of computer technology on manufacturing operations. The changes are not simply replacing old equipment with new, but in most cases it is a new way to plan, produce and handle parts. This can best be illustrated by reviewing several projects presently underway (or in the planning stage) for the following key manufacturing areas: Sheet Metal Detail Parts, Machining Operations, Assembly, Electrical Fabrication, Composites and Supporting Operations (Management Information Systems, Quality and Material Handling).

Annual sheet metal detail parts fabrication of aircraft components at Grumman entails the fabrication of over three million parts made up of over two hundred thousand part numbers. The principal materials are aluminum alloys, titanium and steel. The key manufacturing processes are shearing, routing (profile trim), drilling, forming and finishing.

A specific example of manufacturing in this area, peculiar to the aerospace industry is the drilling and trimming of complex contoured parts (i.e., aircraft skins, fairings and doors). This work is presently accomplished using tedious manual operations, shown in Fig. (6), in conjunction with costly and cumbersome tooling. In fact, most changes over the years (i.e., heavier gages, more difficult to machine materials and tighter tolerances coupled with lower production rates) have had a negative impact on productivity. The advent of large precision robots and low cost micro-processors provided an opportunity to economically automate this section of aircraft manufacturing. The concept developed to address this situation was to design a computer controlled robotic cell, Fig. (7). The robot would be capable of moving from work station to work station, automatically change tools, and perform the programmed functions necessary to produce the part configuration. The cell was designed so the operator's function would be performed outside the range of the robot, on the cell perimeter, to eliminate human/robot interface problems. The function of the operator is to set up the work, load/unload parts from the work station and identify work done to the

Fig. 6. After Form Drill & Trim Operation's

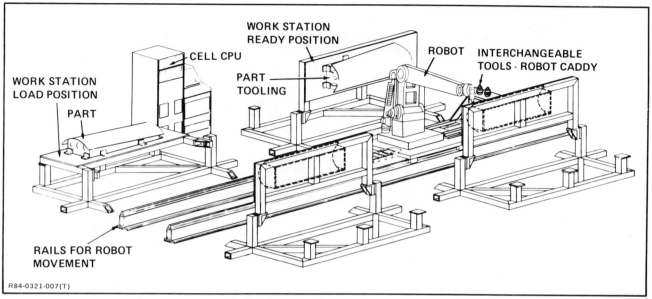

Fig. 7. Robotic After Form Drill & Route Cell

central controller. The system uses a modified Asea Rb 60 robot with a DEC PDP 11-34 mini computer as the cell controller. A summary of the associated automated process needs, with the required support technology, is shown in Fig. (8). The change is significant in several ways. Operator skill is no longer directly related to part quality and with the cell now requiring a significant increase in capital equipment, utilization and planning became key factors for economical success. The soft tooling, tapes and computer data, eliminate the need for hard tooling and provide a technique for quick response at low rate production with minimal cost penalties. Existing hard tooled geometric data must be converted to computer base data on existing programs at a reasonable cost to achieve the desired recurring cost savings. The first cell has been installed on the factory floor and the initial test runs have indicated a sixty percent reduction in fabrication time per part.

CONCEPT	NEEDS	ENABLING TECHNOLOGY
AFTER-FORM ROBOTIC TRIMMING	ELIMINATE OPERATOR INTERFACE WITH ROBOT	AUTOMATIC TOOL PICKUP AND POWER SUPPLY
	HIGH EQUIPMENT USE OF CAPITAL INVESTMENT	MULTI-STATION CELL CONCEPT WITH MOVABLE ROBOT
		CELL CONTROL COMPUTER DATA DISTRIBUTION AND SOFTWARE
	INTERFACE WITH CAD/CAM	CELL CONTROLLER AND INTERFACE SOFTWARE COMPATIBLE WITH CORPORATE CNC PLAN
	ELIMINATION OF HARD TOOLING	DRILL AND TRIM CONTROL WITHOUT TEMPLATE GUIDE
		CAPABILITY TO CALIBRATE PROGRAM ZERO POINT
		UNIVERSAL FIXTURE TO HOLD PARTS CAPABLE OF PRE-PROGRAMMING
	LOW-COST METHOD CONVERTING HARD TOOLING INFORMATION TO COMPUTER DATA BASE	Z-AXIS POSITION CONTROL FOR DIGITIZING
		VISION FOR AUTO-DIGITIZING

Fig. 8. Robotic Cell Technical Requirements

Other projects in the Sheet Metal Fabrication area are the conversion of manual drill and rout operations on flat aluminum parts to numerically controlled Trumf equipment which performs these operations automatically. A similar system, being developed for flat titanium and steel parts, will eliminate a series of manual operations in favor of a numerically controlled punch and nibbling system which will perform these operations automatically. Both operations require the technology for the conversion of existing tooling data to a computer data base, a methodology and planning technique for nesting the parts to reduce material scrap and a support system around the equipment which will ensure maximum utilization of this more capital intense approach to manufacturing.

A similar approach is being taken for extruded shape parts. In this instance an Ingersol partsmaker is used in place of a series of manual operations, see Fig. (9). In addition to the equipment changes, the associated plants are being reconfigured to centralize these operations into one facility to maximize use of the new capital.

In the area of machined parts fabrication, the project's emphasis will be on improving the utilization of the capital equipment required to produce these parts. The principal areas of effort will be a cutting tool regrinding, inventory, a storage and transport system, a mechanized machined part handling and clamping system and the conversion of present existing numerical control to a distributed numerical control system for all major machine tools. To demonstrate the quantitative benefits of these improvements, a flexible machining center will be set up to address small and medium sized machined parts. The technical projects identified above will be used to support implementation of the flexible machining system and, where applicable, be applied to the other machine tools.

Aircraft assembly operations represent approximately fifty percent of total production labor. These operations are typically performed manually using portable tools and expensive fixtures. One of the first projects will be to apply an automated drilling system to the relatively light structure of the E-2C wing outer panel in our Stuart, Florida Assembly Plant. Grumman has been a leader in this work, having ap-

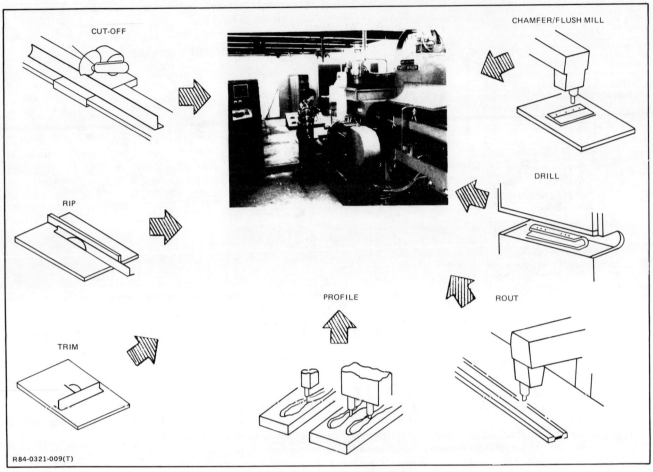

Fig. 9 Numerical Control Parts Maker

plied similar technology to its A-6E aircraft wing drilling operations eight years ago. The success of this program, changes in technology and experience gained, has led to the project presently underway. The system will utilize two drilling gantries, one on each side of the left and right handed assembly fixtures, two computer controllers for the five axes of motion required and four automatic tool changers, as illustrated in Fig. (10). The system will be capable of digitizing the hole pattern directly on the assembly fixture, thereby easing the conversion from hard tooling data to a computer data base. The holes will be finish drilled and countersunk in one shot. Future plans call for the addition of a blind fastener hopper and a delivery and insertion system.

A similar system is planned to address the drilling and fastening operations being applied to aircraft control surfaces. The automated cell concept will use a precision robot and a movable fixture to permit maximum use of the equipment over a variety of structures. This installation is also being planned for the Stuart, Florida facility and should be operational in the second half of 1985.

A fully automated robotic assembly cell is being planned for generic grouping of aircraft structure. The concept of this cell, shown in Fig. (11), will expand from the earlier drill

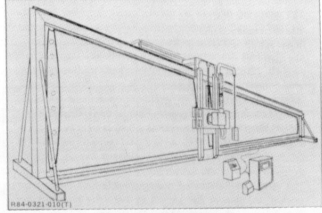

Fig. 10. Automated Wing Drilling

and fasten operations to the development of a system with positioning capability for full structural assembly. Selected elements of technology needed for such a system have been done on a laboratory basis and, with minor product design modifications plus improved vision systems and precision robotics, the system should be sufficiently developed for implementation in 1987 at Grumman's Bethpage, New York Sub-Assembly Operations.

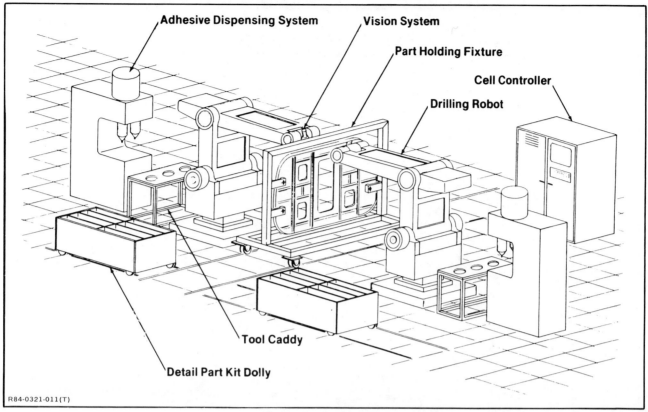

Fig. 11. Robotic Assembly Cell

Other assembly projects address automation of systems for small assemblies on automatic drill and rivet equipment and automated paint and strip operations for aircraft final assembly painting.

The increased reliance on sophisticated electronic systems for new aircraft weapons systems has made manufacturing processes for special wire harnesses and printed wire boards, key elements din manufacturing operations. The program projects, directed at the technical support needs to produce very complex wire harnesses at low production rates, will involve the application of flexible or on-demand manufacturing techniques. At the present time, aircraft wire harnesses are produced by hand using harness boards and "run" sheets as guides, Fig. (12). This method has been customary in the industry because, until recently, equipment was not available to economically build the variety of wire harnesses needed. The recent availability of programmable high speed wire marking equipment has served as a starting point for the development and integration of a variety of devices specifically created for the handling and production of aircraft wire harnesses. The project is directed at changing the production area from essentially manual operations to a significantly automated flexible processing center, as shown in Fig. (13). The project is divided into five principal tasks: high speed ink jet wire marking, automated wire coiling and tying, continuous filament wire handling, mass terminating system and robotic wire routing.

Fig. 12. Wire Harness Fabrication

- *Ink Jet Wire Marking* — Two American Can Company ink jet markers are now in place for high speed wire identification

- *Automated Coiler Tier* — A system has been developed to automatically coil and tie wire being processed by the ink jet marker

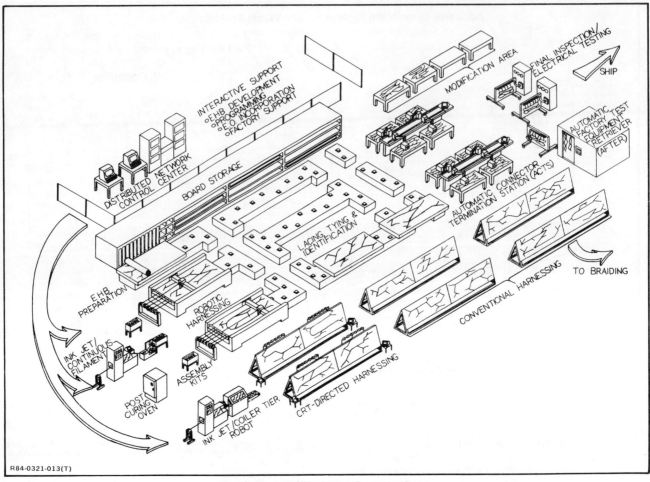

Fig. 13 Conceptual Wire Harness Fabrication Center

- *Continuous Filament Routing* — This system will eliminate the need to cut wires into segments before routing on a harness board. A CRT will be used to direct the operator as to how wire should be routed using a programmable universal harness board to hold the harness in place

- *Mass Termination* — This is a programmable system for automated preparation of the harness wire ends for connectors
- *Robotic Routing* — Application of robotics to the task of automatically routing wires on a wire harness board.

A similar program is underway to institute automated techniques into the fabrication of special printed wire boards.

In the areas of Composites and Bonding, the principal efforts are being directed at operations in the prepreg room. At the present time, approximately fifty percent of the fabrication labor required to build a composite laminate is related to prepreg operations. The principal difference between the production flow for composite structures and that for traditional metal structure is the need to perform the composite prepreg operations within a predetermined time to ensure proper curing of the organic prepreg material in a controlled environment, see Fig. (14). This requires a more disciplined approach to the manufacturing cycle than is needed with metal structures. The capability to apply a programmed flexible tooled system to these operations offers a new opportunity

Fig. 14. Composite Laminate Fabrication Operations

to address the manual labor and material utilization while improving part quality associated with these operations. The project is divided into three key tasks: digitize, nesting and identification techniques to support Gerber cutting operations, develop mechanized ply handling and kitting techniques and automate the laminate bagging operation. Figure (15) illustrates the typical operations and Fig. (16) shows changes planned on typical operations.

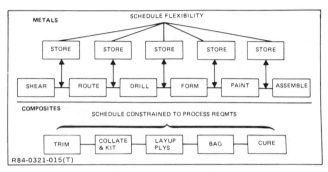

Fig. 15. Metal Part Fabrication

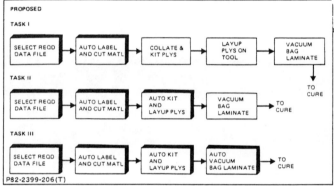

Fig. 16. Prepreg Fabrication Operations Flow

Over one third of the technology efforts being planned in the program are directed at the three key support areas: Management Information Systems, Quality and Material Handling and Storage. The Management Information System's efforts address automating the paper systems for planning and operating the factory. In the area of Quality, the emphasis will be on using the computer as a data base and analysis tool to identify Quality trouble spots and trends in Quality so corrective action can be focused on potential trouble areas prior to a disruption of production. The handling and storage modernization efforts will be directed at incorporating automated storage and retrieval systems which can be linked with an automated inventory capability. The automated retrieval systems will, in many cases, be tied into the production floor operations by using wire guided vehicles. This will permit on-demand material supply and minimize interim marshalling of materials on the production floor.

As we improve our data bases and ability to integrate the information, we plan to use this data to improve the configuration of our plants. The traditional aircraft production line approach has been to group work by aircraft type. The trend toward reduced production rates, larger, more complex assemblies and continued high quality through a reduction of manual tasks is changing this approach. Figure (17) illustrates typical structural assemblies from the five Navy aircraft built at Grumman. These assemblies are re-grouped by geometric class rather than program in Fig. (18). The production rate per assembly class is thus significantly increased over previous aircraft program rates and provides an opportunity to justify the capital investments required to move from manual to mechanized methods. The availability of low cost micro processors, robotics and other flexible systems has made it possible to change assembly methods and significantly reduce costs.

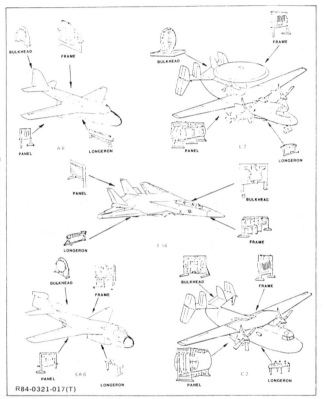

Fig. 17 Multiprogram Aircraft Assembly

Fig. 18. Assembly Grouping by Generic Class

CONCLUSION

The Grumman Tech Mod Program was developed as a result of a total factory view across all responsibility centers. It recognizes the need to integrate technology with new capital equipment. The program has tried to appraise realistically the Department of Defense procurements into the 1990s and to reconfigure the factory and associated methods of producing aircraft to take advantage of the new automation wave based on computer technology and flexible manufacturing systems. We believe them to be especially compatible with our future aircraft production needs.

References:

[1] Application of Composites to Military Aircraft MTAG Non-Metals, R. Hadcock, Grumman Aircraft.

[2] NAVAIR Industrial Modernization Incentive Program (Tech Mod 28th National SAMPE Symposium, Capt. W. Denver Key, USN.

BIOGRAPHICAL SKETCH

Mr. John Huber has over thirty-three years experience with the Grumman Aerospace Corporation in Bethpage, New York. He has a Bachelor of Science Degree in Applied Physics from Hofstra University, a Master of Science in Physics from Adelphi University and is a licensed engineer in the State of New York.

He is presently the Tech Mod Program Manager at Grumman and responsible for the management of the manufacturing technology projects required to support the company's plant modernization effort. Prior to this assignment, he was a section head in the advanced materials and manufacturing development department at Grumman. In this capacity he was responsible for the development and implementation of processes and equipment to improve current manufacturing operations and to meet the new requirements for future aircraft structures.

Some of the development programs conducted under Mr. Huber's direction are an integrated laminating center for composites, automated assembly drilling, ultrasonic machining, laser cutting, application of robotics to aircraft manufacturing and a NASA funded program to develop automated fabrication of structures in space using metal and composite materials. In addition, he has participated in the Air Force Machine Tool Task Force Study and the Air Force seminar on Low Cost Design/Manufacturing Concepts. He has received several patents on automated systems and made numerous presentations on aircraft manufacturing methods.

Pratt & Whitney's $200 Million Factory Showcase

By E.F. Cudworth
Editor & Assistant Publisher

The new $200 million Pratt & Whitney plant in Columbus, GA, represents the largest investment in a facility ever made by the United Technologies Corp. parent company. "The investment," according to UTC chairman and CEO Harry J. Gray, "has created a world class showcase for high technology manufacturing. Our new plant is among the most advanced manufacturing facilities anywhere in the world.... It's truly a factory of the future."

IE involvement

Dedicated last April 19, the P&W plant illustrates some of the latest thinking on automated material handling and allows an excellent picture of industrial engineering involvement in factory automation from first-cut evaluation of the proposal, right through return-on-investment justification, selling top management, material handling system design and vendor selection.

The reasoning behind the initial discussion of the plant in the 1970s was twofold, according to P&W Georgia Operations Manager Eugene A. Demonet, who oversees the Georgia plant and its production of jet engine blades and disks.

"Back in the 1970s, we were considering better ways to build our product and improve its quality. Ours is a business based on high technology. Either we keep advancing the technology or we perish as a business," says Demonet. Reason number one: survival.

Reason number two: "We had developed the precision forging technology to build jet engine compressor blades, but we weren't making them ourselves... we were buying them," Demonet explained. "We had also developed the technology to make the strongest, longest lasting jet engine disks and we wanted to expand our capacity to make them," he added.

While P&W acknowledges about a 75% share of the world's military and commercial jet engine business, its principal competitor, General Electric, has been making its own compressor blades for the past 10 years. Because the disks and blades represent perhaps 25% of the value of the finished engine, the new Georgia P&W facility promises significant cost savings that can be translated into a strong competitive position in the years ahead.

Stanley Barwikowski, materials manager for the Georgia plant, became involved in the early proposal stages of the plant. Barwikowski (BSIE, University of Massachusetts) first became involved in the plant's development when he was an industrial engineering program manager at the East Hartford, CT, Pratt & Whitney headquarters offices. IE involvement in the Columbus, GA, facility included:
☐ Building program.
☐ Plant layout analysis.
☐ Material handling system design and vendor selection.
☐ Manpower forecasting.
☐ Manufacturing plan development.
☐ Part costing.
☐ Selling the program.

"By far," says Barwikowski, "the most significant IE involvement was in selling the program to top management."

Explains Barwikowski, "through the use of state of the art analysis techniques, industrial engineering was able to quantitatively justify the Georgia Plant appropriation. All claims, savings and productivity improvements along with a plan for manufacture were substantiated and documented to top management to show that the Georgia Plant appropriation was sound and well-founded."

The go-ahead on the 700,000-sq-ft plant was given in November 1980, and the facility was completed a few weeks earlier than scheduled. The P&W plant was dedicated this spring, when a host of dignitaries including UTC executives and Georgia state and local officials attended the ribbon-cutting ceremony.

The plant produces jet engine disks and compressor blades for both commercial and military aircraft, including JT8D, JT9D, PW2037 and PW4000 series commercial engines for a wide range of air transport and F-100 engines for F-15 and F-16 fighters. Typical jet engines contain about 21 disks and between 2,200 and 3,200 compressor blades. Two separate production lines at the Georgia plant produce the blades and disks. The two operations are linked through the common computer control network and the automated material handling system.

Both the disks and the compressor blades are very high-value engine parts which must be produced to extremely close tolerances from high-strength metals in the forging and machining process. Finished jet engine prices start in the million-dollar-per-copy range.

Producing the disks

Briefly, the disk production process begins when metal alloys in powder form are consolidated into a mult, which looks like a log. The mult is pre-heated to extremely high temperatures to bring it to a superplastic state. In this highly malleable state, less forging pressure is required to fill the die cavity, thus permitting smaller presses to achieve the critical strength and performance characteristics required of alloys used in jet engines. This proprietary P&W Gatorizing forging process yields less waste of critical raw material, reduces machining and can be performed in a cleaner, quieter environment, P&W says.

Compressor blade process

The compressor blades of varying sizes for the different engine designs, are formed on seven automatic screw presses with capacities ranging from 800 to 2,000 tons. The initial step in the blade production process is the extrusion of titanium or nickel alloy bars which have been cut into short cylinders. At the Georgia plant, all blade transfer operations are handled by either robots or conveyors.

Once loaded into furnaces, the cylinders are heated and later squeezed into preliminary shape by hydraulic presses. After extrusion, the compressor blades are forged by the automatic screw presses which permit the required tolerances.

Following forging, small amounts of excess material are trimmed from the contoured blade, or airfoil, by a human operator. The airfoil portion of the blade is completed by surface finishing and chemical treatment. An automated machining center shapes a large-diameter head left on the preliminary blade form.

The plant is expected to reach design capacity in 1986 when disk production is projected at 12,000 a year and compressor blade production is expected to hit 1.3 million units annually. At full production, about 800 persons will be employed at the plant.

At full production, more than 100 computers, 50 robots and 12 automated guided vehicles will be employed at the Georgia facility.

Computer network

P&W's computer network is hosted by two IBM 4341 CPUs, one for operations and one for backup. Six Digital Equipment Corp. VAX 11/780 minicomputers have dedicated responsibilities including:
☐ Running the combined automated guided vehicle and automated storage and retrieval system developed for Pratt by Eaton-Kenway.
☐ Controlling a flexible manufacturing system in the machining area.
☐ Supervising microprocessor terminals by United Technologies' Mostek or Allen-Bradley.
☐ Controlling an automated blade measuring system.
☐ Controlling an innovative automatic inspection line described later in this report.

"The computer network assures constant tracking of the parts," says Barwikowski. "The computers ensure that the delivery and storage systems won't be a bottleneck," he adds.

Evaluation of the building plan

IE evaluation of the Columbus, GA, venture began with the building program and the operations research group within the P&W East Hartford IE group. Various options were considered, including separate stand-alone facilities (one for engine disk and another for blade production), but a stand-alone combined facility was the final choice.

After analyzing factors such as potential plant population (both direct and indirect labor), capital requirements, size and starting expenses, P&W opted for a fast-track or parallel building and production process program.

IEs provided the financial ROI analysis and relational layouts of the plant in the building program. Plant equipment and department arrangements were developed through the use of CRAFTPWA—a Pratt-enhanced version of the CRAFT (Computerized Relative Allocation of Facilities Technique) simulation package. P&W says the CRAFT-PWA package was developed internally in 1980 to overcome many of the limiting and simplifying assumptions of the CRAFT package. Details of the CRAFTPWA package (developed by staff IEs) were presented at the joint national ORSA/TIMS conference November 10-12, 1980, in Colorado Springs.

Productivity plan

The heart of the productivity gains sought from the plant lies with the manufacturing plan/capacity analysis and the subsequent material handling design and vendor selection process developed by P&W engineers, including Barwikowski and

GEORGIA FACILITY

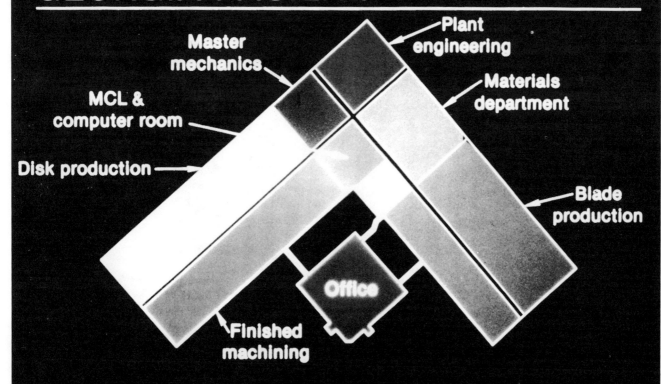

GEORGIA COMPUTER INTEGRATION

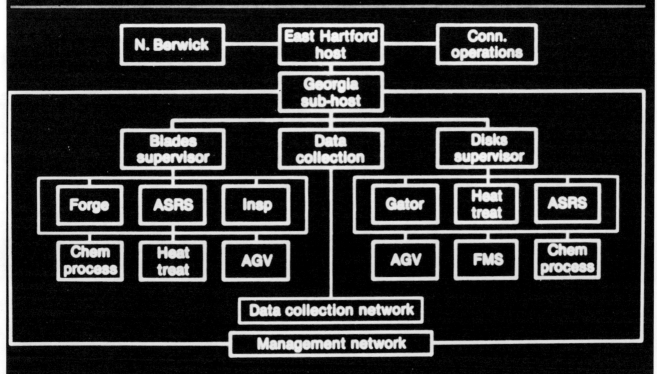

other IEs.

Barwikowski says lot sizes were determined in the manufacturing plan by part-period balancing algorithms. These lot sizes then feed the simulation model for evaluation of lead times and capacity. More than 10,000 lines of GPSS-H (General Purpose Simulation System) code provide details on aspects of the factory and material handling systems. Five-year projections required 1-2 hours of CPU time.

Factory simulation

Varying levels of plant performance and capital mix were entered into the GPSS-H factory simulation model and ECAP (Effective Capacity Analysis Program) to evaluate plant capacity requirements. This permitted analysis of equipment use versus inventory levels, lot flow times and queues.

Lead time estimates were developed for various manufacturing plans using GPSS-H simulation and EOT (Effective Operating Time) and FLOW models. These "estimates," says Barwikowski, "are still the basis for lead times today."

It's important to note that much of the documentation and persuasion of top P&W management to back the Georgia venture required—according to Barwikowski—that "training sessions" be held to prepare the executives to better understand the new systems and technology, much of which goes beyond the conventional wisdom of plant operation of even a few years ago.

Material handling design

P&W IEs worked with various consultants to design the plant's material handling system. Criteria evaluated in the material handling design included cost, production control and how expandable, flexible, adaptable, reliable and easily maintained the system would be. The two top-ranked proposals were for power

Stan Barwikowski, plant materials manager

$12.2 million Tech Mod investment may save U.S. Air Force $100 million

Pratt & Whitney officials expect to save the U.S. Air Force about $100 million in reduced costs and improved productivity in the making of jet engine disks at the company's new Columbus, GA, plant.

As an incentive, the Air Force provided P&W $12.2 million under the Technology Modernization (Tech Mod) program for the integration of three work centers for forging, sonic machining and non-destructive inspection of jet engine disks through a computer network.

"Computer-integrated work centers will improve productivity and process reliability and ensure that the centers will be able to work with one another faster and more accurately," says William W. Bernhart, vice president-manufacturing services for P&W's manufacturing division.

The sonic machining work center is the first application of flexible manufacturing system technology on hard-to-machine materials and parts that must meet close aerospace tolerances, Pratt officials say. The machining center includes vertical turret lathes, automated material handling and electronic probes and gauges for inspecting dimensions and surfaces. The inspection system provides immediate feedback to the machine to ensure quality levels.

This computer-integrated manufacturing effort is expected to yield considerable savings in jet component production, but, says Carl A. Lombard, propulsion Tech Mod program director for the Air Force," the biggest payoff will come in the future, as Tech Mod is expanded. It offers the potential of being a major vehicle in modernizing industry and keeping the United States competitive in the world marketplace."

and free versus a combined automated guided vehicle system and automated storage and retrieval system. The latter won out primarily because of cost advantages and overall flexibility.

For the AGVS and AS/RS, vendors were asked to provide appropriate system designs. P&W industrial engineers "provided all requirements and performance criteria, including sizing estimates," says Barwikowski. Vendor proposals employed three different simulation languages and required some creative approaches by the potential suppliers, including sensitivity analyses, explaining, for example, the impact of a 10% production change, and so forth.

The IE-developed factory simulation model was used to evaluate vendor system performance on the following factors:
☐ Queue times and sizes.
☐ AS/RS location.
☐ AGVS fleet size.
☐ Inventory level.
☐ Product flow times.

While pressed for cost-effective and innovative approaches, P&W engineers say suppliers responded well to the challenges of the new plant's systems designs. Such innovation is evident in the automatic inspection systems employed at the plant.

Advanced inspection system

Final inspection of both engine disks and blades will be accomplished through various automatic systems, as well as through a material control laboratory. Blade contour and dimensions will be verified for surface flaws by undergoing a fluorescent penetrant treatment and inspection system.

The plant's non-destructive inspection center has been integrated into the operation with the help of the U.S. Air Force's Technology Modernization program funding. The system will use computerized ultrasonic, fluorescent penetrant and chemical etch processes. (See sidebar on how P&W is utilizing the $12.2 million Tech Mod funding for the company's automated work centers at the Georgia facility.)

In summing up the new P&W facility, plant manager Demonet, who has already seen company-wide applications of the new systems pioneered at the Georgia plant, explains: "What we're really doing here is showing that, by combining advanced technology and a skilled workforce, we're able to increase productivity dramatically, and that's what it's all about today in the United States." IE

Reprinted from the *1984 Fall Industrial Engineering Conference Proceedings.*

Production Planning and Control at an Aeronautical Depot: An Analysis of Stochastic Scheduling

John S.W. Fargher, Jr.

Depot Operations Directorate
Technical Director
NALC-04A
Naval Aviation Logistics Center
Patuxent River, Maryland 20670

Production Planning
and Control Department Head
Code 500
Naval Air Rework Facility
Cherry Point, North Carolina 28533

ABSTRACT

This paper describes the production planning and workload control systems used at this facility to manage the rework process. The present workload control system and the proposed redesigned system (account for stochastic work content); inventory control system redesign; the automated storage, kitting and retrieval system; and the interfaces between the systems are presented. Finally, "lessons learned" in implementing these systems are discussed.

INTRODUCTION

As Eastern North Carolina's largest employer, the NARF (Naval Air Rework Facility) at Cherry Point has considerable impact on the region's economy. The facility employs approximately 3,200 civilians and generates salaries in excess of $72M (FY 1983) annually. In addition, the NARF does business in excess of $200M (FY 1983) annually. The facility is the only rework facility in the United States for the OV-10 Bronco and the AV-8 Harrier. It is also the East Coast rework point for the F-4 Phantom jet and the CH-46 helicopter. The facility has also aggressively pursued and acquired advanced carbon-fiber/epoxy-resin composite material technology associated with the repair and rework of the latest configuration of the Harrier aircraft, the AV-8B. The "B" model Harrier makes extensive use of the lighter, stronger, more flexible composite material which replaces some of the conventional metal materials in the AV-8A's construction. The facility has also been assigned prospective cognizant field activity/depot overhaul point for the JVX.

The Power Plant Building covers more than five and one-half acres. Within the Power Plant Building, the facility has the capability to overhaul, assemble and completely test a variety of aircraft engines including the T58, T76, T400, F402, J79 and T74. Additionally, the NARF reworks the T58-400 engine that powers the presidential-executive (VH-3) helicopters. These helicopters are frequently used to shuttle the President and other top White House executives. In related engine support, the facility is the overhaul point for ten ground and airborne models of gas turbine compressors, including those used on the SH-60B, AH-64, CH-47D, CH-53 and CH-46 helicopters.

In its Components Program, the facility has the skills, equipment and facilities to rework more than 9,000 different types of aircraft components. This program includes a number of unique features, including a recirculating wind tunnel and a computer-operated hydraulic test facility. In related support, this command is the only DoD activity that reworks the GB-1A, a portable liquid oxygen and nitrogen generating plant. The rework facility offers a customized, rapid-response program which provides a variety of "on-call" services such as field team emergency repairs, field team modifications, "drive-in" modifications, engineering support, fleet training, and customer service. At a moment's notice, our field teams can be deployed to any part of the free world to provide a wide range of aeronautical support services. Because of its productivity record, the NARF was awarded the Naval Material Command Productivity Excellence Award for two years in a row (1982 and 1983), the Navy's Meritorious Unit Commendation Award of 1981, and numerous safety and aviation safety awards (17 years without an aviation accident).

The NARF Production Divisions are organized by product, with cellular, just-in-time production concepts, although some processes are centrally organized because of investment (machine shop) or occupational health and safety (paint stripping, plating and aircraft painting). Production has quarterly schedules to meet involving the full range of aircraft, engines, components, and other support. Induction of workflow is scheduled by Production Planning and Control to assure a smooth, orderly flow to utilize available skills, equipment and plant capacity. As workload is inducted, each item competes for these resources. Level loading is the normal induction mode to spread out resource requirements throughout the quarter; however, batch processing is sometimes used when other work requiring the same or similar resources can be identified and also batch loaded to balance out the quarter. Production is prioritized by scheduled start date, the oldest worked first, with a few exceptions to meet immediate fleet requirements. By prioritizing workload by scheduled start date, the facility is kept from being bogged-down by old work.

Scheduling problems arise because of the nature of the workload. Each item to be reworked is disassembled and repaired according to its material condition. While some units only require minor repair, check, and test to return the unit to a Ready For Issue (RFI) status, other units may require complete disassembly, return to a zero time condition for many components, and overhaul of other components through various metal replacement/metal removal processes (normally in sequence) to return the components to original specifications. The work content within each unit is unknown until the unit is examined and evaluated (E&E'd) by expert artisans, called E&E's, and in some cases until components are routed for repair/overhaul and subsequent E&E. Scheduling requires the allocation, over time, of activities to resources. Typically, these problems may be viewed as the determination of an optimal sequence for the activities on resources and vice-versa and may be formulated as combinational optimization problems.

STATEMENT OF PROBLEM

Most of the literature on combinatorial optimization assumes one of the components (i.e., activities or resources) is fixed and the remaining component is to be scheduled. The most widely known of the "fixed resource" problems concerns the scheduling jobs on machines where machines are considered as special cases of optimum "packing" problems. Here, the jobs are packed into the available machine resources.

The most widely known of the "fixed activities" problems are concerned with workforce scheduling. The fixed activities imply a demand for manpower which is satisfied by scheduling the workforce. In a combinatorial context, such problems can be considered as special cases of optimum "covering" problems.

For these situations where the resources and/or activities must be moved as a part of the allocation, the scheduling problems generalize to include dispatching, routing and location aspects. Typically, these aspects are also combinatorial in nature. These more general problems can be loosely characterized as "distribution" problems.

Experience of depot level maintenance operations in the six Naval Air Rework Facilities, U.S. Air Force Air Logistics Centers, and U.S. Army depots suggests that there is considerable variance of performance at each facility around standard norms for each item. This is due to the unpredictable work content associated with a repair/overhaul operation as illustrated earlier. As a consequence, sometimes it is necessary to use combined approaches which consider the stochastic elements of the system.

Each item that enters a NARF (aircraft, engines, components and other support) can be represented as potential work impact on NARF shop and personnel skills as a CPM-like network. The resources required may represent as many as 50 different labor skills and a multitude of facilities such as machine centers, process shops, disassembly and non-destructive inspection centers, finished parts storerooms, test cells, and finally to packing and preservation. Simply stated, the problem is, "given the current shop status (i.e., the status of completion of each routed component in process as well as components in delay or in the queue awaiting resources), develop a prediction of work backlog for each shop, required resource allocations, and completion time for each component." These predictions need to be reasonably accurate for day-to-day operations up to a reasonable short range planning horizon (one week). Other non-stochastic models can be used for resource allocation beyond this planning horizon to predict the effects of shifts in workload based upon norms and occurrence factors. However, short-range predictions are required to analyze work content generation. The one week planning horizon is chosen because the shortest work determination cycle is the weekly component inductions. Aircraft and engines are inducted on a quarterly schedule while support services are normally a constant level of effort. Because of special probes and bit-and-piece part support or backlog repairable assets not being available, component inductions are subject to constant weekly changes. Component workload normally represents a major portion of the workload of a NARF.

There are several potential approaches. The first approach is to use diffusion approximations. The diffusion approximation approach has the advantage of taking into account the stochastic nature of work content; however, this approach is not computationally efficient. The second is to use time-phased network flow models. While time-phased network flow models are relatively much more computationally efficient, they do not take into account the stochastic nature of work content. A modification of time-phased network models has been selected, however, that loads the entire potential workload into the initial shop, this work content being reduced as the E&E process "buys out" operations that are not required. Additional work content may also be added by individual HWSOs (Hand-Written Shop Orders) specifying the component, process specifications and shops. As work is completed, it is clocked out of a shop by the artisan as he/she completes the task and automatically appears as backlog work for the next shop.

In the development of any model, optimal control rules for the system must be analyzed and related back to the context of real-life depot work flow systems. The decision rule chosen was that of scheduled start date in order that the oldest work, other than higher priority work, be worked first in order to clean out the shops. The system for calculating workload in each shop and tracking parts must be relatively accurate and inexpensive, predictions must be done in real-time, and wasted productive time to clock work must be minimal. On any given day at NARF Cherry Point, there are over 400,000 routed parts, each requiring on the average of seven operations. In a quarter, approximately 1,700 expedites are utilized for parts required back for reassembly that may be back late or with a longer turnaround time than the unit is allowed.

DESCRIPTION OF
THE PRODUCTION PLANNING AND CONTROL SYSTEM

The production planning and control system (Figure 1) begins with the strategic planning process of deciding on philosophies and objectives of the NARF; the resources that are to be committed to obtain these objectives; the opportunities and threats, strengths and weaknesses of the NARF Cherry Point organization; addresses policy issues on how the organization is to be run; strategies; and the specific action plans and milestones to achieve the objectives and strategies. The objectives form the basis for the NARF Cherry Point effort to obtain stable, long-term, high technology workload and the resources to support fleet requirements and depot level maintenance. The strategic plan allows these objectives to be communicated to all levels of our organization.

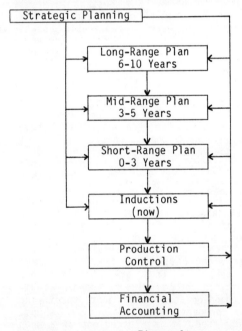

Figure 1.
Framework for Planning and Control

Long, mid and short-range workload plans are maintained to formulate resource (personnel, skills, facilities and equipment, and material support) requirements. The acquisition of facilities is obviously a result of long-range workload plans, as military construction normally has a five to seven year planning horizon. Procurement of major equipment and the skills realized from the apprentice program are products of the mid-range workload plan. Normally, military construction, major equipment acquisitions and development of specific skills are precluded in the short-range plan. The workload projections are normally stated in manhours and by product (aircraft, engines, components, and other support). These projections are broken down into specific manhour requirements and allocated to the various shops and skills required. Facilities, equipment and material support are programmed based upon these requirements. The budget and rate are then developed. Figure 2 illustrates this process.

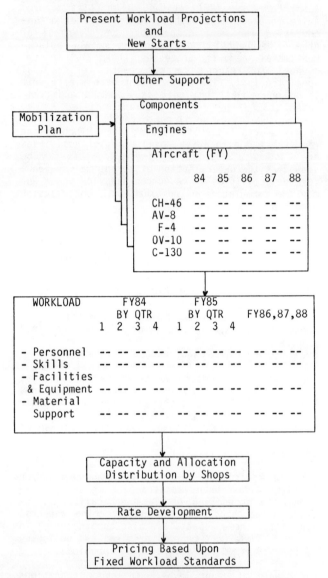

Figure 2.
Workload Planning, Budget and Rate Development

Mobilization planning occurs using this same process. The long, mid- and short-range workload plans allow the NARF to predict employment levels, skill requirements and balance sheet projections (costs and revenues) that are anticipated under present workload conditions that are already in the approved DoD projections.

The Computerized Workload Projection and Budgeting System (CWPABS) provides a standardized workload and budgeting tool that facilitates short-range (0-3 years) detailed planning. The CWPABS produces reports for the funding and operational

budgets. The funding budget is the primary method of acquiring funds for operation of the NARF and, therefore, must accurately project the funding requirements. Because the funding budget is sensitive to changes in workload at any time, the CWPABS was designed for flexibility, minimum turnaround time, accuracy, and user orientation (man-hours of workload). The users of the system enter the required data and run the programs themselves to produce the needed reports. When changes occur, data can be corrected, programs rerun with minimal turnaround time, and new reports printed. CWPABS consists of four subsystems, as illustrated at the bottom of Figure 2:
- Workload subsystem,
- Capabilities and allocation subsystem (also called manpower distribution),
- Rates development subsystem, and
- Pricing subsystem.

Inductions, production control and financial accounting are a result of this planning. The focus here is on individual tasks and transactions. There is continuous feedback and measurement of every action. Management control systems report summaries, aggregates, and totals with specific item information available on an exception basis.

The MIS (Management Information System) for INAS (Industrial Naval and Marine Corps Air Stations) has been designed to provide a standard ADP management information and workload planning system to meet the requirements of the NARF. This uniform system has been implemented at all NARFs. The Workload Control System of the MIS for INAS has been designed primarily as an information retrieval system with the depth of control to the individual operations on each item in the rework process. Information is generated by the use of transactors at each shop to record changes in status and time clocked against the JON (Job Order Number) for that item. This information is processed to obtain management reports in the Feedback segment of the Workload Control System that provide status on each item as well as rolled up to provide cumulative status.

The process by which components of an item are to be reworked, the sequence of operations and shops and engineered time standards are stated on an MDR (Master Data Record). The MDR represents this CPM-type network with all anticipated work flows present. If operations are not required, the E&E "buys out" that process on the MDR, signifying acceptance by an individualized E&E stamp and transacts the appropriate MDR line to kill that operation. If worked by an artisan, the artisan signifies completion with an individualized artisan stamp. If additional operations are required beyond the MDR-specified processes, an HWSO is prepared specifying the sequence of operations and shops.

An MDR is prepared and is provided for all items worked by the facility. Each component, engine, aircraft, fleet and in-house calibration, and preventive maintenance program is represented by a sequence of MDRs for each repairable/routed item. Each operation on the MDR has an associated computer card called Operating Documents (OPDOCs) for clocking time and status for each operation. The OPDOCs are computer-generated from the MDR. Figure 3 shows an example of an MDR and an OPDOC. The MDR must be created before any other part of the workload control system (OPDOCs, Feedback, etc.) can exist.

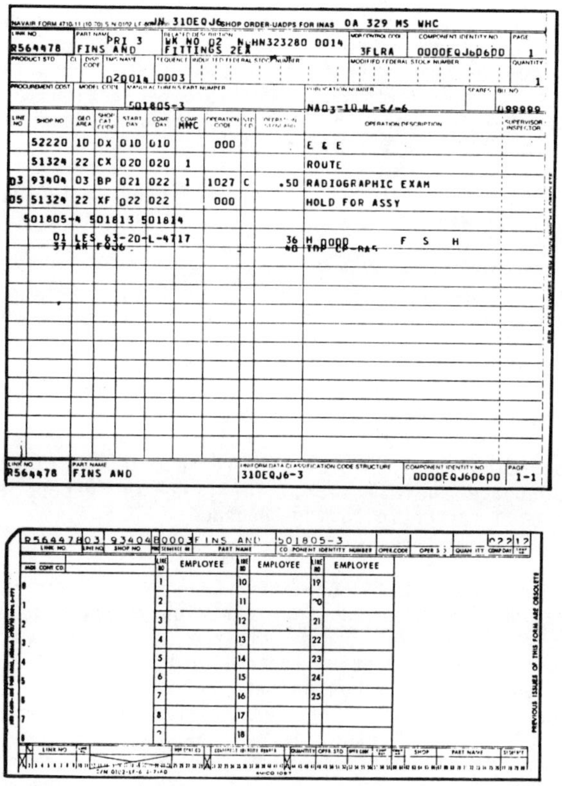

Figure 3.
Sample MDR and OPDOC

During the aircraft, engine and component disassembly stage or calibration and preventive maintenance routing, MDRs and OPDOCs (or HWSOs, if required) are attached to the routed parts for identification and processing. E&Es inspect the parts to determine the depth of rework necessary to return the item to the required specification as well as make determinations for disposition of those parts not requiring rework, removed only for access, or beyond economical repair. Production control personnel are responsible for proper routing of all parts upon completion of E&E.

Parts requiring rework are routed to the proper shops for rework and bit-and-piece part support ordered by production control. Eventually, reworked items arrive at the finished parts storeroom/ASKARS tracked by the workload control system throughout the process. Those items removed for access or not requiring rework are sent directly to the finished parts storeroom/ASKARS. Those items beyond economical repair are sent to supply,

279

or otherwise properly disposed of, and replacement items are ordered by production control from Navy Supply, if needed. From the finished parts storeroom/ASKARS, the items are kitted to assure that everything is readily available to the artisan assembling the aircraft, engine, component, or subassembly.

WORKLOAD CONTROL SYSTEM

The present Workload Control System (WCS) is a batch program that updates the data base, based upon transactions during the workday. The program is run at the end of the second shift, updating the previous day's run. Workload projections are accurate for only the next workday and cannot be forecast by the WCS beyond this horizon.

The reports generated from the WCS program include the following:

- The Daily Rework Requirement Report (DRRR) is a five-part report that is utilized to determine daily priorities at the aircraft level so the aircraft manager can plan the day's work based on actual requirements rather than scheduled requirements, the production controller can load components to the aircraft in a timely, systematic manner, and a concise history and sequence of the operations performed on the aircraft are available to update the MDR deck. The five-part report is printed in the following sequence:

- Part 1: Condition Code Description and Aircraft Movement History contains a description of the condition codes utilized by the aircraft manager and production controller to report status at the Operation Master Data Record (OMDR) labor line level; e.g., "Part Not Available," "Material Not Available," etc., as well as a list of movements of the airframe from one shop to another including the time of day, the date of each movement, and the current shop with a maximum of 19 moves shown.

- Part 2: Completed Operations reflects all OMDR labor lines that have a "Labor Complete" transacted against an operation since the last DRRR was printed (each completed OMDR labor line is reported only one time). Each reported operation will contain the name of every employee who transacted labor against that operation, the man-hours expended by each, an asterisk (*) next to the employee's name who certified completion of the operation, and the last condition code, if any, transacted against the operation.

- Part 3: Past Due and Early Started Operations lists those operations that have not been completed or voided with a schedule start day in the Component Identity Number (CIN) that is lower than the prime schedule day. Also listed are those operations that have had labor transacted against them with a schedule start day higher than the prime schedule day. A summary of operation standard hours is provided to reflect total hours of the operations listed requiring completion by the applicable trade.

- Part 4: Scheduled Operation (Prime Schedule Day), established as the prime schedule day, contains all operations that have not been completed or voided, including HWSOs that have a scheduled start day in their CIN that is equal to the computer calculated prime schedule day. A summary is provided by trade code of the total operation standard hours for all the listed operations.

- Part 5: Forecast is a list of the operations to be done that have a scheduled start day in their CIN that is greater than the prime schedule day. After each forecast day, a summary report is provided listing the total operation standards by trade code that are required for the applicable prime schedule day.

- The Aircraft Position Report shows the progress of each aircraft by listing each checkpoint on the progressing MDR and entering the scheduled completion workday and the actual completion workday, as transacted by the aircraft manager, for each checkpoint. As checkpoints may not be completed in a specific numerical sequence, an asterisk (*) will appear in the field titled "In Checkpoint" when this occurs, denoting an unusual condition. When this condition exists, a "condition code" will be listed under the unfinished checkpoint number, denoting the reason the checkpoint is unfinished.

- Aircraft Status Summary Report shows man-hour information by organizational divisions for the aircraft program, identifying each aircraft separately.

- Transactions Processed monitors all valid and invalid transactions, highlighting the invalid transactions by error legend.

- WCS Error Transaction List contains pertinent data needed to identify the specific OPDOC against which an erroneous transaction was made, record type, error code, and aircraft MDR sequence number.

ASKARS
(AUTOMATED STORAGE, KITTING AND RETRIEVAL SYSTEM)

Each of the six Naval Air Rework Facilities has, or has under construction, an ASKARS. ASKARS is designed to support the production control requirements of the facility to receive, inventory, store, schedule, and retrieve aircraft and pneumatic components. ASKARS at NARF Cherry Point is composed of a central data processor and remote terminals; a storage rack system with stacker cranes, storage/retrieval stations and associated input conveyors; and a driverless tractor system. There are three main racks for Small Item Storage (SIS). Each rack has a storage/retrieval lifting device (stacker crane) that can be operated automatically, semi-automatically or manually. There are approximately 4,140 cubicles in the ASKARS SIS. Each cubicle is capable of holding up to 500 pounds and may be partitioned into many varying configurations. Additionally, there are 176 cubicles for Large Item Storage (LIS) capable of holding up to 1,000 pounds per cubicle, and 60 cubicles for Special Large Item Storage (SLIS) capable of holding up to 2,000 pounds per cubicle. Each rack system (SIS, LIS, SLIS) is equipped with a series of commands to provide the operators with the ability to store, retrieve and inquire. These commands activate the storage/retrieval devices and associated input conveyors. ASKARS is also capable of keeping track of items stored at other locations around the facility.

An ASKARS processor control room is located within the ASKARS enclosure. Located within the control room is the central process controller (Tandem 16) and peripheral equipment responsible for the automatic storage and retrieval of parts. The Tandem processor is physically connected to 15 terminal processors. These remote terminal processors control input/output to the Tandem processor. The processors are limited to basic function requirements based on user location. A master station terminal in the control room is used to perform special and unique system functions. High-speed printers are available to provide daily or as-required reports to users of the system. As a system, ASKARS receives data from both human and electronic input. This input is used to establish the ASKARS Basic Data File. ASKARS is linked to the rework cycle by a driverless tractor. The driverless tractor follows a predetermined route and is equipped with an onboard processor that has limited communication with the ASKARS processor control room. The driverless tractor can tow from one to three trailers with a maximum weight of 15,000 pounds.

Kitting, the unique feature that distinguishes this system from other automatic storage and retrieval systems, is accomplished within ASKARS after the record of parts removed for access, not requiring rework, routed, or requisitioned from Supply to replace parts beyond economical repair, is activated by a transaction. Unless a transaction takes place, ASKARS assumes the item is part of an assembly or not removed. Unless another part is available from another unit with a later scheduled completion date, that part can be diverted. Parts and hardware kits are assembled in advance of the scheduled issue date. ASKARS provides updated reports on the contents of the kits and a list of deficient parts and hardware kits. The completed kits are loaded to production by Production Control as required. Production Control is responsible for expediting parts in the rework process to meet kitting requirements from the uncompleted kit report, to arrive at ASKARS three workdays prior to the scheduled start date to assembly, although parts and hardware kits are retrievable for issue at any time. Parts, hardware kits, and completed kits are stored in ASKARS and logically linked through the various record transactions and the MDRs, and are normally retrieved by the system by a request for a complete kit, although the individual parts to the kits were stored at different time intervals. Kits may also be issued short of parts and ASKARS will provide a listing of all parts missing from the kits. Rotable spare parts are also tracked and stored in ASKARS for diversion to complete kits when a part has a greater turnaround time than the assembly.

NIMMS
(NAVAIR INDUSTRIAL MATERIAL MANAGEMENT SYSTEM)

NIMMS provides the NARFs with a control of financial transactions and receipt and inventory of material used in the rework process. The NIMMS is a data base system with specific programs allowing access to this data base to allow the user to update, add, delete and query the data. The programs operate on UNIVAC-1100/40 computer systems located at the various Navy Regional Data Management Centers (NARDACs) or station data processing (Cherry Point only) colocated with the NARFs. The system, data base and outputs are unclassified; however, users are identified and transactions authorized and tracked to provide user profiles and security of the data base.

NIMMS is an inventory control and management information system. The NIMMS redesign has transformed the system from a serially-processed, tape-oriented system to a data base management system with real-time teleprocessing capabilities, with the user-operated CRT located at the production control and material managment work centers on the production floor as the primary method of data collection and display. Four categories of material records are contained in the NIMMS data base:

1. Naval Industrial Fund (NIF) retail stores,
2. Direct Material Inventory (DMI),
3. Customer Furnished Inventory (CFI), and
4. Specific Requirements Inventory (SRI).

The NIF retail store inventory contains material with repetitive demand patterns (minimum demand frequency of three requisitions semi-annually) and is restricted to consumable material but may also contain non-standard items that are locally manufactured and procured. This inventory data base allows material managers to identify excess inventory quickly and at the earliest possible date. Stocking objectives are reviewed quarterly and automatically established by the system based upon a one-year history. NIF retail store items are tested by a replenishment formula to determine if action is required each time the on-hand, due or stock level quantities are affected. The automatic replenishment formula is:

The automatic replenishment formula feature allows managers to replace the normal formula factor on an individual item or to inhibit replenishment. Excessive and long supply materials (on-hand is greater than stock level) can be transferred to retail stores at other facilities upon physical movement of the material. Material is issued to production at the time the material is actually required and in the appropriate quantities upon receipt of a requisition with a current job order number (JON). On-demand, real-time retail store inventory allows faster reaction to immediate material requirements and greater fiscal control.

The materials established in the Direct Material Inventory are based upon a bill of materials of both consumable and repairable items with the intent of charging progress payments to the customer order as material is identified and issued. The DMI may also contain non-standard items that are locally manufactured or procured. Material is stocked and issued at the procured price with adjustments to the average price for subsequent buys. The risk of ownership of DMI items is totally borne by the NARF, making effective management of this inventory essential. The DMI data base provides the depot managers with the management tools needed to control this inventory and insure minimum waste.

The Customer Furnished Inventory (CFI) is material provided by customers for use on their own jobs. The CFI maintains physical and financial records of this material and provides an accurate tool in developing the total cost of a customer's product. NAVAIR-provided modification kits and interservice and foreign military sales items are normally contained in the CFI.

The Specific Requirements Inventory (SRI) contains those items not carried in the NIF retail store, but required for a specific end-use customer. The SRI contains both consumable and repairable materials.

The redesigned NIMMS provides new capabilities that facilitate requisitioning, receipt, storage, issuance, reconciliation and maintenance of material; provides early location of material sources; computerizes routine operations formerly accomplished manually; provides on-line access to the current status of supply requisitions, inventory balances and history data; presents pertinent on-line information to assist management decisions; and provides output tailored to the users requirements.

Figure 4 is a description of the 8 areas and 22 associated record files that comprise the NIMMS data.

DATE BASE AREA/RECORD	DESCRIPTION OF NIMMS RECORD
(1) Job Order - Activity Control Record	- Record data which are a constant to an individual user activity.
- Job Order Record	- Identify the job order, its associated status, and restrictions in the types of material, and labor and contractural services that may be charged.
- Parameter Control Record	- Provide system access to the NIMMS Parameter Record.
- Parameter Record	- Provide current data prior to report processing functions.
(2) Cross Reference - Part Number Record	- Identify a item by part number and associated manufacturer's code to facilitate the processing of catalog data. The part number is a unique material item identifier.
(3) Store - Store Record	- Provide a detailed description of a store indicating geographical areas for the routing, storage and types of material. Customers for which charges are accumulated are identified. Percentage of adjustments in relation to total dollar value of inventory representing the overall gain or loss for a given year is also indicated.
(4) Material - Material Record	- Provide accurate inventory and funding information, the procurement source, shelf-life span and special material content.
- Store Material Record	- Provide inventory information within a given store on specific material items, identify storage locations for given materials within a given store, indicate total quantity of a material item requested during a quarter, and provide store unit price information.
- Substitute National Item Identification Number (NIIN) Record	- Record an item along with an associated substitute NIIN that is considered a substitute for the ordered part.
- Material Message Record	- Collect user-defined information.

DATE BASE AREA/RECORD	DESCRIPTION OF NIMMS RECORD
(5) Material Due - Material Due Record	- Provide information concerning the requisitioned materials; supply action is indicated (order filled, cancelled, substituted, exchanged materials, etc.).
- Due Status Record	- Provide information regarding the current status of requisitioned materials.
- Due Date Record	- Trigger the automatic follow-up action for a Material Due Record.
(6) Transaction - Transaction Record	- Collect all transactions recorded from the Day Transaction Record for historical purposes including recording all of the financial transactions which are forwarded to the NAVAIR Industrial Financial Management System (NIFMS) and all Material History Records.
- Financial Record	- Collect financial information.
- Day Transaction Record	- Record transactions that occur on a daily basis.
- Material History Record	- Summary of transaction records.
(7) Suspense - Suspense Record	- Collect transactions from the Day Suspense Record that were not processed. These transactions are either processed or purged after correction.
- Day Suspense Record	- Record specific transactions that have not been processed due to errors.
(8) Inventory - Inventory Record	- Collect data generated during a physical inventory.
- Quantity Record	- Contains the on-hand balance of specified

DATE BASE AREA/RECORD	DESCRIPTION OF NIMMS RECORD
	material at the time of a physical inventory request, and the date of the last action registered for the material.
Other - Report Direction Record	- Contain data for the output control of the specific NIMMS reports.
- Error Message Record	- Programmer's aid containing all the error numbers and related error messages used in the NIMMS programs.

Figure 4.
NAVAIR Industrial Material Management System (NIMMS) Program Areas and Associated Files That Comprise the NIMMS Data Base

NIFMS
(NAVAIR INDUSTRIAL FINANCIAL MANAGEMENT SYSTEM)

NIFMS has been designed to provide a standard financial management system in which budget, performance and accounting data are fully integrated. NIFMS captures labor, material, contractual and other costs at the shop order level; accumulates and maintains them in the cost and expense records; and records financial and other information on the customer order records. Customers are billed and all cash receipts and expenditures are processed and accounted for automatically. NIFMS provides required financial data for improving scheduling, expanded labor and material cost information, and a standard data base for interfacing with other systems (NIMMS and WCS) in the Management Information System (MIS) for Industrial Naval Air Stations (INAS). Sixty-two standard management reports are generated for tracking costs at various levels and to provide useful information on direct and indirect programs. NIFMS accomplishes the following specific objectives:

- Provides standard reports that facilitate management by exception.

- Provides on-line query and updating of the data base.

- Simplifies processing of financial accounting data by eliminating routine and repetitive clerical operations and allowing for on-line editing and validation.

- Provides for the integrity of the NIF cash account by ensuring reimbursement billing for work performed and services rendered.

- Provides management the means to exercise greater financial control of disbursements, reimbursements, establishment of liabilities, maintenance of inventories, and other actions having a significant impact on the NARFs' financial position.

- Provides required and prudent controls and audit trail capabilities to meet all internally and externally prescribed requirements.

- Provides for full and ready disclosure of the activity's financial condition and operating results, both through standard reports and on-line query.

NIFMS operates on a UNIVAC 1100/40 computer system. The various components of the 1100/40 system are designed as separate logical units providing maximum modularity.

THE REDESIGNED WORKLOAD CONTROL SYSTEM

Based upon the application of the present WCS, the updating and redesign of the WCS programs is required. The redesign maintains all functions and interfaces of current applications, adds on-line queries and real-time updates to improve service to the users, and simulates rework processes recognizing the stochastic nature of the work. The conceptual approach for the redesign is centered around eight major functional areas:

- Create work plans and document all work processes.

- Mechanize production of MDRs and OPDOCs.

- Automatically generate component induction schedules and integrate component management and reporting.

- Control and maintain status of plant equipment.

- Project work content and cost by rework item.

- Record material usage and costs.

- Defect reporting and control.

- Retain history on the work processes, update occurrence factors on E&E "buy-outs" of MDR lines, and calculate the depth of rework required for subsequent operations.

After several meetings to fully define all of the WCS redesign requirements, redesign began in May 1983. The redesign project involves the design, development, test and implementation of application software. The WCS redesign is being accomplished to run on the TANDEM and UNIVAC systems.

The subsystems of the WCS redesign are:

- Scheduled Inductions -- automated scheduling process for all rework.

- Manage Work -- tracks the disassembly of aircraft, engines, components, etc. Also, it has the capability of reporting on the status of components that are identified as work-in-process.

- Monitor Production -- records plans and schedules, records hours expended by artisans and related costs, updates induction and return data.

- Manage Resources -- provides management, maintenance and tracking of equipment, material and manpower.

- Manage Work Plans -- provides for the creation, management, maintenance and review of work plans and provides the tools to review standards and technical directives used within the work plans.

- Plan Future Workload -- reports actual time for rework vice planned rework time, projects material usage, material costs, labor expenditures and labor costs.

- Record Labor -- produces performance, quality assurance and labor reports, does labor processing, and records leave taken.

The objective of the redesigned WCS is to provide a timely production planning and control interactive data base to assist in planning and executing the rework requirements. To achieve this objective, the WCS is being designed to allow user flexibility in the formatting, presentation and manipulation of the selected data from the data base. UNIVAC MAPPER, an information retrieval and formatting program, allows this flexiblity in an off-line environment. MAPPER is a user friendly interface program that has the capability for outputting charts and graphs as well as formatted information. On-line query is also provided in the redesign effort to reduce the requirement for hard copy reports. Variance parameters can also be predetermined and set to automatically notify the user when the variances are exceeded.

File maintenance is accomplished by the user specifying the changes to be made in specific data bases. The system then applies the changes to all affected data bases. Access for the data is restricted in different ways for different users and other unauthorized access is prevented.

The WCS redesign provides the tools to manage the personnel skills, equipment and facilities, and material required for a dynamically changing workload. Modeling via simulation of various workload levels and mixes can also be accomplished to answer management's what-if questions (impacts on manpower, material, machinery, etc.). The redesigned system also takes into account intervening events such as material delays, work stoppages, and overdue operations. Estimated start and completion days for affected subsequent work operations and higher assemblies are adjusted accordingly. The impact of a material delay, for example, for a particular operation on a sub-route will be readily apparent should the operation become the critical path in the remaining rework processing for the end item. All parts that make up to a next higher assembly will be chained to that assembly. A CPM network is used to compare the anticipated completion date of each part with the date required to support the next higher assembly. Each event is treated as a prioritized queue. Expedites are first in the queue by the scheduled start date, with unexpedited components second, each prioritized. Average (mean) processing times are used. An automatic alert is generated for any schedule in jeopardy.

Projections of material requirements are based on historical accumulations whenever possible. Material personnel monitor, adjust and validate these projections. Material requirements for new work plans are established by the material personnel and then adjusted automatically by the WCS by historical usage. During induction, the WCS will automatically order all 100% piece part materials, unless overridden by the user. Any additional material requirements identified during the E&E evaluation or subsequent rework will be input through the NIMMS as part of the WCS.

The WCS redesign retains a history of all operations performed on an inducted item, a record of technical directives incorporated, and a record of quality certification and/or verification of an operation. The system also provides an audit trail to record information on the employee ID for

WORKLOAD CONTROL SYSTEM INTERFACES

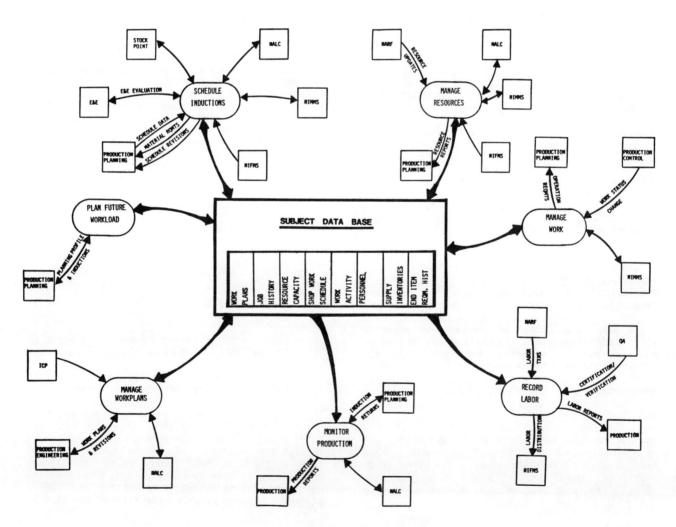

Figure 5.
Workload Control System Interfaces

each input and date, time and location of the transaction.

One integrated telecommunications terminal performs data input and retrieval. This single type of terminal accesses NIFMS, NIMMS, WCS and ASKARS. Data inputs are minimized in volume with editing and validation of point of entry for corrections on-line. Although cards will still be used for non-frequent transactions, bar coding is being considered to expedite data entry. It is envisioned that one of the main uses of bar coding will be to link items in process to the routing and reprocessing of information which supports the rework of the item.

The WCS redesign links the systems described previously (NIMMS, ASKARS, NIFMS and WCS) into a modular, integrated system. Figure 5 describes the interface of the WCS subsystems with NIMMS, NIFMS and ASKARS.

OPERATION OF THE WCS, NIMMS AND ASKARS

The user (operator) initiates execution of a program by entering the appropriate transaction code. The program responds with a formatted screen sent to the user's CRT terminal. The user enters the required information in the appropriate fields and then transmits the screen. The system accepts the user-initiated transactions, edits and validates the input, and performs data base/file maintenance as required. Input is also accepted from selected external interfaces, collected and applied as appropriate. If at any time during the execution of the program run an error occurs, the input screen is sent back to the user with an explanatory message describing the cause(s) of the early termination. System response time, measured from the time the system accepts the input transaction, to the time the output response is available to all major functions, is 2 to 7 seconds, with none greater than 15 seconds.

Whenever possible, data is system-generated based on the data previously collected in the data base and by performing calculations on the data that is available. All systems are capable of both on-demand and batch processing. The batch mode is intended as a back-up to the on-line mode, or utilized where the volume of transactions necessitates mass entry. Data transactions are stored in auxiliary external interfaces between the systems. Output is obtained by calling an output procedure, initiating a report generation, or using output query routines. End-of-day processing provides information for the daily transaction registers and financial summaries.

The systems have been designed for flexibility and ease of adaptation for changing requirements. The concept of each function being a subsystem lends itself to several small transaction programs, in lieu of one large program, thereby making it much easier to accomodate changes.

LESSONS LEARNED

Developing decision support systems to solve the production planning and control problem involves satisfying a multitude of data requirements. A critical aspect of the aggregate planning and control process concerns the dynamic nature of the production process. The present decision or plan is just one of a sequence of decisions and does not establish a permanent production policy. The production process must be continually monitored by production planning and control personnel to ensure a smooth and orderly flow of production within the resources available, inventory managers to maintain the minimum inventory levels to meet demands, and financial managers to keep track of costs to reduce total expenditures of labor and material. Each has a conflicting, yet complimentary, function.

In the design of decision support systems as described previously, the interaction of the real user and manager was essential in developing the complete requirement for the systems. User involvement during design reviews and operational simulations has also been critical in gaining acceptance of the systems as well as providing feedback for later enhancements. The systems must be shown to aid both the practitioner as well as the manager in the decision-making process. These systems were shown individually to support and assist the decision process with the payoff of better knowledge of what was happening or about to happen in the production process, allowing better decisions. The decision support systems all centered around the following design philosophies:

- Definitive and well-formulated user requirements based upon management commitment to better plan and control the complex depot rework process.

- A cooperative effort between all users to support each others' data requirements and a willingness to change for the betterment of the facility.

- Simple user interface with on-line, real-time information to the user to the ultimate goal.

- Flexibility in report format and generation to meet user and manager thought processes.

- Use of a modular approach with interfaces built between current and new systems.

The cooperation from the various groups began at the decision support systems' inception with the development of the requirement and has continued through the implementation of the various systems. This attitude at NARF Cherry Point has allowed each system to develop and mature as the operating personnel gained a better knowledge and understanding of the system's capabilities and limitations.

The downfall of many a successful system, maintenance of the programming software, has also been planned for and is in place. As the system is fully used, enhancements and off-line changes to the models are being developed to improve the systems. Many of the enhancements come from operator personnel. A specialized group is handling the maintenance, update and enhancement of the systems.

BIOGRAPHY

Mr. John S. W. Fargher, Jr., began his career at the Intern Training Center, Red River Army Depot as a Production Design Engineer. He has held positions as a systems engineer for the Small Caliber Ammunition Modernization Program (SCAMP), Frankford Arsenal; Chief of the Production Planning and Control, Hawthorne, NV; Lead Engineer at Rodman Laboratory, Rock Island, IL; Chief Industrial Engineer with the Iranian Aircraft Project Manager's Field Office in Teheran, Iran; Deputy Chief of the Materiel Systems Development Division, U.S. Army Transportation School; Professor of Acquisition/Program Management at the Defense Systems Management College; Deputy Project Manager for the Light Armored Vehicle Directorate, Quantico, VA; and is Head of the Production Planning and Control Department, Naval Air Rework Facility, Cherry Point, NC. He is presently on a temporary assignment as Technical Director of the Depot Operations Directorate at the Naval Aviation Logistics Center.

Mr. Fargher holds a B.S. in Engineering Science from Montana College of Mineral Science and Technology; Master of Engineering in Industrial Engineering from Texas A&M University; and an M.S. in Systems Management from the University of Southern California. He is a member of Alpha Pi Mu and Tau Beta Pi and has authored many articles and several books on program management, integrated logistics support and mathematical modelling.

The Evaluation of a Flexible Manufacturing System — A Case Study

Gary R. Redmond
Senior Staff Manufacturing Engineer
FMC Corporation
Ordnance Division

ABSTRACT

This paper reviews the investigation and analysis pursued by FMC leading to the selection and configuration of a Flexible Manufacturing System (FMS) for random parts machining. Three phases are discussed: (1) initial concept studies that resulted in selecting FMS technology; (2) establishment of key inputs and variables utilized in simulating alternative FMS configurations; and (3) evolution of the system configuration resulting from changing production requirements and system expansion considerations.

BACKGROUND

FMC Ordnance Division, a leader in the design and manufacture of tracked aluminum vehicles, recently introduced the M2/M3 Bradley Fighting Vehicles (BFV) into production. This BFV program required the largest tooling and facilitization effort the division had undertaken since introducing the highly successful M113 program in 1960.

In comparison to the M113, the BFV was a completely new vehicle, with more complex and twice as many manufactured parts. Because of increased part mix and reduced production rates compared to the M113, the BFV parts tended to be processed in the machining department on a batch basis through shops organized by function. Even at lower production rates, the practice of using stand-alone conventional and N/C machine tools for complex workpieces was operationally and economically ineffective. If the BFV's product life cycle, like that of the M113, eventually reached higher production levels than those required by initial contracts, the need for improved manufacturing efficiency would be even more critical.

INITIAL CONCEPT STUDIES

Since the initial BFV production effort, FMC has looked into opportunities for reducing the vehicle cost. As part of this effort, a group of machined parts was identified for detailed analysis. Most of the parts were components in the drive train, weapon positioning or suspension systems and ranged in quantity from one to twelve units per vehicle.

The first concept studies involved special-purpose boring/milling machines and head changers. Their advantages included:
  o Improved part processing to meet close tolerances and improved equipment utilization
  o Reduced labor, lead time, in-process inventory, setup, scrap and rework.

Disadvantages of these concepts included:
o Difficulty in determining machine configuration and tooling costs due to potential variations in production rates
o Equipment lead time
o No backup capabilities in the event of machine failure
o Inflexibility to handle major design or part mix changes

Company decisions and other complications affecting the early development efforts raised several important issues. First, many components on the M113 program were being processed on special purpose machines and a transfer line was being used for hull machining. The inflexibility of the M113 equipment contributed in part to the capital equipment requirements for the BFV program. Secondly, Ordnance Division had received authorization for a parts manufacturing facility that would focus on manufacturing parts on a low-cost basis for the BFV program. To accomplish this objective, the division would have to improve its manufacturing technology and provide flexibility to adapt to increased production rates. The facility was to be located remote from the main complex and limited in its potential size. Thirdly, part of a producibility effort resulted in tolerance changes that offered more choices on the types of machining that could be used to process the parts. An investigation of FMS technology was initiated using the same group of high-cost parts involved in the previous study. Vendor concepts were requested and two were selected as having the highest potential. One concept would require three horizontal machining centers and two vertical turret lathes. The second concept would consist of horizontal turn face boring machines instead of vertical turret lathes. The latter configuration promised reduced part handling and fixture requirements because the part would always be machined in the horizontal attitude; however, the turn face bore machine, expected to be a standard machine tool by the vendor, had yet to be built.

SIMULATION

To properly assess FMS configuration issues, and to comprehensively understand the dynamics of an FMS system, FMC worked in conjunction with Charles Stark Draper Laboratories Incorporated, Cambridge, Massachusetts. Draper Labs is a non-profit corporation involved in research. Their areas of expertise include industrial automation, advanced electronics and advanced inertial navigation systems. Thus, FMC drew upon Draper's expertise in analyzing FMS technolgy relative to its production requirements.

Since the configuration and analysis of an FMS is noninitiative, Draper Labs had developed computer programs which provided capability to simulate many of the dynamics of an FMS machining system. For FMC, a data base was established from which various configurations of an FMS could be created and simulated. Cutting speeds and feeds by material type were defined for the machining processes: rough milling, finish milling, contour milling, drilling; tapping, counter boring, reaming, woodruff key milling, semi-finish boring, finish boring and form tool milling. Operating parameters for each type of machine were also determined for horsepower, thrust, torque, maximum spindle RPM, maximum feed rate, maximum traverse rate, tool change time, rotary table index time, pallet shuttle time and tool storage capacity. Preparing process plans for each part was complicated because, in some cases, a part had multiple processing methods that varied according to the type of machine being used. This complication significantly increased the number of FMS configurations that could be created. The process plan, consisting of machining operations required to complete the part, included the following data: operation number, feature machined, number of features machined, tool type (or machining process), tool diameter and/or number of teeth, length of cut, machine type and material. Fixturing and tooling requirements consisting of fixture concepts, number of fixtures, part load/unload time, number and types of tools and tool set-up time were established for each part.

From the data base, machine time for each part was generated. This data was used with production rate variables [vehicles per period (shipset) and parts per shipset (part mix)] to generate the following information:
o Total machine cycle time by part and shipset
o Cycle time for each machine type by part and shipset
o Distribution of the machining cycle time among the machining processes by part and shipset
o Number of tools utilized in total and in each machining process by part and shipset
o Total number of tool changes per shipset
o Number of unique tools
o Distribution of shipset cycle time among the machine cutting, tool changing, tool positioning, table indexing and pallet shuttling activities
o Tool set-up time by part and shipset

This information was used to determine the approximate number of machines necessary to meet the selected production rate. Configuration of the system was further assisted by a batching and balancing algorithm that allocated parts to specific machines based on the number of tools used by the part, part cycle time and tool capacity of the machine. The following additional information was required for each configuration in order to run the FMS simulator:
o Number of load/unload stations
o Material handling path and nodes
o Routing path of fixtures
o Machine failure frequency
o Due dates used by a scheduling algorithm to feed parts into the system.

The simulation model was able to establish system throughput, as well as utilization of the machine tools, material handling carts and load/unload stations.

Analysis of these simulations significantly increased our insight into the FMS approach. Machine requirements based on part processing methods could be better understood. As the impact of production rate changes were more accurately determined, our confidence in establishing system size and cost increased. Productivity of the system was equal to or greater than dedicated and special-purpose machining alternatives. Opportunities for improving the FMS configuration by changing part processing methods, fixturing and tooling were identified.

SYSTEM CONFIGURATION

Subsequent company decisions affected project development when it was decided that the FMS would be located in the remote parts manufacturing facility. This relocation was considered advantageous because it would make the FMS the cornerstone of advanced manufacturing technology at the new facility. Since the facility was considerably smaller than the main complex, management visibility would be higher, thereby insuring that resources would be available to support the installation, startup and operation of the system. At the same time, some FMS parts were slated to remain at the main complex for machining in dedicated manufacturing cells. A major reconfiguration of the system was required because these parts represented over one-third of the machining load.

Adapting to the part mix changes required FMC and Draper Labs to identify additional parts for machining on the FMS. A parts coding system developed during the planning stages of the parts manufacturing facility was used to search for potential FMS parts. Prismatic and rotational parts were analyzed for suitability on the FMS. In addition to machine run time, the number and lengths of set-up were considered in determining a parts potential.

Once a group of parts was selected, process planning, tooling and fixturing data was prepared. Simulations of the FMS were run with part mix being an additional variable. The results of the simulations showed an imbalance of load between the machine types. There was not enough cycle time to justify more than one machine with turning capability, and it became questionable whether such capability should be included in the FMS. The proposed vertical turret lathe, which required additional fixturing and part handling in order to present the part in both the horizontal and vertical attitudes, was not regarded an optimum method for processing the parts. The turn face boring machine was considered risky since the first unit had not yet been built. The FMS was considered complex enough without the compounded problems of a new machine tool and its associated debugging. Also, with only one turning machine, neither alternative could provide backup within the system. Concurrently, backup was becoming more of an issue because the parts were destined to be machined in the remote parts manufacturing facility, and resources at the main plant would not be available.

Further simulations were performed to determine the impact of removing the turning capability. These simulations involved removing some parts and reprocessing others to perform all the machining operations on horizontal machining centers. The results showed that a system of five identical horizontal machining centers could be configured around twelve parts. The configuration was considered optimum based on all simulations run to date. Productivity would improve significantly over the batch processing method. The deviation from current processing methods would be minimized so that the FMS could benefit from the experience gained on the parts to date. While conventional processing methods required multiple machine/ operations to complete the parts, the new method would maximize the set-up, material handling, scheduling and inspection advantages of the FMS. The number of tools would be minimized to allow incorporation of a significant backup feature into the system. By adding another set of tooling, all parts could be processed on at least two machines. This backup would greatly increase the system's flexibility to produce shipset quantities of parts without reconfiguration or rescheduling in the event of machine downtime.

SYSTEM DESIGN

Having developed confidence in the FMS and its configuration, a vendor specification package was prepared which included the following:
o Part drawings and batch processing documentation
o Production rates
o Input data and results of Draper Lab's simulations for the production rates specified
o Tooling, programming and fixturing specifications
o Software requirements
o Machine and pallet shuttle specifications
o Layout and chip handling specifications
o System simulation requirements
o Material handling system specifications
o System acceptance criteria
o Terms and conditions of purchase

Concurrent with the development of vendor proposals, FMC addressed additional processing required to complete the selected parts. This additional processing meant that certain offline capabilities would be needed. For example, all parts would require some form of deburring. The deburring could easily be accomplished by providing a deburr bench or by conveying the parts to a nearly vibrating deburring machine. Some parts would necessitate the installation of steel inserts between the rough and finish machining operations. Another part required the assembly of a housing and cover prior to the finish machining operation. Finally, certain parts called for drilling and tapping in attitudes that could not be accommodated by fixtures used in N/C part processing. Further simulations of the FMS by Draper Labs determined that offline drilling and tapping operations would not justify 5-axis capability. There was insufficient load to fully utilize the 5-axis machine, so that any additional cost would be high relative to the labor savings. Besides increasing the system's complexity, this particular configuration would preclude the backup of all capabilities in the system.

The ongoing research led to the design of a manufacturing cell that consisted of part queues, deburr bench, oven, freezer, press, assembly table, drilling and tapping machine, and material handling devices. The cell, located in the part staging area of the FMS, enabled the parts to enter and exit the FMS from one location. Parts could then flow from the automated material handling system through the offline workstation with a minimum of handling and ensure that all processing was completed within the FMS.

The vendor proposals, including their system simulation results, provided further insight into the dynamics of the FMS. Previous simulations had not considered the impact of a coordinate measuring machine and an automated parts washer. These workstations tie up a pallet/fixture for a period of time, and thus prevent it from being used elsewhere in the system. By adding these workstations, duplicate pallet/fixtures to maintain machine utilization and system throughput had to be added.

Because no consensus existed among FMS vendors regarding the type of material handling system to use, FMC faced a decision between the tow-line conveyor or the automatic wire guided vehicle systems. Advantages and disadvantages of both alternatives were examined.

Automated Wire Guided Vehicle

Advantages:
o Ease of installation
o Lower purchase cost
o Flexibility to accommodate layout changes late in the project
o Outdoor operation

o   Interface capability with potential AS/RS and manufacturing cells within the same facility
o   Partial backup provided by a spare cart
o   Shorter equipment lead time

Disadvantages:
o   Susceptibility to coolant and chips
o   Drive wheel wear
o   Concrete wear
o   Support area required for battery charging and servicing
o   No buffering of pallets
o   Electric cart complexity
o   Maintenance costs
o   Limited applications in machine shop environment

Towline Conveyor

Advantages:
o   Proven technology and reliability
o   Built-in buffering capability
o   Cart durability
o   Simplicity
o   Lower maintenance cost

Disadvantages:
o   Higher purchase cost
o   Difficulty and cost of installation
o   Longer equipment lead time
o   Reduced flexibility to accommodate layout changes late in the project
o   Limited interface capability with potential AS/RS and manufacturing cells within the same facility
o   Potential limit to the type and number of machines that could be added to the FMS in the future

FMC selected the AWGV because it provided the greatest opportunity to integrate the FMS technology into other manufacturing activities within the parts manufacturing facility. However, this selection still required additional system design changes that addressed the AWGV's inherent disadvantages. Special attention was paid to the control of coolant and chips. Guarding was added to the machine tool that completely encloses the part and fixture during machining. Collection trays were added to the AWGV to prevent coolant and chips from landing on the floor as they fell off the pallet/fixture during transport. A new load/unload station that would confine and transport coolant and chips away from the AWGV path was designed, with a capability to tilt the pallet/fixture 90° and rotate it to ease part loading/ unloading. The resulting design (referred to as an orient/wash station) eliminated the need for the automated parts washer. Specifications for the AWGV were also developed. FMC needed the capability to traverse surfaces not suitable for guide wire installation such as steel gratings and plates. The tire wear problem was not expected to be eliminated so a quick change drive wheel design was required. Computer monitoring of the battery charge on the AWGV assured that a vehicle would be out of service only when a battery change was required. Limit switches on the exterior of the vehicle could be replaced with encapsulated proximity switches. The vehicle could use a rack and pinion drive mechanism for lifting and lowering pallets in and out of pallet receivers. The buffering shortfall could be corrected by adding pallet storage magazines to the FMS.

To justify the system and its increased production rates, major benefits of the system were outlined:
o   Reduced direct, indirect and inspection labor
o   Elimination of set-up
o   Reduced rework and scrap material
o   Lower capital equipment costs to meet increased production rates
o   Reduced inventory
o   Reduced part lead time
o   Reduced material control costs
o   Increased machine utilization
o   Flexibility to adapt to changes in part mix and production rates

The outcome of FMC management's review of the project was a directive to justify the system at lower production levels, which resulted in insufficient load on the system as configured. Further investigation was required to identify additional parts for the FMS application. Parts were selected from aluminum routing as well as machine shop operations. Simulations performed by the FMS vendors resulted in a mix of 17 parts requiring four machining centers.

Reconfiguring the system from five to four machines was a significant change. The system cost could not be reduced at the same level as the benefits. Items in the FMS such as the material handling system, coordinate measuring machine and the computer control system would be required whether there were four or five machines in the system. The four-machine system was considered more complex. As more parts were added, the part debugging effort and the number of unique fixtures and tools in the system increased. Simultaneously, the tool storage capacity was reduced. This prevented running all the parts across any two machines, also reducing the level of backup and complicating the real time scheduling of the system in the event of machine failure.

As the first-phase configuration of the FMS stabilizes, efforts to better understand its dynamics and eventual evolution continue. Further simulations should determine which parts need backup capability given the tool limitations of the system. Alternatice modes of processing subsets of the selected parts in the system will be investigated to address potential raw part shortages. The objective will be to maximize system utilization while minimizing any additional fixturing and tooling required to support a higher throughput of subset parts.

It is expected that the FMS system will expand. Initially this expansion would probably take the form of an additional horizontal machining center to increase production rates of parts selected for the FMS. Parts of rotation could be added to the system by incorporating an N/C chucker, a mini AS/RS and machine loading equipment. Parts could then be batched to the turning machine and randomly processed to the horizontal machining centers. Tool limitations could be reduced by tool control and handling systems. Based on schedule and usage, tools would be transported by AWGV and inserted and removed from the machining centers automatically.

CONCLUSION

FMS technology has shown many advantages in low-volume, high-mix production applications. The evolution of the FMS project was affected most by factors outside the project study as overall division manufacturing plans continued to develop concurrently. These changes would have seriously impeded development of alternative machining approaches consisting of special-purpose machines and head changers. Flexibility of the FMS in the face of change allowed adaptation and continuation of system design efforts. Even after the system design and justification effort had been completed, a major change in production rates and part mix was accomplished in a short period of time. FMC has therefore been able to realize many advantages of an FMS system even before authorization of project funding.

PROPOSED SYSTEM DESCRIPTION

1. Orient Wash Station 2. Automatic Wire Guided Vehicle servicing a Horizontal Machining Center 3. Pallet Storage Magazines 4. Coordinate Measuring Machine 5. Offline Work Station 6. Future Automatic Tool Exchange Equipment 7. Future Horizontal Machining Center 8. Automatic Wire Guided Vehicle Service Area.

The horizontal machining centers are each equipped with a 15-horsepower spindle, a CNC control, an automatic tool changer, a 90-tool storage magazine and an indexing rotary table. The capacity of the four tool storage magazines is greater than the minimum requirements for the selected part mix and provides flexibility to:
o Respond to changes in part processing methods which require additional tooling
o Add parts to the system
o Select part mixes with a greater number of tool requirements
o Process some parts across more than one machine

The operation plan is to incorporate redundant tooling so that critical parts can be processed on more than one machine to minimize the effect of machine downtime on system output and reconfiguration.

The coordinate measuring machine is a gantry type unit incorporating a pallet delivery discharge mechanism and a computer control system. The configuration will allow palletized workpiece fixtures to be delivered to the inspection machine on a sampling basis for part inspection under FMS computer control. The coordinate measuring machine computer control will perform two major functions. First, it will communicate to the FMS computer receiving data on what parts have been delivered to the machine and subsequently what the inspection results are. Secondly, it will support the inspection machine function by providing inspection part programming, part program storage and direct computer control of the inspection machine.

Each load/unload station features a pallet receiver, pallet shuttle and positioning mechanism. Palletized workpiece fixtures will be delivered and picked up by the material handling system. While in the load/unload station, the palletized workpiece fixture will be shuttled in and out of the positioner. The positioner will have two axis of rotation so that the palletized workpiece fixture can be tilted 90 degrees and rotated. Under the positioner will be a chip conveyor tied into a central collection system. This design will facilitate the loading/unloading of heavy parts and the removal of chips from the parts and fixture.

The pallet storage magazine consists of a pallet delivery/discharge mechanism, a ten-pallet storage magazine and pallet service station. The pallet storage magazine will allow palletized workpiece fixtures to be placed in and out of storage by the material handling system under computer control. Two pallet storage magazines with a total storage capacity of twenty pallets is required since there will be more palletized workpiece fixtures than available workstations in the system. The pallet service station will provide the capability to remove a pallet/fixture from the system for checking, repair or changing of fixtures.

The offline workstation is connected to a load/unload station by a roller conveyor. The workstation incorporates into a cell equipment which performs operations that cannot be done on the machining centers. The operations are: deburring, isolated drilling and tapping, assembly/disassembly, sawing, and the pressing of steel rings.

The material handling system consists of three AWGV's, a control computer, a series of wires embedded in the floor and support equipment. The AWGV is an unmanned battery-powered unit which receives its commands from the radio frequency signals in the wires. A stand-alone computer controls the direction and location of travel of the AWGVs, while the FMS computer determines the workstations to be serviced. The support equipment consists of extra batteries, battery chargers, manual cart controls, frequency generator, area controllers and battery carts required to operate and maintain the AWGVs.

The common link among all of the hardware items previously described is the FMS computer. Computer control of complex machining systems requires sophisticated software to obtain high system utilization and to react to planned and unplanned events in the production environment. The functions residing in the FMS computer control system are:
o Traffic Coordination
o Work Order Entry
o Staging
o Simulation
o System Configuration
o Batch Scheduling
o RJE
o Data Distribution

These software functions provide capabilities required to operate an FMS. Among these capabilities are:
o Remote distribution of N/C programs and data files to machine N/C tool controllers.
o Maintenance of NC program libraries.
o N/C part processing at remote facilities via telecommunication links.
o Automated delivery of workpieces to work stations.
o Management of pallet and fixture data.
o Modeling of system activities to assess the impact of changes in production requirements.
o Scheduling, preparation and staging of work orders for workpiece entry into the system and control and monitoring of system resources.
o Diagnostics for maintenance and troubleshooting of system facilities.
o Recording of accounting data and formatting management reports from the data.
o Tracking and management of tools used within the system.

The FMS will be configured to allow expansion and incorporation of new technology. Space is provided within the FMS for a fifth horizontal machining center. Additional growth will be possible adjacent to the coordinate measuring machine and pallet storage magazines. The layout of the machining centers will also allow future addition of automatic tool exchange equipment. This equipment would be part of a tool control and handling system with the following functions:
o Computer generation of tool requirements by machine based on part mix, schedule and usage
o Preparation of tool kits by machine in the preset area
o Automatic loading of tool preset data in the N/C control
o Delivery of tools to the machine tool
o Exchange of tools for spent tools
o Return of spent tools to the preset area for reprocessing

Transportation of tooling between the preset area and the machine tools would be accomplished under computer control utilizing the AWGV.

V. Index

Index

AGV. *See* Robotic vehicles
AML, 11
ARC. *See* Alternate routing combination
ASKARS. *See* Automated storage, kitting, and retrieval system
Accounting system, 30
Aerospace industry, 63, 237
   plant modernization, 262-270
Adaptable-programmable assembly system (APAS), 15
Advanced manufacturing systems
   advantages of, 42
   in aerospace industry, 63
Aircraft assembly, 265-266
Allen-Bradley, 272
Alternate routing combination (ARC), 79-82
American Machinist, 9, 47, 205
Andersson, Edmund, P., 112
Anstead, Edward J., 219
Architecture, of systems, 148-152, 151 *fig. 2*, 174-176
Artley, John W., 104
Assembly operations
   and automated systems, 35
   and robots, 15
Audit
   conducting for CIMS, 246, 248 *fig. 3*
Automated guided vehicle (AGV). *See* Robotic vehicles
Automated storage, kitting, and retrieval system, 280-281, 286
Automation, 8-15, 42, 47-52, 207-208
   achievements in, 11, 12 *fig. 5*
   alternatives to, 49
   benefits of, 50
   using computer-based systems in, 14, 47
   disadvantages of, 50
   equipment features needed for, 11
   fixed, 109-110
   flexible, 108, 110-111, 237-245
   of GE plants, 258-261
   human factors in, 188-193
   islands of, 103, 169, 171, 258, 259
   vs. labor, 48-49
   levels of in decision-making, 189 *tab. 1*
   for low-quantity production, 11, 13
   management systems required in, 50
   of material handling, 92
   motivation for in U.S., 47
   of parts production, 223
   process, 37
   of process planning, 74
   productivity benefits of, 47-52
   programmable, 9
   programming for, 11
   of storage, 264
   trends in, 13, 47
   and unions, 48-49
   and vision systems, 108-111
   impact of work measurement on, 197-201
   effects of on worker motivation, 194-196
Automotive Industry Action Group, 112
Ayres, R.U., 47, 51

Banks, David, 108
Bar-code, 112-119
   code generation in, 115-116
   programs, U.S. Department of Defense, 117-118
   savings with, 118-119
   symbols used in, 114, 116-117
Batch production. *See* Production
Biegel, John E., 252
Biles, William E., 87
Bill of materials
   exploding, 4
   processing program (BOMP), 5
Black, J.T., 88, 120
BOMP, 5
Bradley Fighting Vehicles, 288
Bridges, information, 168-171
Britton, Harley O., 162
BYTE, 193

CAD. *See* Computer-aided design
CAD/CAM. *See* Computer-aided design/Computer-aided manufacturing
CAFI. *See* Computer-aided factory integration
CAM. *See* Computer-aided manufacturing
CAM-I, 7
CAPP. *See* Computer-assisted process planning
CIMS. *See* Computer-integrated manufacturing systems
CNC. *See* Computer numerical control
CWPABS. *See* Computerized workload projection and budgeting system
Calma, 234, 261
Capacity requirements planning, 4
Carter, Charles F., 17
Cell systems. *See* Integrated manufacturing systems
Cellular manufacturing systems, 120-130
   at components facility, 131-134
Changeovers, 28, 30
Chemical process industry, 180
Cincinnati Milacron, 18
   customer requirement proposals, 21
   FM3 philosophy, 20
Coding/classification systems, 126-127, 131
   synergistic benefits of, 135-140
Communication
   between departments, 33
   infrared systems, 101
   with mobile vehicles, 101
Communications oriented production information and control system (COPICS), 5
Computer-aided design (CAD), 57, 58, 108-109, 229
   integrating information sources into, 188
   product types, 58-59
Computer-aided design/Computer-aided manufacturing (CAD/CAM), 7, 13, 49, 229
   developments in, 197-201
   in CIMS, 36, 168-171
   programming routines for, 14
   software costs in CIMS, 39
   work station design in, 190
Computer-aided factory integration (CAFI) system, 238
Computer-aided manufacturing (CAM), 229
   economics of, 42-46
   equipment justification in, 43-44
   systems justification in, 44-45
Computer applications, 3

Index

Computer-assisted process planning (CAPP), 72-73, 224
Computer-integrated manufacturing systems (CIMS), 3, 4, 5, 6, 7, 13, 32-34, 35-41, 72, 154-161, 162-167, 188-193, 209-213, 222-228, 247 *fig.* 1, 248 *fig.* 2
    computer's role in, 36
    and cost reduction, 40
    functions of, 37, 39 *fig.* 3, 154
    hardware and equipment for, 36-37, 38
    implementation costs of, 38
    integrating with CAD/CAM, 168-171
    development team for, 223
    ensuring integrity of, 162-167
    inventory reduction in, 41
    management's role in, 34, 38, 39-40
    modular approach to, 229-236
    and MRP, 180-185
    and operations research, 87-89
    payback time with, 41
    personnel requirements for, 37-38, 39
    philosophy of, 35
    planning, 250
    process planning in, 203-205
    robot costs in, 38
    for small firms, 246-255
    software costs of, 38-39
    subsystems for, 33
    synergy function in, 44
    work measurement in, 203-205
Computer crime, 165
Computer languages. *See* Languages
Computer numerical control, 16, 48. *See also* Numerical control (NC)
    in flexible machining, 69
    and integrated manufacturing systems, 19
    machine tools, 127-128, 209
    self-diagnosis in, 14
Computer technologies, 58 *fig.* 1
Computerization, 136
    components of, 150
    costs of, 57
    hostility toward, 210
    of manufacturing systems, 4-5, 48
    of process planning systems, 72-74
Computerized workload projection and budgeting system (CWPBS), 278-279
Computers. *See also* Minicomputers
    and automation, 106, 188-193
    used in CIMS, 154
    cost of, 108
    DEC, 261, 265
    developments in, 57-58
    IBM PC, 57
    IBM 1620, 3
    IBM 3033U, 34
    in product design, 188-193
    Univac, 281, 284
    user considerations for, 192
    VAX, 272
    and worker attitudes, 191
Computerworld, 150
Control distribution, 4
Control systems, 48
Controller
    aisle, 101
    flexibility of, 158
    work station, 101
    programmable, 19, 37
Controls
    batch, 165-166
    pre-computer, 4
    single-transaction, 166
Cornell, Thomas R., 145
Cost reduction, 3
CRAFT, 60-61
Cross, Kelvin, 192
Cudworth, E.F., 271
Curtin, Frank T., 258

DBA. *See* Data base administrator
DBMS. *See* Data base management systems
DDB. *See* Distributed data base
DSS. *See* Decision support systems
Dar-El, E.M., 78
Data base, 60, 210, 226, 289
    in CAD/CAM, 109 *fig.* 1, 232 *fig.* 3
    in cellular manufacturing, 131-133
    in CIMS, 37, 40, 154-161
    in coding systems, 136
    considerations in CIMS, 231-233
    environment, 164 *fig.* 1
    for facilities information, 59, 60 *tab.* 1
    functional, 154 *fig.* 1
    hierarchical, 155
    interface specification, 156-159
    maintenance, 163-164, 211
    management, 6, 59, 146, 158, 159-165, 230
    manufacturing, 155
    mapping of, 160-161
    network, 155
    selection criteria, 159-160
    subject, 155 *fig.* 2
Data base administrator (DBA), 167, 226
Data base management system, 162-167, 170, 183-184
Data dictionary, 170
Data editing, 164-165
Data processing, 51
Data resource management, 146
Data Resources, Inc., 48
Daugherty, Michael J., 180
Decision support systems (DSS), 57, 147, 286
Deere & Co., 60, 131
Deere Tractor Works, 233
Devaney, C.W., 168
Distributed computing, 173-179
Distributed data base (DDB), 229, 230-231
Draper (Charles Stark) Laboratory, Inc., 288-289
Dumolien, William J., 131
Dunlap, Glenn C., 135
Duriron Company, Inc., 180

Eaton-Kenway, 272
Electronics industry, 52
Employees. *See* Workers
Environment, production, 27-30, 35
Equipment justification in CAM, 43-44

FMC Ordnance Division, 288-293
FMS. *See* Flexible manufacturing systems
Facilities planning, 57-62
    computer-integrated, 61 *fig.* 4
Factory
    automated, 203, 205

Index

of the future, 209, 227, 271
   layout of, 27, 37, fig. 1
Fargher, John S.W., 276
Flexible automation. See Automation
Flexible machining system, 63, 64 fig. 1, 237
   information flow in, 69, 70 fig. 6
   integration of, 68-71
   processor file used in, 70-71
Flexible manufacturing systems (FMS), 18, 35-36, 121, 147, 206-208, 288-293
   annual cost, 255 fig. 3, 257 fig. 4
   applications for, 13
   vs. cellular systems, 124 tab. 2
   dispatching rules in, 77
   energy costs in, 254
   incremental implementation of, 256
   job selection in, 81-82
   labor costs, 252-253, 254
   lead time distribution in, 255 fig. 2
   machine time distribution in, 254 fig. 1
   maintenance costs in, 254
   parts selection in, 76-77
   network analysis in, 77-79
   objectives of, 13
   order planning in, 63-71
   and robots, 11, 13, 15
   scheduling, 76-85
   schematic, 128 fig. 9
   in small-batch manufacturing, 252-257
   trends in, 13
Flexibility
   and cost, 20
   of parts, 12 fig. 6
   and suppliers, 30
Flow-through manufacturing, 35, 38 fig. 2, 40
FORTRAN, 3
French, Robert L., 186, 209
Froehlich, L., 52
FROM-TO analysis, 60
Future
   evolution in automation, 71, 198, 199
   factories of, 22-26, 173-174, 189, 203

Geier, Frederick V., 21
General Dynamics, 263
General Electric, 234, 235 fig. 4, 258-261, 271
Genplan, 73-74
Glenney, Neil, 92
GMS, 60-62
GPSS, 60-61
Graphic input devices, 190-191
Groover, Mikell P., 8, 47
Group technology, 125-127, 135
   at John Deere, 131-134
   at Grumman, 264
   at Lockheed-Georgia, 72-74
   in Russia, 128
Grumman Aerospace Corporation, 260-270

Hales, H. Lee, 57, 246
Hammer, William E., Jr., 162
Handbook of Industrial Engineers, 24
Hardware, computer, 27
   in CIMS, 38, 40, 233
   data errors, 165
   element scoping, 238-239
   failure, 166, 175
   in flexible automation, 242, 244
   link, 293

Harkrider, F.E., 237
Harvard Business Review, 23
Heisley, Michael E., 95
Helander, Martin, G., 188
Hewitt, David G., 173
Hirlinger, Craig R., 135
Huber, John, 262
Hughes, John E., 47
Human factors, in CIMS, 188-193
Human operators. See Workers

IBM, 5, 7, 11, 34
IDBMCS. See Integrated data base management control system
IWC. See Intelligent work center
Industrial engineer, 203
   role in worker motivation, 194-196
   role in strategy development, 24-25, 49-50
Industrial Engineering, 9
Information systems, 148, 162
Ingersol Rand, 82
Ingersoll Milling Machine Co., 34
Inspection programs, 65
   non-contact systems, 16
   programs, 65
Inspection, visual. See Visual inspection
Integrated computer-aided manufacturing (ICAM), 22-23
Integrated data base management control system (IDBMCS), 159
Integrated manufacturing systems, 19-21, 180-185
   cell, 19, 155
   FMS or VMM, 19-20
   factory data automation, 20-21
Integrity, systems, 162-167
Intelligent work center (IWC), 156-160
International Data Corporation, 197
Inventory, 187
   accounting, 5
   carrying cost, 254
   and cell systems, 19
   finished goods, 37
   inprocess, 16
   modeling, 88
   reducing, 28, 41

JD/GTS, 60
Japan, 16, 17, 32, 42, 47, 121-122, 124, 203, 209, 211
Job selection, 65
Job shop operations scheduling, 6

Kearney & Trecker, 13
Kennicott, Thomas C., 199
Klahorst, H. Thomas, 206

Labor force. See Workforce
Languages, computer
   computer, 4, 211
      AML, 11
      COBOL, 169, 170
      FORTRAN, 3, 169
      MCL, 11
      Pascal, 169
      PL/1, 169
      RPG, 169
   simulation
      GPSS, 60-61, 87
      PASAMS, 87

Index

SIMAN, 87
SIMSCRIPT, 87
SLAM-II, 87
Layout
   cellular, 129 fig. 10
   improvement routine, 60
   and material flow, 127
   of plant, 27, 37 fig. 1
Lead time, 29
   critical path, 30
   in FMS, 253-256
Lee, Scott E., 229
Lockheed-Georgia Company, 72-74
Loomis, Russel M., 199

MCS. See Manufacturing control system
MIS. See Management information services (MIS)
McDonnell Douglas, 11, 223-224
Machine control data, 176
Machine time, 17
Management information services (MIS), 38, 57, 58-59
Manchuk, Stan, 222
Manual order point systems, 4
Manufacturing control system (MCS), 219-221
Manufacturing Control Language (MCL), 11
Manufacturing environment, 27-30
Manufacturing management control systems, 186-187
Manufacturing planning, 50, 189 fig. 1
Manufacturing routing, 65
Manufacturing systems, computerized, 6
Mass production. See Production
Master production scheduling (MPS), 28-29, 37
Material handling, 37, 178, 206-207, 259, 293
   in the automated factory, 199
   design, 274-275
   modular integrated system, 92-97
Material requirements planning (MRP), 5-6, 180-185
Mather, Hal, 27
Merchant, Eugene, 18
Meredith, Jack, 42
Metalworking industry, 47
Microprocessors
   and data base systems, 158
   and vision systems, 109
Miller, S.M., 47, 51
Min/max systems, 4
Minicomputers
   Digital Equipment, 4
   growth rate, 5
   Hewlett-Packard, 4
   inventory control with, 94
Modeling
   for material handling, 94-95
   for simulation, 284
Modularity, 230-236
Mosier, Charles T., 229
Motivation, worker, 32, 34

NARF, 276-287
NAVAIR, 263
   industrial material management system, 281-283, 286
   industrial financial management system, 283-284
National Machine Tool Builders Association, 17
Naval Air Systems Command, 262

Nesbitt, John, 108
Nestman, Chadwick H., 154
The New York Times, 106
Nonintegrated manufacturing systems, 6
Numerical control (NC), 8-9, 11, 14, 16, 17, 20.
   See also Computer numerical control
   in cellular manufacturing, 122 fig. 3
   in flexible machining, 69
   and machine tools, 206
   part programs, 63, 225
   parts maker, 266 fig. 9

Object Recognition Systems Inc., 105, 106
Odrey, Nicholas G., 47
Operating management, in CIMS, 209-213
Operations research, in CIMS, 87-89
Order planning, in FMS, 63-71
Orderpicking, 100-101

P:D ratio, 30
Parts
   families, 127
   population, 131
   selection in FMS, 76-77
Payback, 30, 41
Personnel requirements, in CIMS, 39
Philosophy
   business manufacturing, 208
   of CIMS, 36, 154
   of flexible automation, 47
   pallet-loading, 65
   of production management, 32-34
   release, 66-67
   serial production, 65
   tooling, 63-65
Planning and design
   of cellular systems, 122-124
   for computerized facilities, 57-62
   modifications in flexible machining, 69
   top-down, 155
Polacek, Richard, 199
Pratt & Whitney, 271-275
Price/performance curve, 5
Price/performance reduction, 6
Pritsker & Associates Inc., 57
Process control, in CIMS, 37
Process manufacturing, 4
Process planning
   in cellular manufacturing, 133
   computerized, 72-74
   at Grumman, 264
Process technology, 47-48
Product design, 29
Product flow analysis (PFA), 126
Production
   of circuit boards, 29
   economic considerations in, 17
   in FMS, 63-71
   improvements in, 9
   measuring, 30
   planning and control, 65, 276-287
   planning models, 88
   rates, 8
   scheduling, 28
   serial, 69
   small-lot, 120-130
   transfer lines in, 16
   volume, 12 fig. 6

Index

Production techniques, 121 *fig.* 1, 123 *tab.* 1, 211
    batch, 9, 10 *fig.* 1, 12 *fig.* 5
    cellular, 120-130
    continuous process, 120
    flow-through, 35, 38 *fig.* 2, 120
    job shop, 9, 10 *fig.* 1, 12 *fig.* 5, 16, 17, 120-121, 124
    just-in-time, 121 *fig.* 2
    mass, 9, 10 *fig.* 1, 12 *fig.* 5, 16, 17
    project shop, 120
    volume, 211
Productive time, 252-254
Productivity
    increases in, 19
    improving, 39-40, 48
    vs. investment, 17 *fig.* 2
    labor costs, 264
    and material handling, 92
    effect of negative conditions on, 32
    plan for at Pratt & Whitney, 272-274
    statistics, U.S. Department of Labor, 17
    studies, 222
Profits, 32
Programmable flexible automation. *See* Robotics
Programming
    for automation, 11
    dynamic, 89
    geometric, 89
    goal, 89
    integer, 88-89
    linear, 88
    nonlinear, 89
    offline, 11
Project planning, 227
Punched cards, 3, 7
Punched tape, 11

Queueing, in CIMS, 88

Rand Corporation, 18
Random access, 4
Real time access, in MCS, 220
Redmond, Gary R., 288
Reports, in WCS, 280
Return on investment (ROI)
    for automated inspection systems, 105
    in CAM, 44
    at Pratt & Whitney, 272
Riley, Frank J., 199
Robot Institute of America, 51
Robotic cells, 123-124, 129 *fig.* 10
    in aircraft assembly, 266, 267 *fig.* 11
    technical requirements, 265
Robotic vehicles, 98-103. *See also* Robots
Robotics, 110-111, 225
    trends in, 49
Robots, 47, 195, 239
    in aerospace industry, 264-265
    batch production, role in, 13
    and cell systems, 19
    in CIMS, 30
    and flexible manufacturing systems, 11, 13
    in nonsynchronous assembly system, 259
    and product retrieval, 98-103
    and subassembly, 98-103
    and worker displacement, 51
Russia, 128

Sadowski, Randall P., 3
Salomon, Daniel P., 252
Santen, William P., 132
Sarin, S.C., 76
Scanners
    alphanumeric, 113
    bar code, 112-119
    dynamic, 113
    fixed beam, 114
    hand-held, 112
    laser, 37, 95-96, 112, 259
    light-pen, 113, 115 *fig.* 4, 191
    and material handling, 113
Scheduling
    in FMS, 76-85
    in CIMS, 225-226
Schultz, Karl B., 16
Schroer, B.J., 88
Scott, David, 60
Semiconductor industry, 48
Sensing devices, 37
Serial processing, 3
Setup design, 124
Setup time, 11, 16, 28, 199, 209
    in cellular manufacturing, 120-130
    in CIMS, 38, 41
    in FMS, 252, 253
    reductions in, 73
Shell, Richard L., 197
Simulation model, 289-290, 285
Simulation studies, 253
Skinner, Wickham, 203
Smith, Jason L., 32
Smith, Robert E., 98
Software, computer, 5, 18, 27, 73, 147, 169-170, 221
    CAM, 234
    and CIMS, 33, 38
    in coding applications, 136
    data errors, 165
    element scoping, 238-239
    failure, 166
    in flexible automation, 238, 240, 241 *fig.* 2, 242, 244
    functions, 293
    for manufacturing management control systems, 186-187
    mixing and matching, 184
    packages, 211
    for pallet grouping, 65-66
    requirements planning, 5
    for robotic vehicles, 98
    types, 57, 59-60
Sperry Univac, 233, 234
Steelcase Inc., 58
Storage systems, 95-96
Strategy, manufacturing, 23-24, 32, 43
Sullivan, Cornelius, H., 150
Sullivan, Kevin H., 35
Suppliers, 29-30. *See also* Vendors
Synergy, 44, 138
Systems integration, 145-152
Systems Modeling Corporation, 57

Test Engineering, 176
Time allocation, 9, 10 *fig.* 2, 3, 4
Training, 33, 50, 264

Index

vs. capabilities, 194-195
computer-aided, 191-192
of customers, 261
of workers in CIMS, 39, 213
Transfer lines, 16-17. *See also* Production
Tulkoff, Joseph, 72
Turnkey systems, 229-230

U.S. Air Force. *See also* Integrated computer-aided manufacturing (ICAM)
B-1B program, 237, 238
sponsorship of MCL, 11
technology modernization program, 274
U.S. Department of Defense
bar-code estimates, 112
human factors requirements for operating systems, 192
procurement policies, 262
programs for bar codes, 117-118
U.S. Department of Labor
productivity statistics, 17, 48
U.S. Navy, 262
U.S. News & World Report, 48, 51
Unit labor cost, 18, 126 *fig.* 7
Unions, labor, 48-49, 50
United States
and automation, 47, 51, 198 *fig.* 1
free enterprise, interference with, 32
industry leadership position, 16, 203
management system in, 209, 211
management/workforce relationship, 42
technological advantages of, 32
United Technologies, 271
Uplink, 37
Utilization rates, 9, 11, 269
of cells, 132
of equipment, 9, 254

VMM systems. *See* Integrated manufacturing systems
Variable mission manufacturing system. *See* Integrated manufacturing systems
Vendors, 29-30
and facilities planning, 57
of robotic vehicles, 100 *tab.* 1
Vision systems, 15, 108-111, 178
functions of, 105-106
in low-quantity production automation, 13
requirements of, 107
Visual inspection systems, 103-107
Voice-recognition systems
for data input, 14
Vought Aero Products Division, 63, 237-245

The Wall Street Journal, 203
Webster, W. Bruce, 63
Welter, William R., 22
Westinghouse Defense and Electronics Center, 173-179
White, John, 198
Willis, Roger G., 35
Windsor, John C., 154
Work cells
in automation, 109
in flow-through manufacturing, 35
in FMS, 35

Work center loads, 29
Work flow, 34
Work measurement, 197-200, 203-205
Work order scheduling, 69
Work station design, 190
Workers, 8, 97
capabilities of, 194-196
educating, 32
motivating, 194-196
pride, 33
replacement by machines, 48
training of, 35, 39, 50
Workforce
in CIMS, 37
dislocation problems of, 47-52
motivation of, 27
shifts in, 48
Workload control system, 280
interfaces, 285 *fig.* 5
operation of, 286
redesigned, 284, 286
Workpiece, 16-17

Zandin, Kjell B., 203
Zhang, S.X., 88
Zohdi, Magd E., 87

About the editors:

John W. Nazemetz is associate professor, School of Industrial Engineering and Management, Oklahoma State University. Prior to teaching, he was a systems analyst/consultant for the Abex Corporation. He received his B.S. and his Ph.D. in industrial engineering from Lehigh University.

Dr. Nazemetz is a senior member of the Institute of Industrial Engineers. He was the associate program coordinator of the 1984 Fall Conference sponsored by IIE. He is also the director for 1984-85 of IIE's Manufacturing Systems Division.

Dr. Nazemetz has written many articles and lectured extensively on the role of computers in industrial engineering.

William E. Hammer, Jr. is director of information systems for The Duriron Company, Inc. He joined the company in 1967 as senior industrial engineer and was manager of systems before being promoted to his current position. Prior to Duriron, he was a planning engineer for Western Electric. He received his B.I.E. from the University of Dayton and his M.S.I.E. from Ohio State University.

Mr. Hammer has played a major role in the development of manufacturing control systems at Duriron. He is a senior member of the Institute of Industrial Engineers, past president of the Dayton chapter, and past director of the Computer and Information Systems Division.

Mr. Hammer is an Industrial Engineering Visitor for the Accreditation Board for Engineering and Technology (ABET). He is listed in *Who's Who in Engineering* and the *Directory of Top Computer Executives*.

Randall P. Sadowski is associate professor, School of Industrial Engineering, Purdue University. He holds B.S.M.E. and M.S.M.E. degrees from Ohio University and a Ph.D. from Purdue University. Previously he was on the faculty of the University of Massachusetts.

He is a senior member of the Institute of Industrial Engineers. He is a past member of the College-Industry Committee on Materials Handling Education and the American Production and Inventory Control Society.

Dr. Sadowski has published many technical articles and papers, has lectured extensively, and is a consultant for numerous corporations. His teaching and research interests are currently in manufacturing and production systems, with emphasis on control and applied scheduling concepts.